Navigating the FRAM

In the world of safety science, where complex socio-technical systems demand innovative approaches, the Functional Resonance Analysis Method (FRAM) stands as a beacon of insight. Rooted in modern systemic management and Resilience Engineering principles, the FRAM offers a unique perspective, viewing systems through the lens of everyday performance variability.

Developed over decades of meticulous research, the FRAM serves as a powerful tool for modelling intricate systems, encompassing various realms: aviation, healthcare, IT software requirements, crisis management, and more. This book represents a culmination of two decades of research and practical applications of the FRAM. It offers an operational perspective and a useful reference for researchers and practitioners alike, empowering them to tackle a myriad of socio-technical modelling challenges. It blends traditional applications with cutting-edge advancements and looks at case studies in which the FRAM was utilized. Delving into the intricacies of the FRAM's structure and functionality, the book equips readers with operational knowledge essential for navigating diverse domains.

Written for researchers and practitioners in Occupational Health and Safety, Risk Engineering, Safety Engineering, Resilience Engineering, Human Factors, and related fields, *Navigating the FRAM: Mastering the Functional Resonance Analysis Method for Modelling Complex Socio-Technical Systems* resonates with those seeking to deepen their understanding of complex systems.

Riccardo Patriarca is an associate professor in the Department of Mechanical and Aerospace Engineering at Sapienza University of Rome, Italy. He holds a BSc in Aerospace Engineering, an MSc in Aeronautical Engineering, and a PhD in Industrial and Management Engineering. He currently teaches Operations Management, Aviation Safety Management, and Human Factors for the MSc courses in Mechanical Engineering and Aeronautical Engineering at Sapienza. His research is oriented towards modelling complex socio-technical systems – including their cyber dimension – untangling their inherent complexities and advancing risk and resilience-aware solutions for an ever-evolving world.

Navigating the FRAM
Mastering the Functional Resonance Analysis Method for Modelling Complex Socio-Technical Systems

Edited by
Riccardo Patriarca

CRC Press
Taylor & Francis Group
Boca Raton London New York

CRC Press is an imprint of the
Taylor & Francis Group, an **informa** business

Designed cover image: Shutterstock

First edition published 2026
by CRC Press
2385 NW Executive Center Drive, Suite 320, Boca Raton FL 33431

and by CRC Press
4 Park Square, Milton Park, Abingdon, Oxon, OX14 4RN

CRC Press is an imprint of Taylor & Francis Group, LLC

© 2026 selection and editorial matter, Riccardo Patriarca; individual chapters, the contributors

Library of Congress Cataloging-in-Publication Data
Names: Patriarca, Riccardo editor
Title: Navigating the FRAM : mastering the functional resonance analysis
method for modelling complex socio-technical systems / edited by
Riccardo Patriarca.
Description: First edition. | Boca Raton, FL : CRC Press, 2026. |
Includes bibliographical references and index.
Identifiers: LCCN 2025019159 (print) | LCCN 2025019160 (ebook) |
ISBN 9781032854359 hbk | ISBN 9781032850511 pbk |
ISBN 9781003518167 ebk
Subjects: LCSH: System analysis--Simulation methods | Sociotechnical
systems--Simulation methods | Industrial safety--Simulation methods
Classification: LCC T57.62 .N364 2026 (print) | LCC T57.62 (ebook)
LC record available at https://lccn.loc.gov/2025019159
LC ebook record available at https://lccn.loc.gov/2025019160

ISBN: 978-1-032-85435-9 (hbk)
ISBN: 978-1-032-85051-1 (pbk)
ISBN: 978-1-003-51816-7 (ebk)

DOI: 10.1201/9781003518167

Typeset in Times
by SPi Technologies India Pvt Ltd (Straive)

Dedication

---

to Emanuele F.T., our morning star

Contents

PROLOGUE   Introducing the FRAM

SECTION I   Step 1: To Identify and Describe the System's Functions

SECTION II   Step 2: To Identify the Variability

SECTION III   Step 3: Aggregate Variability, Actual, and Potential

SECTION IV Step 4: Assess the Consequences of the Analysis

SECTION V Implementation and Reflections on the FRAM

SECTION VI   Final Thoughts

Contributors

Arie Adriaensen
Faculty of Technology, Policy and
  Management
Department of Multi-Actor Systems
TU Delft
Delft, The Netherlands

Michel Anzanello
Universidade Federal do Rio Grande
  do Sul
Porto Alegre, Brazil

Antonio De Nicola
Agenzia nazionale per le nuove
  tecnologie, l'energia e lo sviluppo
  sostenibile (ENEA)
Rome, Italy

Ivenio Teixeira de Souza
Universidade Federal do Rio de
  Janeiro (UFRJ)
Rio de Janeiro, Brazil

Andrea Falegnami
Uninettuno International Telematic
  University
Rome, Italy

Pedro Ferreira
CENTEC
University of Lisbon
Lisboa, Portugal

Flávio Sanson Fogliatto
Universidade Federal do Rio Grande
  do Sul
Porto Alegre, Brazil

Josué E. Maia França
Petrobras and KTH
Rio de Janeiro, Brazil

Leonardo Bertolin Furstenau
Universidade Federal do Rio Grande
  do Sul
Porto Alegre, Brazil

Niklas Grabbe
TUM School of Engineering and Design
Chair of Ergonomics
Technical University of Munich
Munich, Germany

Assed Naked Haddad
Universidade Federal do Rio de Janeiro
  (UFRJ)
Rio de Janeiro, Brazil

Rees Hill
Zerprize Limited
Christchurch, New Zealand

Erik Hollnagel
Professor Emeritus at Linköping
  University - LiU (Sweden), Mines
  Paristech (France), and Syddansk
  Universitet (Denmark)
and
Honorary professor at Macquarie
  University
Sydney, New South Wales, Australia

Jeanette Hounsgaard
Senior Consultant (retired) at The
  Region of Southern Denmark
Denmark

Olivia Lounsbury
University of Oxford
Oxford, United Kingdom

Ehsan Mahmoodi
Department of Intelligent Production
  Systems
University of Skövde
Skövde, Sweden

Hideki Nomoto
Advance Technology Research Center
  of Japan Manned Space Systems
  Corporation
Tokyo, Japan

James Norman
Independent researcher
USA

Riccardo Patriarca
Sapienza University of Rome
Rome, Italy

Gesa Praetorius
Swedish National Road and Transport
  Research Institute (VTI)
Linköping University
Linköping, Sweden
and
Western Norway University of Applied
  Sciences
Bergen, Norway

Tarcisio Abreu Saurin
Universidade Federal do Rio Grande
  do Sul
Porto Alegre, Brazil

Amit Sharma
University of Bergen
Bergen, Norway

Francesco Simone
Sapienza University of Rome
Rome, Italy

David Slater
Cardiff University
Cardiff, United Kingdom

Doug Smith
Memorial University of Newfoundland
St. Johns, Newfoundland, Canada

Mark Sujan
University of York
York, United Kingdom

Riana Steen
BI Norwegian Business School
Oslo, Norway

Natália Maciel Tocchetto
Universidade Federal do Rio Grande
  do Sul
Porto Alegre, Brazil

Andrea Tomassi
Uninettuno International Telematic
  University
Rome, Italy

Maria Luisa Villani
Agenzia nazionale per le nuove
  tecnologie, l'energia e lo sviluppo sostenibile (ENEA)
Rome, Italy

Naruki Yasue
Department of Micro Engineering
Graduate School of Engineering
Kyoto University
Kyoto, Japan

Enrique Ruiz Zúñiga
Department of Intelligent Automation
University of Skövde
Skövde, Sweden

Paulina Zurawska
Faculty of Technology, Policy and
  Management
Department of Multi-Actor Systems
TU Delft
Delft, The Netherlands

Foreword

Rogier Woltjer
Lund University, Sweden

Ivonne Herrera
Studio Apertura, NTNU Social Research AS, Norway

Around 2007, when the Functional Resonance Accident Model – as it was called at that time – and its associated method were in their infancy, the FRAM's potential to address complexity and emergence already created considerable curiosity. This was the year of the first FRAMily meeting. Developments from FRAM as a model to the FRAM as a practical method, that is, the Functional Resonance Analysis Method, aimed to meet the growing need for systemic analysis approaches.

Initially, the method was rather underspecified with no detailed definitions, instructions, or software, giving room to diverse interpretations of its implementation. Scientists and practitioners used Erik Hollnagel's presentations, the FRAM chapter from his 2004 book *Barriers and Accident Prevention* and, later, course materials, to analyze complex socio-technical systems. Back then, FRAMily meetings and early projects using the FRAM discussed evolutions of the method intertwined with real world application, which we recall as a process characterized by learning-by-doing. Examples include how to visualize variability and resonance, design the first software tools, handle large models, and communicate findings. Over the years, sharing studies and experiences across domains during FRAMily meetings and conferences contributed to the evolving method, reinforced by Erik Hollnagel's 2012 landmark FRAM handbook.

This edited book *Navigating the FRAM: Mastering the Functional Resonance Analysis Method for Modelling Complex Socio-Technical Systems* stands as a statement of the latest FRAM progress. It shows that substantial headway has been made since the early days of the FRAM, in terms of both maturity and varieties of the method and its application in diverse domains. We wish to congratulate Riccardo Patriarca and all of the chapter authors for their efforts in consolidating this valuable resource showcasing the current state of FRAM in terms of both guidance and application.

In the spirit of the FRAM perspective of studying why things go right, this book presents success stories and progress towards everyday use in various domains, not only by researchers but in close cooperation with practitioners. Moreover, this book illustrates the successful application of the FRAM providing useful insights with a number of analyses and adaptations of the method to various purposes and needs.

This book will be a useful resource and facilitate further application by exploring the method's steps with increased granularity and detailed guidance, for readers interested in both qualitative and quantitative analyses, through small- and larger-scale models. The FRAM's evolution is closely linked with its practical application with the FRAM Model Visualizer, and other tools are in the making. An important contribution of the book is the initial formalization of the FRAM steps and foundational principles, which may prove useful towards engineering resilient systems. A related development that this book highlights in several chapters is the use of various Artificial Intelligence techniques, from fuzzy logic to machine learning and Large Language Models. This provides a fresh perspective on how various analytical tools may be combined to enhance our understanding of complex work domains. All of this progress provides a diverse set of answers to questions already raised during the early FRAMily meetings, and raises new ones.

The authors invite the reader to learn from their latest results from learning-by-doing and sharing their thoughtful reflections. With the helpful tailoring of content to different reader profiles, the editor and the author team have managed to produce a book that is relevant to a wide audience, pushing the state of the art, inspiring application, and encouraging cross-domain learning. We are confident that this book will be a valuable resource for understanding everyday work, emergent system behaviour, and performance variability via the FRAM.

This book thereby revisits what FRAM has become, and we look forward to seeing its impact on the FRAM becoming an everyday method. The book represents an invitation for opportunities for further applications and improvement, and we are keen to follow the results of the various new research endeavours and industry applications that it may inspire, in safety management and beyond.

Preface

Riccardo Patriarca

Before opening this book, you have been staring at a sketched futuristic honeycomb on the front page.

Since the book is about the Functional Resonance Analysis Method (FRAM), it *has* to involve hexagons. If you are new to the field, hexagons represent graphically the building elements of the FRAM itself, that is, a function, as you will see in the coming chapters.

Or perhaps the honeycomb is meant to resemble a beehive? Yes, a beehive: a marvel of complexity. A place where thousands of bees work tirelessly, each with their own role. On the surface, it appears as a system functioning effortlessly. But a closer look reveals a delicate balance, where constant adaptations, intricate communications, and shared awareness ensure the hive's survival in an unpredictable world.

Both perspectives hold true.

Much like a beehive, modern socio-technical systems thrive – or falter – based on their ability to adapt to ever-changing conditions. The FRAM was conceived to explore this dynamic interplay, offering a lens to understand how everyday actions resonate within a system. It moves beyond the traditional focus on failures and single causes, instead appreciating the variability and interdependence that drive performance everyday.

Additionally, I chose a digital, futuristic beehive to reflect the book's forward-looking perspective. It goes beyond reiterating existing cases by introducing novel methodological advancements, and offering guidance for future applications.

I also have a second metaphor for you. This one stays in the book's title *Navigating the FRAM*, which invokes the image of the Norwegian polar research ship called "the FRAM", used to navigate the uncertain and risky water of the Artic and Antarctic in the 20th century. Once you acknowledge system's complexity, you might find beneficial to explore the nuanced landscape of the FRAM as a tool, and mastering it to uncover the resonances within your own systems, ensuring they not only survive but flourish in the face of uncertainty.

The inspiration to edit this book came from my own experiences navigating the complexity of explaining socio-technical modelling (and, perhaps, of beekeeping alike). I realized that while the FRAM has proven valuable across diverse domains, its richness often remains inaccessible to those outside an inner circle of experts.

Consequently, this book represents a guide for practitioners, researchers, and students to venture inside a hive, or to piloting a ship. It showcases the FRAM's principles, steps, and applications, with a particular focus on practical use cases, revealing how this method can uncover the subtle interactions that shape resilient systems.

Whether you are discovering the FRAM for the first time or seeking deeper insights, I hope these pages will inspire you to appreciate the underlying dynamics of the method and transfer these lessons to your own domain.

Acknowledgements

From its inception, this book was designed as a collaborative endeavour – an edited volume bringing together diverse authors, each contributing their unique perspectives. It has, in essence, been a socio-technical writing project involving multiple stakeholders, various technological tools, and numerous organizational dimensions. With an internationally diverse team of contributors and reviewers spanning Europe, Brazil, Canada, Japan, New Zealand, and the United States, the writing process has been both dynamic and enriching.

Bringing this book to your hands required an adaptable system where individual efforts came together to create a whole far greater than the sum of its parts.

This holds true, also from my perspectives as an editor, a husband, and a father.

I am deeply grateful to my family and my colleagues who embraced this journey and devoted their time to support my vision for this book.

To some extent, these pages are a testament to the resilience of such a diverse joint cognitive system!

Prologue

Introducing the FRAM

1 The FRAM for Socio-Technical Safety Management and Beyond

Riccardo Patriarca

A (SHORT) HISTORIC OVERVIEW

A core idea behind the Functional Resonance Analysis Method (FRAM) development was recognizing that safety is not just about avoiding harm, it is about understanding how systems actually work, day in and day out. Safety science as a field has indeed shifted over time from focusing on purely technical faults to exploring human decisions, organizational structures, and the messy reality of how things get done.

Traditionally, we often think of safety as the opposite of risk. We reduce it to figures and probabilities, trying to calculate where things might go wrong, leveraging a long-standing cultural heritage (de Moivre, 1711). Yet the very word "safe," while being connected to risk, traces back to the Latin "salvus," meaning "intact" or "whole." Safety, then, is more than just minimizing bad outcomes. It is about preserving the integrity of people, property, and the environment. This perspective acknowledges that uncertainty, complexity, and adaptation are normal parts of working life, no matter what domain.

If on the one hand, risk is understood as any uncertain situation where something of value is at stake, on the other hand, safety means keeping that uncertainty within a range we find tolerable. But this is not a simple numbers game, or at least it should not be. Safety science cannot be restricted to calculating odds, it also involves concepts, theories, and methods that guide how we think about, discuss, and manage risks.

By the 1980s, an initial wave of regulations and major accident investigations pushed the focus of analysis beyond hardware and engineering. Otherwise, inexplainable events forced us to ask tougher questions: How do people's decisions and actions matter? What about the role of management, maintenance practices, or even how information flows through an organization? Reports of accidents no longer blamed technology alone; they pointed to human, organizational, and cultural factors. The recent injection of automation and even artificial intelligence (AI) introduces new layers of complexity, raising fresh questions about trust, accountability, ethical boundaries, and the delicate balance between human judgment and machine-driven decision-making.

DOI: 10.1201/9781003518167-2

3

We realize that stable "safe states" are rare and that everyday operations depend on people constantly adjusting to shifting conditions. Oppositely, early safety models assumed accidents followed a simple chain of causes, often pinned on a single human error. But as our understanding deepened, we saw that complex systems rarely fail because of one bad decision or faulty part. Instead, accidents can emerge from subtle interactions, unpredictable feedback loops, and small glitches that happen constantly and yet, under certain circumstances, lead to larger outcomes, often adverse. As such, over time, models became richer, reflecting that control and coordination, rather than isolated barriers, keep systems running.

A question naturally arises: How do we really capture the interplay of people, tools, tasks, and environments for both expected and unexpected events?

This is where Systems Theory, and by extension, Resilience Engineering, come into play. Instead of treating variability as a nuisance, they recognize that constant adaptation keeps systems functioning, and the meaningful unit of analysis should lie in the system as a whole, not just as a sum of its constituents. This perspective is particularly necessary to handle the full breadth of complexity in modern socio-technical systems defined by intricate technologies, diverse stakeholders, and constant change.

The FRAM builds on these premises, being framed as a method to model variability without forcing it into a neat cause-and-effect chain. It is designed to help us see how everyday adjustments might lead to stable performance, or drift towards failure, depending on how they interact and "resonate" within the system.

In the chapters that follow, we will explore the FRAM's principles, steps, methodological extensions, and practical applications. We will show how this method can give us new insights into navigating the inherent complexity of socio-technical systems. Rather than chasing a tidy list of causes for incidents or hazard identification, the various chapters offer insights on how to appreciate the patterns that emerge when multiple factors combine. With the FRAM, the goal is not just to prevent a specific failure but to understand how systems work well most of the time, how to strengthen those success factors, and how to detect adverse underlying mechanisms.

This book aims to offer a representation of the more recent developments of research and practical applications of the FRAM.

THE TIME OF THE FRAM

The FRAM is a systems-thinking approach that models complex socio-technical systems by examining how performance variability can resonate to produce both intended and unintended outcomes under expected and unexpected conditions.

Rooted in Resilience Engineering principles, the FRAM systemic view focuses on understanding how everyday work unfolds in dynamic environments. Its key contribution lies in acknowledging and modelling the underlying behaviours of performance variability, rather than only looking at isolated failures.

The FRAM's development began in the early 2000s, culminating in its first formal description in 2004. Initially, the acronym stood for "Functional Resonance Accident Model," a reflection of its early focus on accident analysis (Hollnagel & Goteman, 2004).

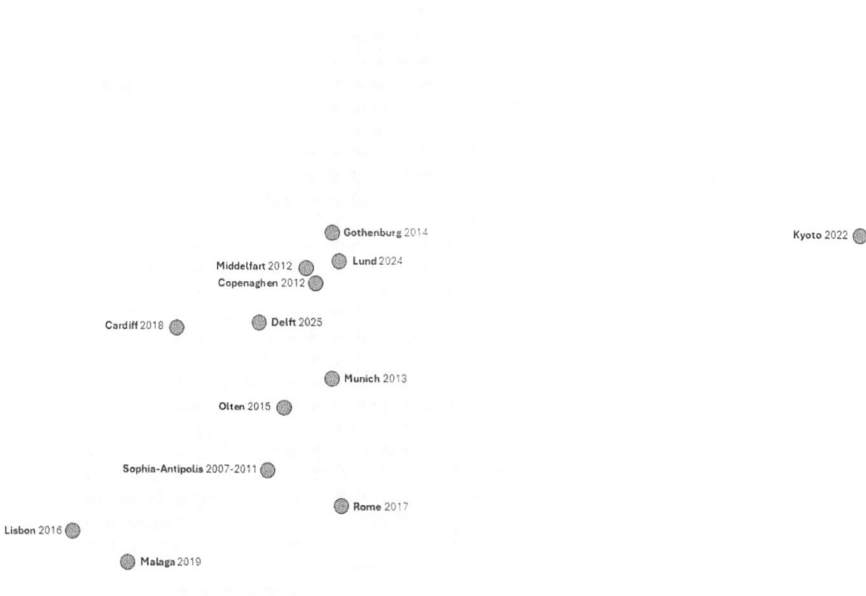

FIGURE 1.1 FRAMily meetings over time (2011–2025). The 2020 meeting was cancelled and postponed to 2021 due to pandemic restrictions and was eventually held in 2022.

Over time, intensive discussions within the FRAM community of experts and users, known as the "FRAMily" (cf. Figure 1.1), highlighted that the FRAM is, in fact, a method rather than a mere model. Moreover, it can be applied far beyond accident analysis, extending its reach to any scenario where people and technology interact closely. This perspective is discussed extensively in the FRAM handbook by Hollnagel (2012).

By redefining the FRAM as the "Functional Resonance Analysis Method," its potential broadened significantly. Indeed, the FRAM fundamentally allows building a model of a selected work reality with minimal underlying modelling assumptions. Today, the method serves as a versatile modelling tool, extending well beyond pure safety management purposes. The FRAM can even be used to model how the FRAM itself works!

In this perspective, it is better described as a *method-sine-model* rather than a *model-cum-method*. Reflecting this concept, models developed using the FRAM should be referred to as "FRAM-built models" rather than just "FRAM models." For the sake of brevity, these terms will be used interchangeably throughout this book even though, formally, the term "FRAM model" should be used for specifically referring either to the model of the FRAM itself—cf. Patriarca et al. (2020) and Chapter 20 of this book—or to the plywood reconstruction of the renowned Norwegian ship "the FRAM"[1] used for Arctic and Antarctic expeditions between 1893 and 1912.

WHY DID I DECIDE TO EDIT THIS BOOK?

It was a hot summer day close to the Mediterranean Sea. That's when I decided to edit this book. The decision came when I tried to explain to patient non-experts what has been one of my main research activities over the past 10 years, that is, modelling socio-technical systems. It was at that moment that I vividly realized the challenges of explaining and describing what this type of modelling means. It was not just the temperature that made it hard; the struggle was all too familiar. I had faced similar moments before, explaining the concept to engineers in different fields.

In my experience, discussions about the FRAM often suffer from misunderstood terms, overlooked fundamentals, and valuable insights scattered across too many sources. Although the FRAM is widely regarded as useful for a range of purposes, its key ideas can be buried under specialist jargon or locked away in research papers that assume a common, advanced background.

By drawing on a diverse set of expert contributors, I aimed to produce a resource that not only clarifies the FRAM's underlying principles and methods but also demonstrates its practical applications in a coherent and accessible manner. I believe the FRAM provides a crucial lens for understanding modern socio-technical systems, acknowledging the variability and complexity of everyday work, and offering a structured framework to model and appreciate these dynamics. In doing so, I hope this book will equip practitioners, researchers, and students with the understanding and confidence needed to apply the FRAM effectively in their own contexts with sufficient rigour.

Ultimately, my decision to edit this book came from a desire to make the FRAM more accessible, ensuring that its nuanced perspectives on complexity, performance variability, and resilience could reach a large audience.

My aim is to provide readers with a solid, methodically sound foundation and tangible examples in diverse contexts that they can rely on to apply the FRAM effectively and confidently, no matter what their domain or level of expertise.

WHY SHOULD YOU READ THIS BOOK?

This book offers more than an introduction to the FRAM. It provides a path to deeper understanding, enabling you to work on the method effectively. Here are the key benefits you can expect:

- Acquire timely expertise: This book equips you to understand and apply the basics of the FRAM—one of the most influential safety science methods, as it has been applied in various domains for diverse purposes and its recent adoptions in diverse contexts.
- Gain a balanced perspective: This book attempts to go beyond surface-level explanations, exploring the FRAM's principles, building steps, and practical implications. Guided by top scholars and seasoned practitioners, you will learn how and when to harness the FRAM's potential for obtaining meaningful results and what challenges you will most likely face.

- Draw inspiration from real-world applications: Through real-world examples and success stories, you will see the FRAM in action across different sectors and scenarios. On purpose, every chapter offers a detailed case study in a different domain. These practical insights are expected to guide you in adapting the FRAM to your unique environment, enhancing both system understanding and resilience.
- Bridge domain-specific expertise with cross-domain insights: The FRAM's true strength lies in its versatility. Nonetheless, every chapter attempts to follow a structured approach to encourage you to look beyond your own domain and learn from other fields. By discovering patterns and lessons drawn from diverse industries, you will expand your perspective and refine your strategies for tackling complex challenges in socio-technical modelling.

WHAT YOU WILL FIND IN THE BOOK

The book is organized into seven sections. It opens with a Prologue (which includes this chapter) to set the stage, followed by four core sections (Sections 1–4), each dedicated to one of the four building steps of the FRAM. These sections unfold as a compendium, intended as a single resource gathering the concepts needed to understand and apply the FRAM with its novel dimensions of analysis.

Next comes Section 5, which offers more general perspectives related to software and learning mechanisms within the FRAM framework. The book concludes with an Epilogue (Section 6), reflecting on the benefits and limitations of using the FRAM, essentially addressing the question, "To FRAM or not to FRAM?". A Glossary, curated by Patriarca and Hollnagel, closes the book.

OVERVIEW OF THE PROLOGUE

Besides this chapter, the Prologue provides an unofficial history of the FRAM as written by Prof. Hollnagel, the expert behind the method's conception. It then presents a brief yet formal description of FRAM principles and building steps, later referred to as the traditional steps. Finally, a scoping overview of the last five years of literature offers readers a status update on recent contributions in the field.

OVERVIEW OF SECTION 1

In Section 1, the focus is on identifying and describing the system's functions.

Hounsgaard (Chapter 5) opens this section by presenting their strategy for creating effective, simple models with a small number of functions and links through a hospital case.

Later, Steen and Norman (Chapter 6) show that Viable System Modelling can offer a practical lens through which interpreting FRAM functions could become easier, particularly when these functions are derived from diverse sources. Their work involves analyzing documents using Natural Language Processing related to an aviation accident.

Slater and Franca (Chapter 7) advance the discussion by leveraging the latest developments in generative AI to build models that can inform preliminary investigations of system properties. Their case study in the Oil and Gas industry highlights the critical role of the relationships with the operators in ensuring model significance.

Afterwards, De Nicola and Villani (Chapter 8) use AI to compare the semantic similarity of functions in models derived from different investigative approaches, specifically contrasting work-as-imagined with work-as-done.

OVERVIEW OF SECTION 2

In Section 2, about the identification of variability, Praetorius and Sharma (Chapter 9) start exploring how to interpret variability in the FRAM using qualitative data, applying their methodology to maritime pilotage.

In a more semi-quantitative approach, De Souza et al. (Chapter 10) introduce Fuzzy Set Theory to formalize linguistic assessments of variability into measurable indicators, demonstrated through a dam construction case.

Zuinga et al. (Chapter 11) further this exploration by employing Discrete Event Simulation as a quantitative method to describe variability for operations management.

OVERVIEW OF SECTION 3

Section 3, about the aggregation of variability, opens with Falegnami and Tomassi (Chapter 12), who offer a comprehensive investigation of such aggregation via Monte Carlo simulation, a method early discussed in literature for such a purpose, with a case study in dairy production, a domain where the FRAM is not yet widely applied.

Grabbe (Chapter 13) combines a similar Monte Carlo simulative approach with network assessment metrics in an automated driving scenario to clarify critical points.

Nomoto (Chapter 14) pushes the quantification of aggregated variability further by developing a dedicated FRAM-based Machine Learning algorithm to optimize hospital bed allocation.

Adriaensen and Zurawska (Chapter 15) follow a more qualitative approach, formalizing the study of aggregated variability through Interdependence Analysis, using the principles of Observability, Predictability, and Directability.

OVERVIEW OF SECTION 4

Section 4 is about the management of variability, which begins with Smith (Chapter 16), who introduces the concept of functional signatures to identify variability paths and define targeted interventions.

Saurin et al. (Chapter 17) integrate lean principles into the FRAM, applying their approach to waste management in a medical sterile unit.

Finally, Sujan and Lounsbury (Chapter 18) share a set of lessons learned from their extensive experience with the FRAM, providing practical guidance for analysts seeking to draw conclusions from their models.

OVERVIEW OF SECTION 5

Section 5 comprises two chapters. Rees Hill (Chapter 19) provides an overview of the traditional software used for FRAM analysis, the FRAM Model Visualizer (FMV), while Ferreira (Chapter 20) offers a critical reflection on the complete process of building a FRAM model.

OVERVIEW OF THE FINAL SECTIONS: EPILOGUE AND GLOSSARY

In the Epilogue (Section 6), Patriarca and Praetorius summarize the key concepts behind the applicability of the FRAM, offering guidance on when it is, or is not, a reasonable method to use.

The book concludes with a Glossary curated by Patriarca and Hollnagel, which covers all the core concepts discussed throughout the text and is intended to serve as a reference for future inquiries about the FRAM itself.

HOW TO READ THIS BOOK?

I envisioned three primary reader profiles, or approaches you may want to embrace to read this book, that is, the *curious reader*, the *domain expert, and* the *method geek*. Here is an overview of how the three profiles are conceived:

- *The curious reader*: If you are new to the FRAM, think of this as the most natural way to go through this book. You can simply read from start to finish, allowing the chapters to guide you from foundational principles towards more intricate aspects of the method. With each step, the complexity and depth of the FRAM are expected to become clearer. By the end, you will have a solid understanding of how the FRAM works and why it matters to get a solid basic of its fundamentals before starting graphing whatever number of hexagons. No special preparation is fundamentally required, especially if you focus only on the main text in each chapter.
- *The domain expert*: For readers who already possess deep knowledge of a particular field, each chapter will contain focused sections that highlight how the FRAM can be applied within specific domains. These "in-focus" boxes present detailed analyses, case studies, and lessons learned, all tailored to a specific professional case. By paying special attention to these highlighted parts, domain experts can discover how to integrate the FRAM's principles into their own operational contexts, making the method's implications more tangible and relevant.
- *The method geek*: If you already know the FRAM fundamentals and are eager to dive directly into the finer details - be it algorithms, qualitative observation protocols, or data analysis techniques - this book will accommodate you! Rather than following the standard chapter sequence, you are invited to dip into individual chapters and locate the specific methodological elements that interest you most. Specialized content for the "method geek" is clearly signposted, enabling you to focus on technical discussions without wading through introductory material.

Table 1.1 summarizes the various chapters and their intended focus on a specific methodological step, domain and technical aspect.

Through this strategy, I hope the book remains both accessible and technically rich. The main text offers a coherent narrative for the curious reader, while domain experts and method enthusiasts find additional layers of detail integrated.

Since my aim is to provide a resource that both introduces the FRAM to newcomers and expands the horizons of experienced practitioners, I hope this presentation might ultimately make the FRAM more applicable and more adaptable to a wide range of challenges in complex socio-technical systems.

I invite you to join us on this journey, to embrace new perspectives, and perhaps to bring greater resilience to the systems you help shape.

TABLE 1.1

Overview of the Book Chapters and Their Respective Focus

Book Section	Chapter, Authors	Core Method Focus	Domain Expert - Focus	Method Geek - Focus
Prologue	2—Hollnagel	Introduction to the FRAM	/	/
Prologue	3—Patriarca, Simone	Principles and building steps	/	Formal representation of each building step
Prologue	4—Falegnami, Tomassi	Literature review of the FRAM	/	Steps for the review process
1	5—Hounsgaard	Step 1	Hospital operations (management of pressure-relieving-mattresses)	Interview and debriefing guide
1	6—Steen, Norman	Step 1	Aviation operations (runway movements coordination)	Viable System Model, Natural Language Processing
1	7—Slater, Franca	Step 1	Drilling in an Oil and Gas facility	Artificial Intelligence for model generation
1	8—De Nicola, Villani	Step 1	Chemical experimental laboratory operations	Semantic similarity via Artificial Intelligence
2	9—Praetorius, Sharma	Step 2	Maritime pilotage	Hierarchical Task Analysis, Applied Cognitive Task Analysis
2	10—De Souza et al.	Step 2	Dam construction	Fuzzy Set Theory
2	11—Zuniga et al.	Step 2	Manufacturing	Discrete Event Simulation
3	12—Falegnami, Tomassi	Step 3	Dairy production	Monte Carlo simulation

(Continued)

TABLE 1.1 (Continued)
Overview of the Book Chapters and Their Respective Focus

Book Section	Chapter, Authors	Core Method Focus	Domain Expert - Focus	Method Geek - Focus
3	13—Grabbe	Step 3	Autonomous driving	Monte Carlo simulation, network metrics
3	14—Nomoto	Step 3	Hospital operations (bed allocation)	Machine learning
3	15—Adriaensen, Zurawska	Step 3	Industrial cobot; Automation in aviation	Interdependence Analysis
4	16—Smith	Step 4	Arctic shipping operations, home transition hospitalization	Functional signatures
4	17—Saurin et al.	Step 4	Waste management in medical sterile unit	Lean principles
4	18—Sujan, Lounsbury	Step 4	Post-surgical patient deterioration	Lessons learnt for managing variability
5	19—Hill	Software	/	Metadata
5	20—Ferreira	Implications for the method	/	Critical reflections on building a FRAM model
6	21—Patriarca, Praetorius	Epilogue	/	Critical reflections on FRAM usability

NOTE

1 Interestingly, the word "FRAM" in Norwegian translates to "forward," which beautifully reflects the method's inherent focus on looking ahead and proactively understanding complex systems.

REFERENCES

De Moivre, A. (1711). De mensura sortis. *Philosophical Transactions*, vol. 27, pp. 213–264.

Hollnagel, E. (2012). FRAM: The functional resonance analysis method: Modelling complex socio-technical systems. In *FRAM: The Functional Resonance Analysis Method: Modelling Complex Socio-technical Systems*. Ashgate.

Hollnagel, E., & Goteman, Ö. (2004). *The Functional Resonance Accident Model*. Cognitive Systems Engineering in Process Control.

Patriarca, R., Di Gravio, G., Woltjer, R., Costantino, F., Praetorius, G., Ferreira, P., & Hollnagel, E. (2020). Framing the FRAM: A literature review on the functional resonance analysis method. *Safety Science*, 129. https://doi.org/10.1016/j.ssci.2020.104827

2 A Brief and Unofficial History of the FRAM

Erik Hollnagel

I am asked every now and then about the origin of the FRAM, and here is an account to the best of my recollection, given advancing age and the temptations of hindsight (Fischhoff, 1975). Like so many other products of the human mind, the FRAM owes its existence to a persistent problem. In this case, the problem was the difficulty in accounting for the occurrence of disproportionate (non-linear) outcomes, which seemed insurmountable for conventional accident models that are all based on simple linear cause–effect reasoning and the four symmetries:

- Magnitude: large consequences cannot have small causes,
- Valence: negative outcomes must have negative causes,
- Time: events will take place in the same way in the future as they took place in the past, and
- Comprehensibility: complex outcomes have complex causes, and vice versa.

In the domino model, for instance, the force by which the last domino piece (injury) falls is the same magnitude as the force by which the first domino piece (social injury) falls, although the model does not even try to explain how the first piece falls!

The English philosopher George Henry Lewes had, fortunately, recognised the problem already in 1877 in the context of discussions following Darwin's theory of evolution and had proposed a distinction between resultant and emergent effects. Lewes juxtaposed the two types of effects and noted that the latter are neither additive nor predictable from knowledge of the system's components nor decomposable into such components. For Lewes, emergent effects were the outcomes of natural selection, as Darwin had originally proposed. Since an accident theory that can only account for the occurrence of linear and proportional outcomes of unexpected combinations or clusters of hazardous conditions is of limited value in the face of what we think actually happens, for instance, the "normal" accidents that Perrow (1984) had drawn attention to, it was therefore necessary to look for a solution based on an alternative accident theory (cf. Dekker et al., 2011). Epidemiological models such as the Swiss cheese model took a step in the right direction with their combination of active failures and weakened defenses. The most obvious example of a non-linear event is probably an avalanche, where the outcome is clearly out of proportion to

DOI: 10.1201/9781003518167-3

whatever the underlying cause may have been. For most avalanches, the triggering event is usually something of limited magnitude, such as a skier mistakenly entering an unstable zone or a small animal that suddenly jumped in the wrong direction and thereby moved into the zone that could start an avalanche. Avalanches may even happen spontaneously without any known or observable triggering event.

When I started to think about the FRAM, I referred to it in my own notes as a snowflake model of accidents.[1]

RESONANCE

An avalanche is a good example of a non-linear outcome, but so is resonance. A well-known example of resonance is the collapse of the Tacoma Narrows Bridge in 1940 (Gazzola et al., 2023). In this case, the resonance was due to a gentle wind blowing along the strait, which made the bridge oscillate and eventually collapse. The collapse was fortunately caught on film (today it would have been recorded by multiple smartphones). Resonance can be due to the effects of an external force that coincides with the natural frequency of the system, as when an adult pushes a child on a swing to increase the amplitude. But resonance may also happen for other reasons, for instance, as stochastic resonance (Gammaitoni et al., 1998), which caught my attention when it was described in a New Scientist article that I just happened to read. Yet to explain how adverse events (accidents) can happen, it is necessary to go beyond stochastic resonance. Here, the Millennium footbridge in London serendipitously provided an outstanding example. The Millennium Bridge, also known as the London Millennium Footbridge or, among Londoners, the wobbly bridge, is a horizontal suspension steel bridge for pedestrians crossing the River Thames, between St. Paul's Cathedral and the Tate Gallery. This is a very unorthodox construction: in contrast to a vertical suspension bridge, the deck of the Millennium Bridge could oscillate horizontally, due to lateral forces, rather than vertically, as we are used to and as the designers had imagined.

When many people cross a suspension bridge at the same time, either a group of wanderers/hikers or a contingent of soldiers, they take care not to walk in step in order not to introduce resonance. The bridge initially opened on June 10th, 2000. Unexpected lateral vibration due to resonant structural response caused the bridge to wobble so much that it was closed on June 12th and remained closed for almost 2 years while modifications were made. Due to unanticipated lateral forces when people walked across the bridge (the designers had only considered the vertical forces, which would have been adequate for a usual suspension bridge), it started to wobble. People instinctively respond by synchronising their steps with the lateral movements of the bridge. Videos clearly show how people spontaneously began to walk in step, just as they do when they walk on the deck of a ship rolling in the waves. But this synchronisation only amplified the lateral forces and, in this way, increased the amplitude of the oscillation. The phenomenon was clearly not stochastic but systematic and predictable and was for that reason named functional resonance. If it had been stochastic, it would be of limited value for safety-related efforts where the ability to predict is essential.

RECURSIVE GOALS–MEANS ANALYSIS

Besides the concept of resonance, the idea that a function in FRAM could have one or more preconditions called aspects came from earlier work I had done on a Goals Means Task Analysis (GMTA), described, for instance, in CREAM (see Hollnagel, 1998). A GMTA is essentially a recursive Goals–Means analysis principle, which was also the basis for the General Problem Solver (GPS) (Newell & Simon, 1963). The Goals–Means analysis can be illustrated by this simple example:

- Goal: get to the office in the morning.
- Means: get into your car and drive.

However, if your car does not start (it might have been a cold night), this creates a new goal, namely "To start the engine of the car," and there may be several means by which this can be achieved. If it is not possible to start the engine, then you could also consider an alternative solution or look for different means that will bring about the same goal, for instance, "to take a local train to the station close to your office." But this possibility leads to two new problems (that in turn become two new goals): first, how to get to your local train station, and second, how to get from your arrival station to the office. The GMTA is indeed recursive. And so is the FRAM, which can be defined as a recursive analysis method. The recursiveness allows the FRAM to begin the analysis with any function. But the FRAM uses six different aspects, rather than just preconditions, namely Input, Output, Precondition, Resource, Control, and Time (Hollnagel, 2012).

The GMTA defines the following check rules, which are also applicable to the FRAM:

- For each goal, except the top goal(s), which are called entry functions, there must be a higher-level or "parent" goal. For each goal, there must be a goal description, which in most cases corresponds to the Output of the function (because the function serves to establish the condition defined by the goal).
- For each goal, except the top goal (or entry function), there must be a precondition from which the goal is derived.
- For each task step, there must be a task step description. Task steps are defined for goals. The task steps basically describe what the functions do.
- For each precondition, there must be a precondition description. Preconditions are defined for task steps (functions) only.
- For each postcondition, there must be a postcondition description. Postconditions, including side effects, are defined for task steps.

Due to the near universality of the Goals–Means decomposition principle, it is found in other places, for example, in the Functional Analysis Systems Technique (FAST) (Bartolomei & Miller, 2001), which also was an inspiration for the FRAM.

Another source of inspiration for the names and the idea of multiple aspects came from the Structured Analysis and Design Technique (SADT) (Marca & McGowan, 1987).

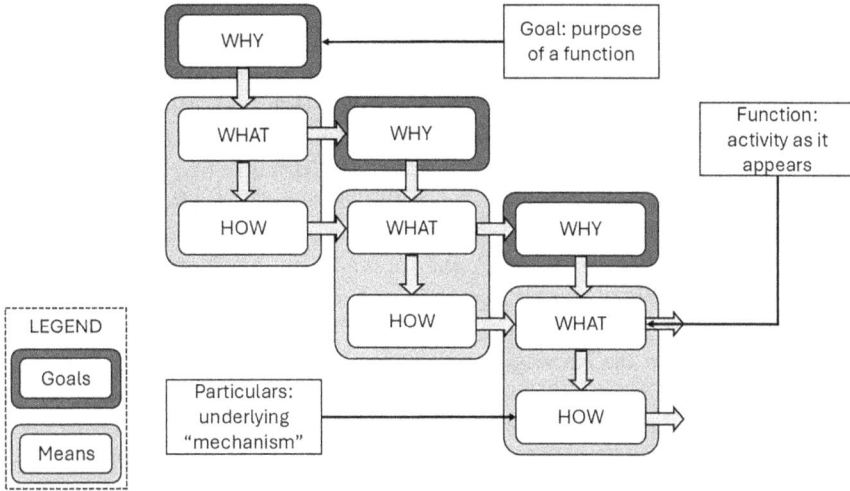

FIGURE 2.1   The recursive nature of Goals–Means decomposition.

Yet another contribution to the FRAM was the concept of couplings for the links between functions as proposed by Perrow (1984). And finally, the idea that any function, via its couplings, can affect other functions came from my interest in object-oriented programming and futile attempts to master the SMALLTALK programming language.

Recursiveness represents an important feature of both object-oriented programming and GMTA. The means, the WHAT of one level, becomes the goal, the WHY at the more detailed level. In the same way, the HOW and the WHAT exchange roles (see Figure 2.1). This can clearly be continued ad infinitum, and because of this recursiveness, it does not matter by which function an analysis using the FRAM begins. By applying the method consistently, all necessary functions will eventually be identified.

Even when briefly explained, as in the above, all the major pieces (or conceptual building blocks) of the FRAM were in place, and they all came together on a day in December 2003. I was attending the conference of the Australian Aviation Psychology Association in Darling Harbour in Sydney, where I sat in the audience waiting for a presentation entitled "Error 'molecules' and their implications for system safety." The author shall remain anonymous; suffice it to say that he was from NASA, Ames Research Center. I actually never heard the presentation, because as soon as I saw the title, I quietly left the auditorium. Instead, I slowly wandered down to a small outdoor cafe on Circular Quay in Sydney Harbor in front of the Danish architect Jørgen Utzon's opera house, where I had a rather nice ciabatta sandwich for lunch (as far as I recall), and still puzzled by the title of the presentation I had left, I started to think about ways to explain accidents as emergent outcomes without needing the existence of either error atoms or error molecules. In hindsight, I must have tried to combine the idea of non-linear effects or emergent outcomes with resonance as a possible "cause," in order to overcome the simple linear thinking about errors, to say nothing of error molecules or, even worse, error atoms, as causes. To be useful as a kind of theory of why non-linear outcomes could appear, it was clear that resonance had to be systematic, hence predictable in some way rather than stochastic. This led to the idea of functional resonance, for which the London Millennium Bridge mentioned above provided a perfect illustration. All that remained was to find a good acronym. I had thought of functional resonance, hence the first letters F and R, but was in those days before Resilience Engineering and Safety-II was still thinking in terms of accident models, hence the last letters A and M, which put together with FR became FRAM, an easily pronounceable acronym – then meaning Functional Resonance Accident Model and first described as such in Hollnagel (2004). Having lived in Norway, I also knew the Norwegian word "fram" that means "forward" (it has the same meaning in Swedish but alas not in Danish), and I also knew about the famous Norwegian Polar explorer Fridtjof Nansen (1861–1930) and the FRAM expedition (1898–1902). The vessel FRAM was designed by the Norwegian naval architect Colin Archer to withstand the enormous pressures of the Arctic ice as Nansen planned to reach the North Pole. It was Nansen's plan that FRAM should get stuck in the ice and then slowly drift over the North Pole, from Northern Siberia to Greenland. But the currents unfortunately did not move as he had expected, and the FRAM simply became stuck in the ice. After two winters onboard, Nansen decided instead to walk on skis to the North Pole. Decades later, the FRAM was also used by another Norwegian explorer, Roald Amundsen, in an expedition to the Antarctic, where he famously reached the South Pole in 1911, five weeks before the English explorer Robert Scott arrived. The FRAM is to date the only wooden explorer vessel that has been to both the Arctic and the Antarctic. It therefore seemed an appropriate name for a hopefully sturdy method.

Several years later, when the first small FRAM workshops were held in Sophia Antipolis, the discussions made it clear that the FRAM was actually a method and not a model, but the acronym was fortunately able to accommodate both interpretations. So, the FRAM now stands for the Functional Resonance Analysis Method. There was no epiphany or Heureka moment involved, just plenty of serendipity; in some sense, the FRAM, true to its nature, emerged out of many unsystematic readings and speculations.

TERMINOLOGY: MODEL OR METHOD

As described above, the fourth letter of the acronym FRAM at first meant model, but this was revised after a while to mean method. The revision may seem trifling, but it is actually significant. One consequence is that it is linguistically improper to refer to a FRAM model; the correct expression is "a model built by the FRAM" (or a FRAM-built model, to be less circumlocutions) because the M in FRAM acronym means Method and not Model. The FRAM is in this way different from other analysis methods, such as the Swiss cheese (Reason, Hollnagel, & Pariès, 2006) and the Bowtie (de Ruijter & Guldenmund, 2016), which both are models or analogies that purport to explain how accidents happen, as the domino model also did before them (Heinrich, 1931). All three rely on conspicuous and powerful physical analogies, and none of them are actually models of how accidents happen. The same goes for STAMP (System-Theoretic Accident Model and Processes) and the associated STPA (System Theoretic Process Analysis), where STAMP is unique by not referring to a model, despite the M in STAMP, unless Jens Rasmussen's abstraction hierarchy, that also is the basis for ACCiMap, can be considered a model, which is a questionable assumption (cf., Lind, 2003). ACCiMap does represent how certain structures are interrelated, in this case, so-called control structures, which merely are named boxes with connecting lines in a conventional flow chart format, hence by no means a physical analogy or a model as defined by Coombs et al. (1970).

The FRAM does, of course, not refer to a physical analogy either, but instead relies on four clearly articulated principles, simply listed below and described in the FRAM literature, as well as later in this book.

- The principle of equivalence of successes and failures
- The principle of approximate adjustments
- The principle of emergence
- The principle of functional resonance

Models are ubiquitous in the scientific literature as well as in many other places, for instance, to support political or financial decisions (Borner, 2021). Most researchers also find it irresistible to propose models as part of their work. But what is a model, actually? A model is essentially a simplified representation of a real-world phenomenon or a set of salient aspects of the real world, a set of physical or functional characteristics, or more formally: "The basic defining characteristics of all models are the representation of some aspects of the world by a more abstract system. In applying a model, the investigator identifies objects and relations in the world with some elements and relations in the formal system" (Coombs et al., 1970, p. 2).

The problem with models of accidents is that we do not know how accidents actually happen in the real world. We know, of course, about the outcomes of accidents, the harm and the injuries, but we do not actually know how an accident as such happens. This is possibly because causation is metaphysical while causes and effects are physical, hence observable, as David Hume argued centuries ago (Hume, 1739). Contrariwise, we can and do have models of physical phenomena such as superconductivity and turbulent diffusion of crystallisation and coagulation, and of both standing and freak waves, because we have strong theories to explain what happens and because we can actually experiment with the phenomena under controlled conditions and verify that the models and the hypotheses are correct. None of this is possible for models or hypotheses of accident causation, with the possible exception of the domino model, where you can actually put a set of domino pieces in a row and then push the first to watch the rest fall (there are elaborate examples of that available on the WWW). No one would dare to do a controlled experiment of the Swiss cheese accident hypothesis, except that you could buy a chunk of Emmental cheese (or another hard cheese with holes) and cut it into slices, but that would not be as convincing a demonstration as the domino pieces falling, and although it would be easier to dispose of the evidence afterwards. Instead, we have to wait for a suitable natural experiment to take place (Leatherdale, 2018). Carrying out realistic experiments with accidents would be impossible in a democratic country because it is highly unethical. What we traditionally call accident models are therefore no more than plausible hypotheses of how accidents happen, as diverse as the domino hypothesis (simple linear causality), the Swiss cheese hypothesis (an epidemiological analogy), and the bow-tie (simple linear thinking once more). The hypotheses and the so-called models are therefore just rough and easily understandable analogies that provide a convenient way of rendering the description of an event in terms of the concepts provided by the respective models/analogies, and the resulting accounts, that we call traditionally explanations, are necessarily as simple as the hypothetical accounts they are based on.

"Many factors contribute to incidents and disasters. Processes of causal attribution influence which of these many factors we focus on and identify as causal. Causal attribution depends on who we are communicating to, on the assumed contrast cases or causal background for that exchange, on the purposes of the inquiry, and on knowledge of the outcome" (Woods et al., 2010, p.33). Each model should therefore more properly be named a model-cum-method (a model with a method since the method is inseparable and indistinguishable from the model). In contrast to these approaches, the FRAM is a pure method, or a method-sine-model (a method without a model). Instead of being based on a simple analogy or hypothesis, the Functional Resonance Analysis Method relies on four clearly articulated principles listed above that provide a basis for analysing and representing how something has happened, happens or should happen, leading to what is commonly, but incorrectly, called a FRAM model. Of the four principles, the fourth principle of functional resonance comes closest to being a model or an analogy since it actually refers to the physical phenomenon of stochastic resonance (Gammaitoni et al., 1998).

THE FRAM IS A GENUINE SYSTEMS THEORETIC METHOD

The FRAM is a genuine systemic method. It describes a system as a set of mutually coupled functions, rather than as a physical or control structure composed of parts that can be analysed individually and later combined to prove an overall result.

- In the FRAM, such decomposition would prevent and distort the overall understanding of how the system functions as a whole.
- The FRAM, therefore, does not assume that functions operate independently. Indeed, the main purpose of a FRAM analysis is to account for the ways in which functions depend on each other and how such mutual dependencies can give rise to unexpected outcomes.
- The FRAM neither assumes that functions perform in the same way singly as when they play their part in the whole. The FRAM thus strongly disagrees with the substitution myth (Bradshaw et al., 2013).

- In a FRAM-built model, all functions are coupled (upstream and down-stream). A FRAM-built model, therefore, includes feedback loops and non-linear interactions as essential features. And the FRAM fully acknowledges that the systems being analysed and modelled are non-trivial as defined by the great cybernetian Heinz von Förster (von Förster & Pörksen, 2002).

NOTE

1 Indeed, a very (2012) early draft of a FRAM glossary from includes the following entry: "Snowflakes: 'Snowflakes' are a nickname for FRAM hexagons. The name is based on the similarity between the graphical representations. The idea of a non-linear development was also initially communicated by using an avalanche as an analogy. Today, the 'snowflake' reference is, however, considered obsolete and potentially misleading, and should therefore be avoided. Due to this analogy, the hexagon representing a snowflake became the icon for a FRAM function.

REFERENCES

Bartolomei, J. E., & Miller, T. (2001). Functional analysis systems technique (FAST) as a group knowledge elicitation method for model building. In: *Proceeding of The 19th International Conference of the System Dynamics Society*.

Borner, K. (2021). *Atlas of Forecasts: Modeling and Mapping Desirable Futures*. MIT Press.

Bradshaw, J. M. et al. (2013). The seven deadly myths of "autonomous systems". *IEEE Intelligent Systems*, 28(3), 54–61.

Coombs, C. H., Dawes, R. M., & Tversky, A. (1970). *Mathematical Psychology*. Prentice Hall, Inc.

de Ruijter, A. & Guldenmund, F. (2016). The bowtie method: A review. *Safety science*, 88, 211–218.

Dekker, S., Cilliers, P., Hofmeyr, J.-H. (2011). The complexity of failure: Implications of complexity theory for safety investigations. *Safety Science*, 49(6), 939–945.

Fischhoff, B. (1975). Hindsight $\neq$ Foresight: The effect of outcome knowledge on judgment under uncertainty. *Journal of Experimental Psychology, Human Perception and Performance* 1(3), 288–299.

Gammaitoni, L., et al. (1998). Stochastic resonance. *Reviews of Modern Physics*, 70(1), 223.

Gazzola, F. et al. (2023). A new detailed explanation of the Tacoma collapse and some optimization problems to improve the stability of suspension bridges. *Mathematics in Engineering*, 5(2), 1–35.

Heinrich, H. W. (1931). *Industrial Accident Prevention*. McGraw-Hill Insurance Series.

Hollnagel, E. (1998). *Cognitive Reliability and Error Analysis Method (CREAM)*. Elsevier Science Ltd.

Hollnagel, E. (2004). *Barriers and Accident Prevention*. Ashgate Publishing.

Hollnagel, E. (2012). *FRAM: The Functional Resonance Analysis Method: Modelling Complex Socio-technical Systems*. Ashgate Publishing.

Hume, D. (1739). *A Treatise of Human Nature*. Oxford University Press.

Leatherdale, S. (2018). Natural experiment methodology for research: a review of how different methods can support real-world research. *International Journal of Social Research Methodology*. 22. 1–17. https://doi.org/10.1080/13645579.2018.1488449

Lind, M. (2003). Making sense of the abstraction hierarchy in the power plant domain. *Cognition, Technology & Work*, 5, 67–81.

Marca, D. A., & McGowan, C. L. (1987). *SADT: Structured Analysis and Design Technique*. McGraw-Hill, Inc.

Newell, A. & Simon, H. A. (1963). GPS, a program that simulates human thought. In E. A. Feigenbaum & J. Feldman (Eds.), *Computers and Thought*. McGraw-Hill.

Perrow, C. (1984). *Normal Accidents: Living With High-Risk Technologies*. Princeton University Press.

Reason, J., Hollnagel, E., Paries, J. (2006). Revisiting the "Swiss Cheese" model of accidents. EEC Note No. 13/06, EUROCONTROL, Brussels (Belgium). Available at: https://www.eurocontrol.int/sites/default/files/library/017_Swiss_Cheese_Model.pdf

Von Förster, H., & Pörksen, B. (2002). *Understanding Systems: Conversations on Epistemology and Ethics*. Carl-Auer-Systeme

Woods, D. D. et al. (2010). *Behind Human Error* (2nd ed.). CRC Press.

3 The FRAM Essentials
Principles and Building Steps

Riccardo Patriarca and Francesco Simone

NAVIGATING UNCERTAINTY AND THE WEIGHT OF DECISION-MAKING

In a rural district just beyond the castle walls, barter looks an attractive practice especially for families who live far from the city markets. Yet, people have little to trade. After each harvest, the lord's officials arrive to collect tributes in grain and other goods, leaving little for the farmers to exchange.

Once the collectors depart, Aldus, a humble farmer, takes a bold step. He reveals a small jar of honey he managed to hide beneath the straw in his granary. In that moment, he has become an outlaw, and he is not alone. Nearby, another farmer has secretly kept a bag of wool from his flock, and he is now seeking a discrete buyer willing to trade.

It looks like a fair trade: the winter this year is much colder than anything Aldus can remember, endangering his young son. He decides to barter the honey. He prepares his libra, ready to exchange it for enough wool to keep his child warm. The farmer is accustomed to bartering, and he has measures on his side. He knows how to quantify weights, volumes, lengths, and even time. But a thought intrudes in his mind: How much wool will be enough for these icy nights? What weight, volume, or length would finally stop his son from shivering? Is the quantity of wool he can afford enough for the coming nights, or will he need to uproot his family and seek shelter elsewhere? What if the officials discover his hidden stores?

For the first time in his life, Aldus finds himself trying to weigh more than physical goods. He is attempting to weigh complexity. As he struggles to finalize the trade, Aldus realizes that his decision cannot be simplified to a single figure, and managing it requires a rather large mixture of factors: fabric thickness, body heat, the shelter of stone walls, the chill in the air, the frequency and magnitude of the imperial reprisals, and countless other intertwines factors. The simple objective of providing comfort for his son requires understanding the interplay of multiple elements, not just tallying up weights and volumes.

Something that, needless to say, it is quite complex to do. It demands a model that transcend raw numbers, one that acknowledges how various resources, environment, humans interact dynamically over time. Something that once acknowledged, does generates a discomfort you cannot simply let go.

DOI: 10.1201/9781003518167-4

In that moment, Aldus learns a lesson far more profound than any routine action could teach. If we rely solely on numbers, vital nuances slip away unseen. To find answers that satisfy, not only for trade and farming, but for all domains analyzed deeply in this book, we must appreciate the complexity and context from which mean-ingful outcomes emerge. Flattening problems to simple solutions will generate ridic-ulously biased outcomes. Only by recognizing and navigating the full richness of interdependent factors can one discover strategies that truly endure. Real progress comes from genuine understanding, achieved through the challenging exercise of grappling with every layer of complexity of a work process.

Beyond Aldus' metaphorical tale, the greatest challenge in managing socio-technical systems lies in recognizing that certain aspects of reality are simply too complex to be explained through straightforward cause-and-effect modelling. Aspects such as culture, situational judgements, trade-offs, experience, and policies cannot necessarily be reduced to measurements without losing the richness of their inner meaning. If reduced, their nuances could become obscure in a way that com-promises a meaningful decision-making process.

While this reduction is often necessary, what is the meaningful boundary to make it useful?

If you are approaching this book is mainly because you are aware that over the last two decades, safety science has moved gently (or, for some scholars, abruptly, look-ing at what Kuhn would call a "paradigm shift") beyond simple cause–effect reason-ing, embracing a resilience-oriented perspective to study both successful and failed outcomes. One should indeed focus on understanding a work process and modelling the activities that define it, allowing its success, as well as, occasionally, its failure. This focus is required to give visibility to the tensions that happen every day, and it should be at the basis for any thoughtful decision-making process.

The FRAM views system performance as emerging from dynamic, variable inter-actions among components. It promotes "functional resonance" as an emergent phe-nomenon that can intensify or diminish individual functional variations, guiding the system's behaviour in certain directions. Measuring these interactions and their influ-ence remains challenging, as acknowledged in the literature (Patriarca et al., 2020).

This introductory chapter summarizes the FRAM principles, connecting them to the traditional building steps of a FRAM model. The term "traditional" is here used to refer to the statements provided in the FRAM handbook by Hollnagel (2012). This distinction is of interest because the remainder of the book will see several different integrations and extensions of the same steps to solve detailed challenges faced by experienced FRAM analysts.

PRINCIPLES

Being the FRAM an approach rooted in resilience engineering, it focuses on under-standing the variability of performance and ultimately addresses the mechanisms that guarantee (or fail to) resilient performance. Within this context, a system can be con-sidered resilient if it can adjust its functioning prior to, during, or following events (changes, disturbances, and opportunities) and thereby sustain required operations under both expected and unexpected conditions (Hollnagel, 2006).

THE PRINCIPLE OF EQUIVALENCE OF SUCCESSES AND FAILURES

This first principle challenges the historical view that successes and failures have distinct and separate causes. A view where failures are seen as the outcome of a non-compliance event (mainly in one of the categories of technical failure or human error). The same view suggests that successful events (i.e., work in processes where there are no reported events) are only possible when everything works as imagined. However, a FRAM analyst knows that both successful and unsuccessful outcomes arise from the same underlying mechanisms, that is, adaptability in everyday work. The very same types of adaptations, under different circumstances, may contribute to an unintended and undesirable result, or they are at the basis for what sustains operations in the large majority of circumstances.

Acknowledging this equivalence helps move away from the seduction of finding a root cause, encouraging a neutralized approach. It shows that learning from how things go right is as important as learning from how they go wrong, giving rise to studying everyday performance. Note that we tend to avoid using the term "normal" work, because this could become a trap of deciding what is normal, eventually making small nuances unnoticed. In the FRAM, such subtle adaptations, nuanced fluctuations in performance, and interconnected changes are considered valid and essential units of analysis.

THE PRINCIPLE OF APPROXIMATE ADJUSTMENTS

This principle acknowledges that socio-technical systems are inherently intractable: conditions rarely match the tidy prescriptions found in procedures or plans. Faced with limited resources, time, information, manpower, and assets, individuals, teams, and organizations are continually forced to adapt their performance to guarantee systems operability. This capacity for manoeuvre displays the system's adaptability, and it is usually effective (systems do not fail on a daily basis!), yet it cannot be perfectly accurate. They are approximate solutions shaped by uncertainties, constraints, and conflicting goals. Such approximations ensure that tasks are completed in changing environments, but they inevitably introduce variability into the system's functioning. This performance variability is the same entity that guarantees both successful and unsuccessful outcomes.

Understanding this principle encourages analysts and decision-makers to focus on how people cope with complexity, rather than simply trying to enforce compliance with ideal procedures.

THE PRINCIPLE OF EMERGENCE

The principle of emergence highlights that performance variability from multiple interacting functions can combine in ways that are neither predictable nor directly traceable to individual elements. Minor deviations, each insignificant on its own, may align and amplify one another, resulting in disproportionately large consequences. This non-linear effect means that outcomes cannot be fully explained by referencing isolated parts of the system. While traditional cause-and-effect analyses often assume that failures stem from singular malfunctions or identifiable errors in specific components, the FRAM acknowledges that effective and ineffective outcomes emerge from the dynamic interplay of activities, resources, and operating conditions.

As such, this principle claims that decomposing the system and studying parts in isolation is of little benefit, rather recommending a system-wide investigation. Recognizing emergent phenomena shifts attention towards understanding how everyday variability coalesces, enabling a comprehensive approach to understanding complexity.

THE PRINCIPLE OF FUNCTIONAL RESONANCE

The principle of functional resonance explains how everyday performance variability may occasionally resonate and escalate into significant, system-wide impacts. Unlike straightforward cause–effect chains, resonance occurs when fluctuations in several functions reinforce one another, culminating in unexpectedly high levels of variability. This can lead to outcomes that exceed what any single variability could produce on its own. The analogy to mechanical resonance highlights that this phenomenon is dynamic, context-dependent, and sensitive to timing and interactions.

This principle encourages analysts to examine the complexity of interactions (i.e., couplings' variability) rather than searching for simple causes (i.e., non-compliant functions). It underscores the importance of modelling and understanding systemic patterns to understand what the relationships are that sustain daily operations and to anticipate and mitigate potential adverse outcomes.

BUILDING STEPS, IN A NUTSHELL

The following paragraphs detail each building step of the FRAM, enhanced through a formal analytical formulation. The formulation is intended as a complement to other ontological representations of the FRAM, such as the ones already available in literature (Lališ et al., 2019; De Nicola et al., 2023).

While an ontology is deemed necessary for clarifying FRAM concepts, the proposed analytical representation provides an unambiguous picture of what the FRAM steps entail, articulated in a systematic language. The result is a FRAM model—or more precisely, a FRAM-built model—that supports a more nuanced management of variability. This distinction is significant because the FRAM is described as a method-sine-model: it is not a model in itself but a method used to describe systems, limiting the bias of interpretation. This enables both prospective and retrospective analysis while minimizing the influence of modelling assumptions, as would occur with a model-cum-method. See Figure 3.1 for a simple FRAM model and the first of many presented in this book.

STEP 0: TO RECOGNIZE THE PURPOSE OF THE ANALYSIS

Every FRAM analysis starts with a precise definition of its purpose. Indeed, a system is a social construct that exists only with respect to some boundaries. This very first step is meant to set such an operating domain. The definition of the system's

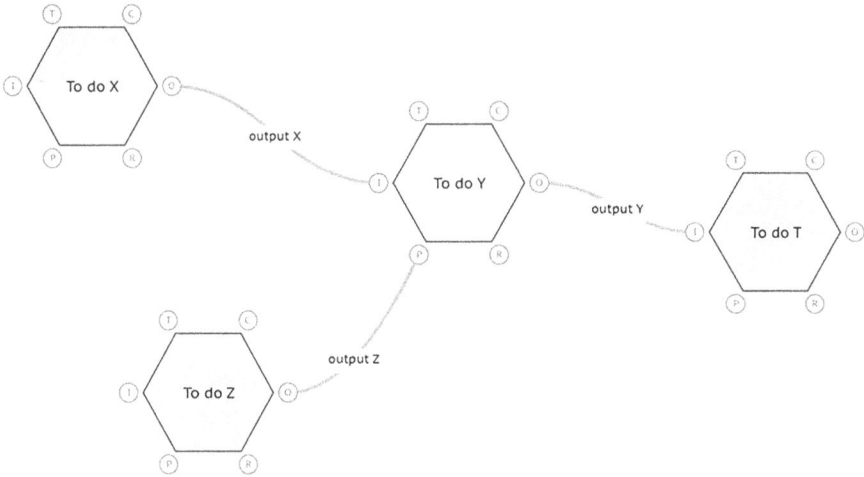

FIGURE 3.1 Exemplary FRAM model.

boundaries guides the following building steps, starting from the initial functions to be identified. As a result, at this stage, the analyst is aware of what could roughly be the scope of their FRAM model. The term "roughly" is consciously used here, as it happens frequently that the initial expectations are exceeded once getting closer to the process, thus pushing to reconsider the initial boundary.

Method Geek

To formalize this preliminary step, let U represent a utopistic complete (which cannot ever be complete!) model of a process to be studied. To render the analysis viable, we define a subset $S \subseteq U$ as the maximal domain within which all the FRAM analysis will occur. By establishing S, we impose a strict analytical boundary: no assertions, estimations, or conclusions will be drawn for any elements outside of S. In practice, the scope S is something that is relevant for the stakeholders of the analysis, and all outcomes are contingent on this pre-defined, limited domain.

STEP 1: TO IDENTIFY AND DESCRIBE THE SYSTEM'S FUNCTIONS

Once the scope is set, the first step of a FRAM analysis requires the actual identification of the system's functions and of their aspects. Noticeably, the type of each aspect can be one among Input, Output, Precondition, Control, Resource, or Time. While not all of them must be present at the same time for every function, it is always recommended to reflect on each possible richness for the analysis. An interested reader is invited to check in the glossary of this book for the nuances that support the labelling of items as one of the aspects.

This step corresponds to determining the **I** functions and the $\mathbf{J_i}$ aspects that are capable of defining the scope S meaningfully:

$$\mathbf{I} := \left\{F_1,\ldots,F_i,\ldots,F_I\right\} \quad ; \quad \mathbf{J_i} := \left\{a_{i,1},\ldots,a_{i,j},\ldots,a_{i,J}\right\} \tag{3.1}$$

where **I** is the set of all functions F_i to be potentially included in the FRAM model, and $\mathbf{J_i}$ is the set of all the aspects $a_{i,j}$ of such functions to be potentially included in the FRAM model.

$$F_i : F_i \in \mathbf{I} \tag{3.2}$$

where F_i is a function of the FRAM model (i.e., the i-th one), while **I** is the set of all model's functions. It is also true that:

$$a_{i,j}^k : a_{i,j}^k \in \mathbf{J_i} \quad ; \quad k \in \left\{I,O,P,C,R,T\right\} \tag{3.3}$$

where $a_{i,j}^k$ is an aspect (i.e., the j-th one) of type k of the i-th function, while $\mathbf{J_i}$ is the set of all the i-th function's aspects. The superscript k can indicate one among Input (i.e., I), Output (i.e., O), Precondition (i.e., P), Control (i.e., C), Resource (i.e., R), or Time (i.e., T). The definition of F_i and $a_{i,j}^k$ through **Eqs. (3.2–3.3)** concludes the first building step of FRAM.

STEP 2: TO IDENTIFY THE VARIABILITY

Subsequently, the method involves the characterization of variability for each identified function. Within the FRAM, three types of variability can be isolated:

- *Exogenous variability*. It refers to the variability that comes from the external environment of a specific function. It may be related to factors such as changes in the surrounding environment, changes in the regulations the system must comply with, or—more generally—whatever unexpected event the system could face while performing the function. The exogenous variability introduces unforeseen challenges—and opportunities—to the system's normal operating conditions, which may affect its vary. The good news is that an exogenous dimension of variability is the observable one at the function's Output.
- *Endogenous variability*. It refers to the variability arising within a function itself. It may be related to factors such as dynamics within the system executing the function, the individual's skills, technical specifications of the equipment involved, or, more generally, the way the function is executed. The endogenous variability reflects how the system operates during its normal conditions, that is, how its inherent characteristics influence its behaviours.

- *Cross-function variability*. It refers to the variability related to the function being executed before and after other functions. It highlights how dependencies and interactions between functions can let the system's performance variate. The cross-functional variability is the basis of functional resonance taking place within the socio-technical system operations.

On this basis, every aspect of type Output implies the existence of an exogenous variability, and its provision is affected by its generating function and the upstream functions connected to its generating function.

Method Geek

Formally, the exogenous variability is formalized as follows:

$$\forall a^o_{i,j} \in \mathbf{J_i} \ \exists \ V^{exo}_{a^o_{i,j}} = \mathbb{F}\left(V^{end}_{a^o_{i,j}}, V^{cross}_{F_i, a^o_{h,j'}} \right) \tag{3.4}$$

where $V^{exo}_{a^o_{i,j}}$ is the exogenous variability related to the j-th aspect of type Output of the i-th function of the model, $\mathbb{F}$ represent a morphism (either logical or analytical) between the endogenous and the cross-function variability, $V^{end}_{a^o_{i,j}}$ is the endogenous variability of the function F_i with respect to its j-th aspect, and $V^{cross}_{F_i, a^o_{h,j'}}$ is the cross-function variability induced on F_i by the j'-th aspect of type Output of the h-th function of the model F_h.

Thus, the definition of the exogenous variability involves and depends on the definition of both the endogenous and the cross-function variabilities. This latter represents a propagation of variability among upstream and downstream functions, and it is discussed in Step 3, where functions' variabilities are aggregated. The endogenous variability, instead, is characteristic of the very single function under analysis.

Method Geek

To this end, it is formalized as follows:

$$V^{end}_{F_i, j} = \mathbb{G}\left(PI_l \right) \ ; \ l \in \mathbf{L}_{a^o_{i,j}} \tag{3.5}$$

where PI_l is a performance indicator to be related to the j-th Output of the i-th function of the FRAM model, $\mathbf{L}_{a^o_{i,j}}$ is the set of all performance indicators related to the aspect $a^o_{i,j}$, and $\mathbb{G}$ is a morphism (either logical or analytical) between these indicators. A single Output can be described by more than one performance indicator, but its endogenous variability $V^{end}_{a^o_{i,j}}$ considers them all simultaneously.

Here lies the first spot where performance indicators are involved in the FRAM: the way in which the Output is generated by the function can be referenced by one (or a set of) indicator(s), which can be either quantitative or qualitative.

The formalization of the characterization of the endogenous variability can be extended via additional considerations depending on the nature of the performance indicators being included in **Eq. (3.5)**. Specifically, for each i-th function:

$$\forall F_i \in \mathbf{I}; a^o_{i,j} \in \mathbf{J}_i; l \in \mathbf{L}_{a^o_{i,j}} \quad \exists \quad PI_l := PI_l^* \lor PI_l := PI_l' \tag{3.6}$$

where PI_l^* denotes a meaningful and measurable performance indicator within $\mathbf{L}_{a^o_{i,j}}$, and PI_l' denotes a meaningful and non-measurable performance indicator within $\mathbf{L}_{a^o_{i,j}}$. Without forcing the analysis towards quantitative or qualitative reasoning, for a determined l, the corresponding performance indicator could be quantifiable or not, exclusively.

Then if:

$$\forall l \in \mathbf{L}_{a^o_{i,j}} \quad \exists \quad PI_l := PI_l^* \tag{3.7}$$

where every performance indicator, being identified to characterize the endogenous variability of the j-th output of the i-th function, is meaningful and measurable. Thus:

$$\mathbb{G} := \mathbb{G}^* \tag{3.8}$$

Detailing the possibility for a completely quantitative assessment of the endogenous variability of the function with respect to the output.

If instead:

$$\forall l \in \mathbf{L}_{a^o_{i,j}} \quad \exists \quad PI_l := PI_l' \tag{3.9}$$

where every performance indicator, being identified to characterize the endogenous variability of the j-th output of the i-th function, is meaningful and non-measurable. Thus:

$$\mathbb{G} := \mathbb{G}' \tag{3.10}$$

Detailing the need for a completely qualitative assessment of the endogenous variability of the function with respect to the output.

STEP 3: TO AGGREGATE THE VARIABILITY

Variability is not an end in itself but rather dependent on its other manifestations within the FRAM model, as they verify in a selected instantiation. The exogenous variability of a function is the result of its endogenous variability and the cross-function variability affecting the provision of its Output. To this extent, an instantiation that locks specific couplings allows modelling cross-functional variability. Indeed, in an instantiation, Outputs become inputs for other functions in a determined way. Variability can thus propagate when the Output of a function depends on the Output of another function. This propagation can acquire various forms depending on the nature of the function and its nested parameters, that is, analytically deterministic, Bayesian stochastic, and rule-based.

Method Geek

So, let F_h be a function of the model, such as at least one of its Output $a_{h,j'}^o$ equals an aspect $a_{i,j}^k$ of type Input, Precondition, Control, Resource, or Time in the function F_i:

$$F_h, F_i \in \mathbf{I} : a_{h,j'}^o = a_{i,j}^k \quad ; \quad k \in \mathbf{K} := \{I, P, C, R, T\} \tag{3.11}$$

The cross-function variability $V_{F_i, a_{h,j'}^o}^{cross}$ is formalized as follows:

$$V_{F_i, a_{h,j'}^o}^{cross} = \mathbb{H}(PI_m) \quad ; \quad m \in \mathbf{M}_{a_{h,j'}^o} \wedge m : PI_m = PI_l \tag{3.12}$$

where PI_m is a performance indicator to be related to the j'-th Output of the h-th function of the FRAM model, $\mathbf{M}_{a_{h,j'}^o}$ is the set of all performance indicators related to the aspect $a_{h,j'}^o$ and influencing the performance indicators PI_l of the i-th function, and $\mathbb{H}$ is a morphism (either logical or analytical) between these indicators. Please note again that the same Output can be described by more than one performance indicator influencing a performance indicator PI_l of the downstream function F_i, but its cross-function variability $V_{F_i, a_{h,j'}^o}^{cross}$ considers them all, simultaneously.

The way in which a downstream function is affected by the Output of an upstream function can be referenced by one (or a set of) indicator(s) to describe the upstream function to be related to one (or a set of) indicator(s) to describe the behaviours of the downstream function. Such indicators can be either quantitative or qualitative, but they must, or forced to, be relatable.

Method Geek

The formalization of the cross-function variability could be further detailed: considerations could be made once more on the nature of the performance indicators being included in **Eq. (3.12)**. Specifically, for each i-th and h-th function being related:

$$\forall F_i, F_h \in \mathbf{I}; a_{i,j}^k, a_{h,j'}^o \in \mathbf{J_i} ; l \in \mathbf{L}_{a_{i,j}^o} ; m \in \mathbf{M}_{a_{h,j'}^o} \ \exists \ PI_m := PI_m^* \vee PI_m := PI_m' \tag{3.13}$$

where PI_m^* denotes a meaningful and measurable performance indicator within $\mathbf{M}_{a_{i,j}^o}$ to be related to a meaningful and measurable performance indicator PI_l^* within $\mathbf{L}_{a_{i,j}^o}$, and PI_m' denotes a meaningful and non-measurable performance indicator within $\mathbf{M}_{a_{i,j}^o}$ to be related to a meaningful and non-measurable performance indicator PI_l' within $\mathbf{L}_{a_{i,j}^o}$. Again, to avoid forcing the analysis towards quantitative or qualitative reasoning, for a determined m (and a corresponding l), the performance indicator could be quantifiable or not, exclusively.

Then if **Eq. (3.7)** holds true and:

$$\forall m \in \mathbf{M}_{a_{h,j}^o} \quad \exists \ PI_m := PI_m^* \tag{3.14}$$

where every performance indicator is identified to characterize the endogenous and the cross-function variabilities, and, in turn, the exogenous variability, is meaningful and measurable. Thus, the tree morphisms for the characterization of variability become:

$$\{\mathbb{F}, \mathbb{G}, \mathbb{H}\} := \{\mathbb{F}^*, \mathbb{G}^*, \mathbb{H}^*\} \tag{3.15}$$

Depicting the possibility for a completely quantitative assessment of the variability being modelled in the FRAM.

If instead **Eq. (3.9)** holds true and:

$$\forall m \in \mathbf{M}_{a_{h,j}^o} \quad \exists \ PI_m := PI_m' \tag{3.16}$$

where every performance indicator, being identified to characterize the endogenous and the cross-function variabilities, and, subsequently, the exogenous variability, is meaningful and non-measurable. Thus, the tree morphisms for the characterization of variability become:

$$\{\mathbb{F}, \mathbb{G}, \mathbb{H}\} := \{\mathbb{F}', \mathbb{G}', \mathbb{H}'\} \tag{3.17}$$

Stressing the need for a purely qualitative assessment of the variability being modelled in the FRAM.

The existence of $\{\mathbb{F}^*, \mathbb{G}^*, \mathbb{H}^*\}$ and $\{\mathbb{F}', \mathbb{G}', \mathbb{H}'\}$ is deemed exclusive because of the impossibility of adhering to **Eq. (3.13)**. There is no way to combine numerical (e.g., PI_m^*) and non-numerical (e.g., PI_l') indicators without any loss of generality. Accordingly, a choice must be made at this stage: to force the evaluation of variability through $\{\mathbb{F}^*, \mathbb{G}^*, \mathbb{H}^*\}$ or, conversely, through $\{\mathbb{F}', \mathbb{G}', \mathbb{H}'\}$. Please note how this latter sentence translates in the transformation of quantitative performance indicators into qualitative ones, and vice versa.

The aggregation of variability could be performed at the model level, too. By "model level," we mean the possibility to generate an aggregated assessment that encompasses multiple instantiations at a time. Indeed, the previous lines provided indications to aggregate variability within and among functions, but took for granted the existence of causal and temporal relationships. Such relationships do not exist in an FRAM model but only in a specific instantiation at a time.

The endogenous, the cross-function, and—consequently—the exogenous variability of a function cannot be statically defined but in a specific instantiation. Accordingly, the variability of a function shall be evaluated for each instantiation of interest. Then, instantiation-dependent variabilities shall be evaluated aggregately to reckon the variability of the function at the model level.

Method Geek

To this end, each s-th instantiation of the model defines:

$$\forall s \in S \; ; F_i, \in I \; ; a_{i,j}^k \in J_i \; ; k \in K \; \exists \; PI_1, PI_m, \{\mathbb{F}, \mathbb{G}, \mathbb{H}\} \tag{3.18}$$

where S is the set of all models' instantiations.

The adherence to rather than Eqs. (3.15) or (3.17) ensures the identification of the exogenous variability of each j-th Output of the i-th function of the model being considered in the s-th instantiation, namely $V_{a_{i,j}^o,s}^{exo}$. Each function could be thus characterized by more than one $V_{a_{i,j}^o,s}^{exo}$ depending on the s-th instantiation being analyzed.

The variability of the i-th function at the model level (envelope of multiple instantiations) is formalized as follows:

$$\mathbb{V}_{i,j} = v_{i,j}\left(V_{a_{i,j}^o,s}^{exo}\right) \; ; \; s \in S \tag{3.19}$$

where $v_{i,j}$ represents an envelope rule for the j-th Output of the i-th function to aggregate each $V_{a_{i,j}^o,s}^{exo}$ among different instantiations, and $\mathbb{V}_{i,j}$ is a morphism (either logical or analytical) to characterize the variability at the model level of the j-th Output of the i-th function.

Please note how the envelope rule. $v_{i,j}$ refers to $V_{a_{i,j}^o,s}^{exo}$, in which the performance indicators have already been aggregated by means of the $\{\mathbb{F}, \mathbb{G}, \mathbb{H}\}$ morphisms. Thus, there is no relation to be developed among different performance indicators, but $v_{i,j}$ rather refers to a pre-determined (set of) performance indicator(s) that changes (either quantitatively or qualitatively) over the FRAM model's instantiations.

STEP 4: TO MANAGE THE VARIABILITY

No one here should enter the merit on how good we are at doing variability (cf. principles of equivalence of successes and failures). Indeed, the FRAM involves the understanding of when and how such variability emerges, understanding system behaviours holistically.

In mechanics, resonance refers to the phenomenon that occurs when a (technological) system is driven at its natural internal frequency. When moving at this frequency, the system absorbs energy more efficiently, and resonance leads to oscillations of larger and larger amplitude. On this premise, every mechanical system has an inborn natural resonant frequency at which it tends to vibrate when disturbed. In mechanics, this frequency depends on the system's properties, like its mass, its stiffness, or its shape. An infamous example is the breakdown of the Tacoma Narrows Bridge: the frequency of wind gusts matched the natural resonant frequency of the bridge, causing large oscillations that led to its collapse. Accordingly, pure mechanical resonance occurs exactly when the natural resonant frequency of a system matches the external frequency induced by another system, ultimately leading to uncontrollable behaviours.

However, in real mechanical systems, a dampening factor is present (e.g., friction or resistance), and energy is somehow always dissipated. The dampening factor reduces the amplitude of oscillations over time, preventing the unbounded increase favoured by the resonance: there is hope! When the decrease prevails over the increase, such a situation is in its extreme. Accordingly, in an opposite way to resonance, antiresonance refers to the phenomenon where a mechanical system exhibits vibrations of very small, or even zero, amplitude when subjected to an external driving force applied at a particular frequency (i.e., the system's natural antiresonant frequency). In other words, when a system is driven at its natural antiresonant frequency, the antiresonance leads to oscillations of smaller and smaller amplitude, minimizing or even cancelling them.

Method Geek

To be formal, resonance verifies when:

$$\omega_0^+ = \omega_{ext} \tag{3.20}$$

where ω_0^+ is the natural (internal) resonant frequency of the mechanical system, and ω_{ext} is the frequency at which the mechanical system is driven by an external system.

And antiresonance verifies when:

$$\omega_0^- = \omega_{ext} \tag{3.21}$$

where ω_0^- is the natural (internal) antiresonant frequency of the mechanical system, and ω_{ext} is the frequency at which the mechanical system is driven by an external system.

However, functional resonance includes both the resonance and the antiresonance notions, and it conceptually extends them beyond pure mechanics. In the FRAM, systems—whether mechanical or not—are modelled through a set of functions that relate in such a way that their variability resonates (i.e., it is amplified) or antiresonates (i.e., it is dampened). However, while such a relationship is quite straightforward for mechanical systems, it is not easily generalizable to any system. To this end, three issues arise.

The first one regards the definition of an equivalent for frequency. What is the frequency of an FRAM function? Unluckily, there is no easy answer. Nevertheless, reconnecting to the ideas presented above, the frequency of a function surely depends on the aggregated variability of its Outputs, which is in turn evaluated through the performance indicators related to the variability of its Output in different model instantiations. While the term "frequency" is juxtaposed to quantitative performance indicators with ease, it is not true for qualitative ones. Accordingly, and more generally, the frequencies of a function can be thought of as patterns of interactions and the function's states in various instantiations.

Method Geek

To formalize this definition, the frequency of the j-th aspect of type Output of the i-th function is:

$$\omega_{i,j} := \mathbb{V}_{i,j} \to \mathbb{V}_{i,j} \tag{3.22}$$

namely, a morphism (either logical or analytical) from $\mathbb{V}_{i,j}$ to $\mathbb{V}_{i,j}$ itself.

The second issue refers to recognizing whether a function's "frequency" is internal or external. Consider a FRAM model made up of only two functions. The patterns of interactions and the function's states of the first function represent its internal frequencies. On the contrary, the patterns of the second function are the external frequencies of the first one. Such reasoning applies in reverse, too. Now, let's generalize this approach to a FRAM model made up of more than two functions. The patterns of interactions and function states of a specific function represent its internal frequencies, while the patterns of interactions and function states of all the other functions in the FRAM model represent its external frequencies.

When such frequencies generate functional resonance (both resonance and antiresonance), they correspond to natural frequencies (both resonant and antiresonant). To this end, the third issue regards the identification, among all their frequencies, of the FRAM functions' natural frequencies. Indeed, not all patterns are natural frequencies: only some of them have the potential to generate functional resonance. While in traditional mechanics this step is trivial, and there are well-established means (equations) to retrieve the natural frequency of a system, this is not true for a socio-technical model obtained via the FRAM.

The identification of natural frequency is dependent on external frequencies. Functional resonance exists (i.e., manifests) because of the complexity arising from the interactions of functions, rather than being present a priori and taken for granted (as for mechanical resonance). The suggestion here is to evaluate each internal frequency of a function with all the external frequencies of all other functions in the model, or, to some extent, to those that pertain to the selected function.

Method Geek

In other words, the frequency of the j-th Output aspect of the i-th function is a natural frequency when there exists a morphism between the aggregated variability of the j-th aspect of type Output of the i-th function and the aggregated variability of a generic j'-th aspect of a generic h-th function:

$$\omega_{0i,j} := \omega_{i,j} \quad ; \quad \omega_{i,j} : \exists \mathbb{V}_{h,j'} \to \mathbb{V}_{i,j} \tag{3.23}$$

where $\omega_{0i,j}$ denotes a natural frequency of the j-th aspect of type Output of the i-th function.

Such evaluation leads to the identification of natural frequencies, ultimately supporting the management of the system's variability by understanding how functions relate for the generation of functional resonance.

CONCLUDING REMARKS

Managing socio-technical systems requires moving from the comfort of straightforward metrics to embracing the intricate dynamics of socio-technical systems. Aldus' decisions at the beginning of this chapter underscore the importance of understanding how performance variability, human judgement, and contextual factors interact to shape outcomes. These insights remind us that meaningful progress does not arise from oversimplifying complexity but from engaging with it thoughtfully and holistically.

By integrating both qualitative insights and quantitative measures, we can model systems that reflect their true dynamic, adaptive, and interdependent nature. This chapter offered a summary of the essentials at the core of the FRAM and attempted to derive the foundational principles in an analytical manner which can guide diverse integration and extensions of the method itself (cf., e.g., Patriarca et al., 2025). An interested reader may also find it valuable to use the same formalization to reproduce other existing approaches used to extend the FRAM (e.g., the Abstraction Agency, Monte Carlo simulation, and functional signatures). Some of these perspectives will be explored from diverse perspectives and facets in the remainder of the book to ultimately understand their true potential.

REFERENCES

De Nicola, A., Villani, M.L., Sujan, M., Watt, J., Costantino, F., Falegnami, A., & Patriarca, R. (2023). Development and measurement of a resilience indicator for cyber-socio-technical systems: The allostatic load, *Journal of Industrial Information Integration*, vol. 35, 2023, https://doi.org/10.1016/j.jii.2023.100489

Hollnagel, E. (2006). Prologue: Resilience engineering concepts. In: *Resilience engineering: Concepts and precepts*. Ashgate Publishing, edited by: Woods, D. D., & Leveson, N. https://doi.org/10.1136/qshc.2006.018390

Hollnagel, E. (2012). *FRAM: The Functional Resonance Analysis Method: Modelling Complex Socio-technical Systems*. Ashgate Publishing.

Lališ, A., Patriarca, R., Ahmad, J., Di Gravio, G., & Kostov, B. (2019). Functional modeling in safety by means of foundational ontologies, *Transportation Research Procedia*, vol. 43, pp. 290–299. https://doi.org/10.1016/j.trpro.2019.12.044

Patriarca, R., Di Gravio, G., Woltjer, R., Costantino, F., Praetorius, G., Ferreira, P., & Hollnagel, E. (2020). Framing the FRAM: A literature review on the functional resonance analysis method, *Safety Science*, 129. https://doi.org/10.1016/j.ssci.2020.104827

Patriarca R., Lovaglio L., & Simone F. (2025). Functional resonance analysis via a genetic algorithm to ensure cost-effective maintenance planning, *International Journal of Production Economics*, 281, 109516. https://doi.org/10.1016/j.ijpe.2025.109516

4 The Last Five Years (2019–2024) of FRAM Literature

Andrea Falegnami and Andrea Tomassi

CONTEXT: SETTING THE STAGE FOR THE REVIEW

Conducting a review on the Functional Resonance Analysis Method (FRAM) requires initially distinguishing relevant documents from those associated with completely unrelated terms yet identical acronyms (e.g., Fram sea strait or Ferromagnetic Random-Access Memories). In this endeavour, the published research "Framing the FRAM" by Patriarca et al. (2020) is used here as a fundamental work. Such a review not only catalogued emerging concepts up to 2019 but also mapped the network of scientists engaging with these ideas. The co-citation network analysis within that study highlighted the close connections within the FRAM community, often revealing less than five degrees of separation among researchers, including those less actively involved. This interconnectedness underscores the collaborative nature of the FRAM research community and its commitment to advancing the field. In addition, "Framing the FRAM" extensively examined the FRAM's application across various fields, analyzing over 1,700 documents via a structured protocol to highlight its evolutionary traits and increasing academic recognition. The review differentiated between retrospective analyses of past incidents and prospective modelling for risk management, noting both qualitative and quantitative advancements in the FRAM. This chapter seeks to provide up-to-date insights on a literature analysis of FRAM-related literature, continuing the rigorous examination set by "Framing the FRAM", and complementing it with a thematic overview. An interested reader is indeed invited to go first through such a journal article, which is available as open access, and come back to this chapter for an up-to-date narrative to actually navigate the current FRAM panorama.

METHODOLOGY

Method Geek

This chapter iterates the search strategy and the analyses originally undertaken in "Framing the FRAM" and strives to adopt the structured and systematic process as the leading methodology proposed by Moher et al. (2009), that is, PRISMA. The PRISMA protocol, an acronym for Preferred Reporting Items for Systematic Reviews and Meta-Analyses, offers a comprehensive framework to systematically review and synthesize findings from numerous studies on a particular topic.

DOI: 10.1201/9781003518167-5

PRISMA provides a checklist that guides researchers through this process, covering essential aspects such as defining precise objectives, detailing their search strategies for relevant studies, and explaining the criteria for selecting and analyzing the data. The PRISMA protocol thus functions as a quality assurance tool for literature research review, ensuring that the conclusions drawn are both thorough and reliable.

The PRISMA protocol execution has been paired with the data analytics platform KNIME (Berthold et al., 2009). KNIME (Konstanz Information Miner) is a powerful open-source platform designed for data analytics, reporting, and integration. KNIME users can visually assemble workflows by dragging and dropping nodes, each representing a different data processing step. The platform supports a wide array of data manipulation and analysis techniques, from basic data cleaning and transformation to complex machine learning and predictive analytics. Users can seamlessly integrate data from diverse sources, perform advanced statistical analysis, and visualize results using an extensive library of built-in tools and third-party extensions. One of the key strengths of KNIME lies in its extensibility, allowing for user customization and capabilities extension through scripting in languages like Python, R, and Java. With its robust, scalable architecture, KNIME can handle large datasets and complex computations, making it a versatile solution for data analysis.

Mimicking the original "Framing the FRAM", every document in the English language mentioning "Functional Resonance Analysis Method" or "Functional Resonance Accident Model" in the title, abstract, or keywords was included for a first scan. Especially the addition of the second acronym (Accident Model) was provided only to ensure a wider literature review, yet it should not have been necessary considering the evolution of the terminology already discussed in the previous chapters of this book. As for the original research, the search spanned multiple scientific repositories, including Scopus, Web of Science, EBSCOhost, and Google Scholar, ensuring a broad and inclusive dataset. The query was launched on April 27th, 2024, and included any documents from October 2019 to April 2024. The total number of documents resulting from the databases' queries was 2771 (Scopus: 250; Web of Science: 160; EBSCOhost: 141; Google Scholar: 2220). For the purposes of coverage comparisons, only exact duplicates within each database had to be removed, leaving overlaps and items discarded for other reasons untouched by the join operations.

Method Geek

The Scopus database is often deemed a reliable and comprehensive source of articles, which includes over 76 million records from more than 40,000 publishers worldwide. Estimates indicate that Scopus's documents' coverage exceeds 86% (Tomassi, Falegnami, & Romano, 2024).

The first outcome of this exercise is to confirm that the FRAM-specific research field coverage and relevance are well-established in Scopus, indicating that the amount of potentially missing research is minimal. Comparatively, while EBSCOhost offers some complement to Scopus, Web of Science (WoS) might be disregarded in FRAM literature overviews since its documents are almost fully included in both Scopus and EBSCOhost. This could be due to its slower indexing pace, and such discrepancy is particularly notable since our search query targeted very recent publications (2019–2024). On the other hand, Google Scholar returns a significantly higher number of (grey) documents. However, a range of sources, including dissertations, HTML articles, white papers, and articles in unclear peer-reviewed journals and magazines, are often not indexed in standardized institutional databases. Despite the high volume, many obtained sources only mentioned the FRAM, without any real empirical or theoretical work on it. Accordingly, the information retrieved from Google Scholar is shown to be generally poor for the purpose of this work, especially limitedly suitable for bibliometric purposes. To verify this claim, a limited set of bibliometric information (e.g., authors, document title, abstract, DOI, document type, source title, and year of publication; these seven variables were chosen because they are theoretically available from all databases, and because they form the basis of any structured literature review) was selected from all data provided by the four databases and the number of missing data per document and per database was then counted (e.g., while WoS and Scopus consistently reported all seven essential variables, EBSCOh provided only partial data in the majority of cases). WoS demonstrated 100% accuracy, while Scopus exhibited 94%, and EBSCOh demonstrated 54%. Scholar exhibited the lowest level of completeness, reporting only 51 results – ~2% of the total – with the full seven pieces of information. In 1780 cases, the DOI was absent; in 479 cases, the year of publication was missing; and in 1429 cases, the document type was not reported or not applicable as data. We have thus verified that, for a literature review on the FRAM, Scopus, EBSCOh, and WoS are reliable sources of information. Omissions of information are rather rare. Google Scholar, on the other hand, shows the opposite behaviour, thus confirming our initial feeling.

At this stage, the 2771 documents resulting from the queries were imported into KNIME for processing, where tasks such as merging the databases, removing duplicates (562 documents), nonpertinent items (1129) and documents not written in English (846) were visually inspected to ensure consistency and reduce the potential for automation error. Finally, 234 documents were included in the subsequent analyses.

The research team complemented the analysis of the contribution typology with another documentary exploration method. This methodology has been applied in various scholarly projects (Garito et al., 2023; Tomassi, Caforio, et al., 2024), yet it marks its debut in extracting FRAM-related themes from a document collection. More specifically, the documents in the corpus were treated as nodes in a specific bipartite graph. In addition to *document nodes*, such a graph also has *term nodes*. If a document contains a particular word, the corresponding nodes are connected by a link: the more frequently a word occurs in a document, the stronger the link.

Once the Document-Terms matrix, representing raw term frequencies within documents, is transformed into a TF-IDF matrix, the methodology of the systematic review can be further improved.

Method Geek

TF-IDF, an acronym for Term Frequency-Inverse Document Frequency, is a statistical measure that facilitates the comprehension and quantification of the significance of words within a corpus of documents. It assigns a higher weight to words that are exclusive to a specific document and a lower weight to words that are ubiquitous across all documents in a given corpus. The concept is of central importance in the fields of information retrieval and text mining, as well as in clustering and document classification. In a TF-IDF matrix, each row represents a document, each column represents a term (or word), and each cell contains the TF-IDF score of a word in a specific document. The value of each cell is comprised of two elements: the term frequency (TF) and the inverse document frequency (IDF). The initial component, TF, represents the frequency with which a particular word (or term) is encountered within a given document. It is typically normalized by dividing by the total number of words in the document, thereby accounting for the variable length of documents. This component provides the answer to the question of the frequency of occurrence of the given word in the document. The second component, IDF, is used to ascertain the importance of a given word within the context of the entire document corpus. The rationale is that certain words, such as "the" or "is" (referred to as "stopwords"), are prevalent in a multitude of documents and may not be particularly useful in differentiating one document from another. The IDF is calculated as the natural logarithm of the total number of documents in the corpus, divided by the number of documents in which the word in question is present. A word that appears in many documents will have a low IDF, indicating that it is less informative. The combination of TF and IDF for each word in each document gives the TF-IDF score. The TF-IDF score increases in proportion to the number of times a word appears in a document but is offset by the frequency of the word in all documents in the corpus.

When investigating a specialized domain like the FRAM, the TF-IDF matrix becomes instrumental. It enables the researchers to identify key terms that are both frequently used and critically significant to the topic under review. The TF-IDF matrix helps focus the review process on the most salient aspects of the available literature, not only streamlining the identification of relevant literature by eliciting knowledge concealed in the documents but also enhancing the overall quality of the analysis, leading to more precise and insightful conclusions in the systematic review process. The process is sketched in Figure 4.1 to improve clarity.

a) Initially the documental corpus can be thought as a fully disconnected graph

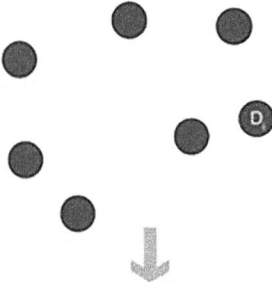

f) Once the bipartite graph has been connected, it is straightforward to obtain its adjacency matrix whose submatrices have zero entries except for the Doc-Terms one. DxD and TxT are Zero matrices since the graph is bipartite (i.e., no link exists between nodes of the same set); the Terms-Doc is actually zero because the connections are oriented and they can go only from the document-nodes to the term-nodes. Therefore, only the Doc-Terms possesses values greater than zero

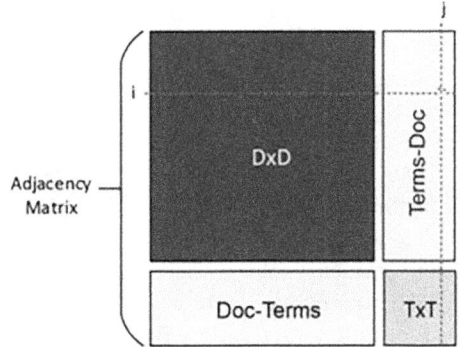

b) T terms were identified as relevant for the review

c) Actually any of these T terms are associated with several words pertaing the same semantic field (i.e., list of related words)

d) Now we have two kinds of nodes in our fully disconnected graph: Document-nodes (the dark circles) and Term-nodes (the grey rounded rectangles), i.e., the graph is bipartite

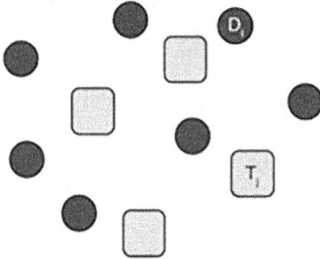

e) If the term T_j is present in a document D_i a connection is built; the strenght of the connection is proportional to the times T_j occurres within the document D_i

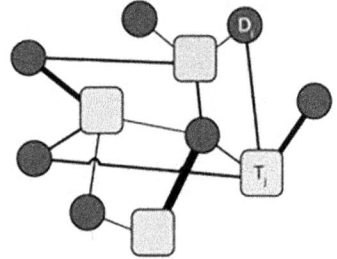

Adjacency Matrix

FIGURE 4.1 Illustration of a bipartite graph representation for a systematic review of the FRAM. This diagram depicts the initial development and subsequent linkage within a bipartite graph: starting as a fully disconnected graph, relevant terms are identified and associated with related semantic fields, leading to the construction of a graph with documents as nodes and terms as nodes. Connections are established based on the occurrence of terms within documents, with the strength of these connections reflected in the adjacency matrix, highlighting the non-zero submatrix where document-term relationships are quantified. This structured approach facilitates the systematic analysis of term relevance and distribution across the document corpus.

The adjacency matrix is a tabular representation of a graph, whose element *ij* is zero if the link from node *i* to node *j* is missing; otherwise, its value represents the strength of the link. Representing a graph through its adjacency matrix allows for leveraging linear algebra for the associated graph's information management. Moreover, by representing the bipartite graph as an adjacency matrix, it is straightforward to derive the documents-terms matrix, which is commonly used in information retrieval, document classification, topic modelling, and other computational linguistics tasks. The two matrices have a subtle but important difference: the first is a square matrix having the same nodes both on the rows and the columns; the document-term absolute matrix, often referred to simply as the document-term matrix (DTM), is a fundamental representation of the bipartite graph where each row corresponds to a document and each column to a term. The value within this matrix, for example, at a specific row and column, denotes the raw count of how often a particular term appears in that given document. For example, if the term "Resonance" is found five times in a document, the corresponding entry in the matrix will reflect this number, creating a direct link between the document and the term in the graph, with the magnitude of the entry showcasing the strength of this connection. Moving beyond the raw frequencies captured in the DTM, the TF-IDF matrix offers a shrewder view by transforming the document-term matrix. This transformation is achieved by adjusting term frequencies to account for the length of each document through normalization and by modifying the importance of terms based on their rarity across all documents using inverse document frequency. The resulting TF-IDF score for each term in each document is thus a product of these two measures, which reduces the influence of terms common across many documents and boosts the significance of rarer terms, helping highlight their unique contribution to the meaning of a document. This progression from the DTM to the TF-IDF matrix illustrates a shift from simply mapping the structure of a document-term bipartite graph to understanding the relative importance of each term within a broader corpus context. While the adjacency matrix and the DTM provide a straightforward numerical representation of the graph's structure, the TF-IDF matrix refines this by weighing the terms. This method ensures that the significance of terms is adequately represented, reflecting not just their frequency but also their informational value across the set of documents.

The review at hand was focused on determining which documents discuss specific methodologies and, at the same time, how certain domains (e.g., healthcare) were represented differently in the document selection. To further validate our findings, topic modelling was performed as a confirmatory step.

The network analysis package *igraph* was used to generate the graph adjacency matrix, from which the document-term matrix (doc-terms) can be extracted (Ju et al., 2016). The analysis was carried out using KNIME and the Python libraries Matplotlib and Seaborn for data visualization (Stancin & Jovic, 2019).

FINDINGS AND THEIR SIGNIFICANCE

The descriptive analysis identifies the characteristics of the FRAM research sector and its diachronic evolution.

PUBLICATION FORA

The term "fora" – derived from the Latin term "forum" – is used in this book to reference platforms where scholars and practitioners gather to exchange ideas and advance knowledge. These fora take various forms, mainly conferences and journals. Conferences serve as dynamic, interactive venues where participants present research findings, discuss emerging trends, and build collaborative networks. Journals, on the other hand, provide a more formal medium for disseminating peer-reviewed research, ensuring that scholarly work undergoes rigorous evaluation before publication. Both types of fora are crucial in fostering academic discourse and innovation, facilitating a vibrant exchange of ideas across disciplines. As shown in Table 4.1, the sample of documents explored shows a trend of increasing scientific production, with journal articles and conference proceedings being the largest contributors (169 and 42, respectively).

The analysis of the document distribution from 2019 to 2024 reveals little. Table 4.2 shows the trend for the top 5 journals. *Safety Science* demonstrates significant variability, with a notable peak of 10 articles in 2020, followed by a sharp decline to one in 2021, and fluctuations in the following years. *Reliability Engineering and System Safety* shows a gradual increase in article count until 2021 where it reaches a peak of 4 articles. After this peak, the journal's contributions stabilize, averaging between 1 and 3 articles per year.

TABLE 4.1
The Distribution of Documents by Year and Document Type within the FRAM Dataset from 2019 to 2024

Document Type	Year						Total
	2019	2020	2021	2022	2023	2024	
Article	5	30	29	39	46	16	**165**
Book chapter		2	3	2	4	2	**13**
Conference paper	6	11	7	10	7	1	**42**
Review		3	4	2	2	3	**14**
Total	**11**	**46**	**43**	**53**	**59**	**22**	**234**

Note: The document types include Articles, Conference Papers, Reviews, and Book Chapters.

TABLE 4.2

The Distribution of Articles in the Top 5 Journals from October 2019 to April 2024

Journal	Year					
	2019	2020	2021	2022	2023	2024
Safety Science	3	10	1	2	4	2
Reliability Engineering and System Safety		2	4	1	3	3
Applied Sciences (Switzerland)		2	2	2	3	
Applied Ergonomics		4	2	2		
Cognition, Technology, and Work		2	3	2		

Note: The data highlights "Safety Science" as having the most variability, while the other journals display more stable trends over the years.

Applied Ergonomics maintains a relatively consistent output over the years, peaking at 4 articles in 2020. Similarly, *Cognition, Technology and Work* exhibits a low but steady contribution, with peaks of 3 articles in 2021. Lastly, *Applied Sciences* shows minor fluctuations in its document count, peaking at 3 articles in 2023.

KEY VOICES

The distribution of the number of authors involved in documents reveals insightful trends in collaborative efforts within the dataset (see Figure 4.2). The data highlights a strong preference for collaborative research.

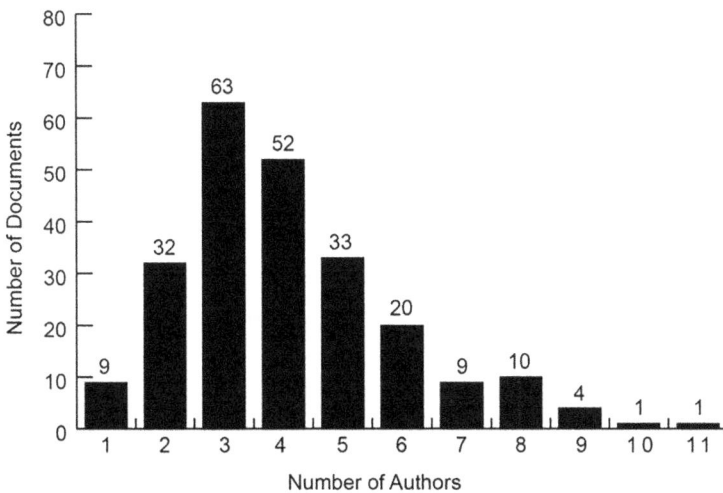

FIGURE 4.2 The distribution of documents based on the number of authors involved. The most common team size is three authors, with 63 documents, followed by four authors with 52 documents. Smaller teams, such as single-author and two-author documents, are less frequent. Large writing teams are equally rare.

Specifically, documents authored by three individuals are the most prevalent, with a notable count of 63, suggesting this as the optimal team size for the majority of research outputs. Documents with four, five, and two authors follow in frequency, with 52, 33, and 32 documents, respectively, underscoring a significant preference for small to medium-sized research teams. The number of documents decreases as the number of authors increases. Slightly larger teams are occasionally formed, potentially for more complex or interdisciplinary research projects. Indeed, instances of documents with nine or more authors are particularly rare, with counts dropping significantly. Single-author documents are relatively rare, with only 9 instances, highlighting that individual research efforts are less common in the FRAM-related research, which usually benefits from different views.

The network visualization built with VOSviewer further elucidates the collaborative dynamics previously described, showcasing the intricate web of interactions among researchers (Van Eck & Waltman, 2011) (Figure 4.3).

Method Geek

VOSviewer is a software tool for creating and visualizing bibliometric maps. It excels at building networks such as co-citation, co-authorship, and co-occurrence, revealing patterns and structures within large datasets.

VOSviewer uses the Leiden clustering algorithm as the default algorithm for analyzing co-authorship networks, like the one at hand. The unit of analysis is set to "authors" and the "full count" method is used; that is, each author is fully credited for each publication to which she or he has contributed. The visualization produced is a co-citation network, which maps the connections between authors based on shared citations, providing insights into research collaborations and intellectual communities. The chosen layout method is "association strength", which arranges the network based on the strength of the links between authors, allowing for a more accurate representation of the relationships. The clustering resolution is set to 1, a parameter that controls the granularity of the clusters identified; a value of 1 provides a balanced level of detail, detecting both larger groups and smaller, more specific clusters. In addition, the minimum cluster size is set to 6, meaning that only the clusters with at least 6 items are retained; therefore, too-small clusters have been omitted to reduce visual clutter.

Central to this visualization are the highly connected scholars such as Patriarca, Hollnagel, Saurin, and Sujan, who play crucial roles in fostering extensive collaborations. These central figures highlight the field's reliance on key researchers to drive large, interconnected research efforts. Surrounding these central nodes are distinct clusters representing subgroups of researchers with frequent collaborations. For instance, the group around Hirose includes researchers like Rubio-Romero and Zúñiga, indicating a tightly knit connection likely focused on specialized research topics. The same occurs for the group around Smith. This pattern reinforces the earlier observation of a predominant culture of collaboration, particularly in smaller, focused teams. The varying thickness of the connecting edges in the network

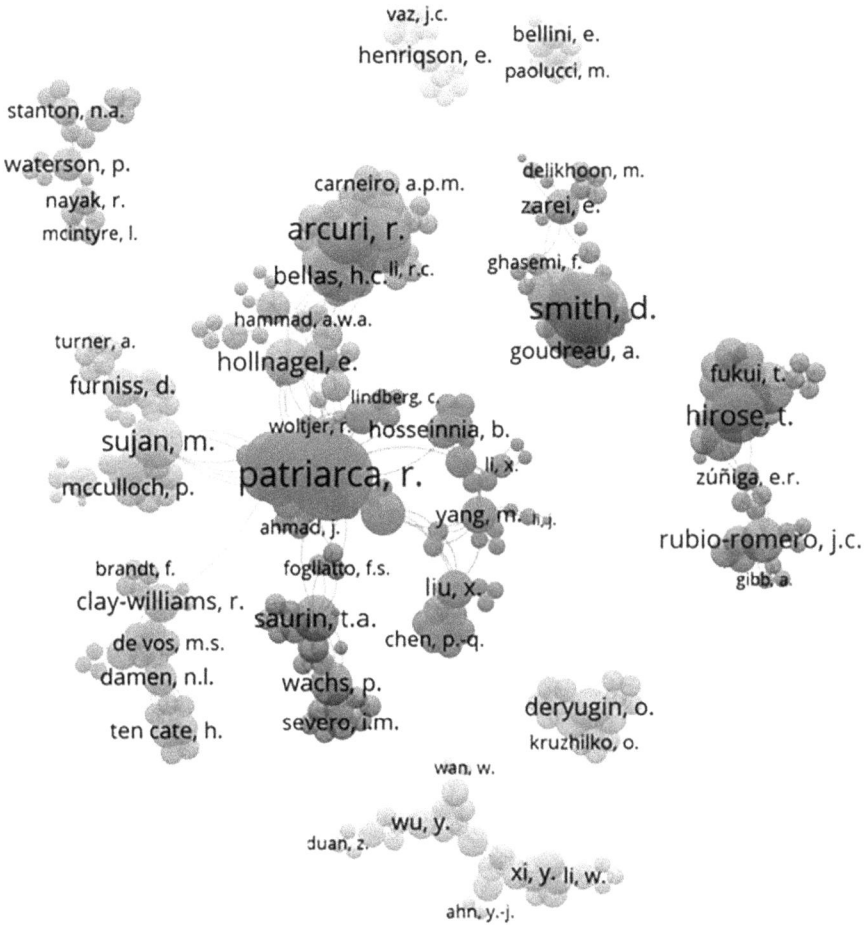

FIGURE 4.3 Co-authorship network illustrating the extensive and intricate collaborations among researchers in the FRAM literature. Central figures are highly connected, indicating their pivotal role in the community. Distinct clusters represent frequent collaborations among smaller subgroups, while peripheral connections highlight interdisciplinary and international collaborations. The varying edge thickness signifies the strength of these collaborative ties, emphasizing the dynamic and interconnected nature of research within this field.

illustrates the strength and frequency of these collaborative ties. Thicker edges signify robust, frequent interactions, often seen among core team members, while thinner edges represent more occasional or peripheral collaborations. This variation aligns with the data's indication of flexible collaboration approaches, with researchers engaging in both strong, frequent partnerships and broader, less frequent ones depending on the project's needs. Peripheral nodes, such as those involving Wu and Xi, connected through fewer and thinner edges, give an indication of

TABLE 4.3

The Top 20 Authors by the Number of Documents They Have Contributed to within the FRAM Dataset

Author	Article	Book Chapter	Review	Conference Paper	Total
Patriarca R.	6	1	1	3	11
Smith D.	5		2	1	8
Veitch B.	5		2	1	8
Di Gravio G.	4		1	3	8
Costantino F.	3		1	3	7
Hollnagel E.	4	1		1	6
Jatobá A.	4			2	6
Hirose T.	3			3	6
Sawaragi T.	3			3	6
Praetorius G.	4	1			5
Salehi V.	4		1		5
Kaya G.K.	4			1	5
Sujan M.	4			1	5
McCloskey R.	3		2		5
Saurin T.A.	3			2	5
Rubio-Romero J.C.	2	2	1		5
Falegnami A.	2			3	5
Wagner C.	4				4
Clay-Williams R.	3	1			4
Adriaensen A.	3		1		4

Note: The chart highlights a range of authors, from highly prolific to moderately active, reflecting the diverse and collaborative nature of the research community. This distribution underscores the central role of key researchers in driving collaborative efforts while also showcasing the valuable contributions of a broader group of authors.

interdisciplinary or less frequent collaborations. These peripheral connections underscore the diversity within the research community, reflecting contributions from various fields and institutions, thus enhancing the collaborative nature of the field.

The distribution of document counts among the top 20 authors further exemplifies the same observations (Table 4.3).

Leading figures such as Patriarca, R., Smith, D., and Veitch, B., with their substantial document counts, embody the robust, frequent interactions indicated by the thicker edges in the network visualization depicted in Figure 4.3. Their prolific contributions highlight their central roles in fostering and sustaining research collaborations, aligning with the network's depiction of strong core team members. The presence of numerous researchers with document counts ranging from four to seven underscores the diversity of the collaborative efforts within the community. These researchers likely engage in both frequent partnerships within their primary networks and broader collaborations across different fields. The chart also reflects the significant contributions from authors with lower but still notable document counts.

TABLE 4.4
Distribution of Research Perspectives

Perspective	N. of Documents
Prospective	112
Both	48
Not Applicable	40
Retrospective	34
Total	**234**

Note: It exhibits a strong preference for prospective studies, representing 48% of the total. Retrospective studies account for 14%, while 21% of the documents integrate both. The "Not Applicable" value a ccounts for those documents discussing methodological issues and for reviews.

The research team investigated two distinct dimensions of research methodologies applied within the dataset: quantification (qualitative *versus* quantitative) and perspective (prospective *versus* retrospective).

Prospective versus Retrospective

Table 4.4 reveals an overwhelming inclination towards prospective studies, comprising 48% of the total. Prospective research designs are forward-looking and primarily involve the analysis of risks or safety concerns, collecting data over a future period. This orientation involves inquiries into potential developments and outcomes, assessing preventative measures, and predicting future trends. The dominance of prospective studies highlights proactivity among scholars, who use the FRAM to understand work and draw conclusions on how adaptive practices might be managed to ensure the system operates safely. Retrospective (14%) studies involve the analysis of existing data, mostly linked to incident/accident analyses. Although fewer in number, in line with the shift of perspective of the FRAM since 2004 to today, retrospective studies via the FRAM allow evaluating historical data to understand drivers and patterns. The usage of the FRAM in the most recent studies reflects a significant evolution in the approach to analyze complex socio-technical systems and their adaptive practices. A detailed examination of the methodologies listed in the recent articles reveals a shift towards more advanced and integrated techniques compared to those used in earlier studies. This trend is reflected in the number of documents comprising both perspectives (21%).

Qualitative, Quantitative, or Hybrid?

The majority of the studies (47%) employ qualitative methodologies (see Table 4.5). This predominance of qualitative research underscores a focus on understanding phenomena through detailed, contextual, and in-depth data collection techniques

TABLE 4.5
Quantification Degree in the FRAM Literature at Hand

Quantification Degree	N. of Documents
Qualitative	109
Semi-/quantitative	79
Hybrid	46
Total	**234**

Note: Hybrid items collect either reviews or studies in which the quantification does not directly affect the FRAM methodology.

such as interviews, observations, and case studies. These methodologies are particularly effective in exploring complex, multifaceted issues where numerical data alone
might not capture the nuanced dynamics at play. The FRAM is indeed not meant to
analyze simply numerical data alone, where more traditional approaches can be generally considered more established. The quantitative studies emphasize the measurement and analysis of variables through statistical, mathematical, or computational
techniques. This approach is crucial for establishing patterns, testing hypotheses, and
generalizing findings across larger populations.

Fuzzy logic has gained prominence in FRAM analyses, offering enhanced capabilities for handling uncertainties and providing a more nuanced understanding of
system behaviours. Hirose and Sawaragi integrated fuzzy logic with the FRAM using
a cellular automaton to dynamically simulate and analyze socio-technical system
behaviours, reflecting this trend (Hirose & Sawaragi, 2020). This integration allows
for a more detailed exploration of the variability in system performance and the
potential risks associated with it. Another notable research study proposes a novel
integrated framework combining rough sets and fuzzy logic with the FRAM to tackle
the complexity of systemic performance variability in aircraft deicing operations.
This integration boosts the FRAM by employing rough sets to streamline and minimize the rule base and using fuzzy logic to quantify data's inherent vagueness. This
dual approach offers a solid framework for predictive modelling, improving simulations of operational conditions and enhancing decision-making and risk management
in aviation safety under varied and uncertain environments (Slim & Nadeau, 2020).

Numerical algorithms have also been increasingly integrated with the FRAM,
enabling real-time data analysis and predictive modelling. This represents a significant advancement over earlier methodologies, providing dynamic and adaptive
solutions for risk management. For example, Bellini et al. (2020) introduce a
numerical extension to the FRAM called Q-FRAM, designed to handle complex
system behaviours by quantifying functional variability and resilience. This
approach develops indicators based on performance measures that reflect the system's ability to anticipate, respond, monitor, and learn from changes, thus enhancing decision-making in resilience management. The method is validated through a
numerical example, illustrating its potential to guide system improvements and risk
management.

A similar study proposes a risk assessment procedure for cloud migration processes that integrates the FRAM with tools for defining specific performance requirements for service suppliers. Using the Critical-To-Quality (CTQ) method to determine IT platform customers' quality drivers and applying technical standards to define security management system requirements, the approach was verified through a real-life case study on cloud migration supply chain risks. The findings contribute to enhancing knowledge in IT systems' risk assessment and provide a foundation for further research (Fargnoli & Murgianu, 2023). This approach is claimed to allow improving both the efficiency of data collection and analysis and the facilitation of the development of more effective and responsive safety strategies.

Recent literature indicates a consistent growing use of Bayesian networks (Delikhoon et al., 2024; Z. Guo et al., 2024; Zarei et al., 2022). These methodologies offer robust frameworks for quantifying risks and understanding the stochastic relationships between different system components. This trend is exemplified by integrating the FRAM with probabilistic approaches in their safety assessments, showcasing the shift towards data-driven safety analyses. Another recent study integrates the FRAM and dynamic Bayesian Network for quantitative resilience assessment of chemical process systems, focusing on technical-human-organizational interactions (Zinetullina, 2020). The method is claimed to allow enhancing resilience assessment by analyzing accidents to develop attributes for withstanding or recovering from disruptions. In some cases, these studies highlight a tendency to reductionism, which should be carefully considered so as not to reduce the FRAM to another linear accident modelling exercise.

Hybrid methodologies are continuing to grow: the combination of the FRAM with other approaches, such as HCR-analysis (human collaborative relationships), allows for a more holistic analysis of system safety. Such integrations leverage the strengths of different methodologies, providing a more detailed and nuanced understanding of system interactions and potential failure modes. Lee, Yoon, and Chung demonstrated this by integrating the FRAM with HCR to analyze maritime operations, illustrating the benefits of hybrid methodologies (Lee et al., 2020).

Even though his work lies just before the time interval investigated, it is worth mentioning the work by Pardo-Ferreira et al., who thoroughly examined methods combined with the FRAM (e.g., Monte Carlo simulation, Fuzzy Logic, Model Checking, Hierarchical Analytical Process, and Abstraction Hierarchy). Highlighting a shift from traditional safety models to resilience engineering focused on adaptability, the research ended up with a systematic review of 22 publications, mainly oriented towards aviation and maritime. This work shows that despite the frequent, yet not consistent, preservation of the theoretical basis of the FRAM, there is no dominant combined approach, indicating large diversity (Pardo-Ferreira & Martínez-Rojas, 2019). This work was the only review available in the time span investigated in 'Framing the FRAM'. In the current corpus of the present analysis, however, there are 14 review papers. Tierra-Arévalo et al.'s work on the construction sector, for example, is distinctive (Tierra-Arévalo et al., 2024). The evolution of the FRAM from a predominantly qualitative approach to one that incorporates a balanced mix of both qualitative and quantitative techniques, including supplementary methods such as Bayesian networks and Monte Carlo simulations, underscores its growing robustness, as demonstrated in Salehi et al. (2020). This integration of advanced computational methods not only enhances the utility of the FRAM but also reflects its increasing alignment with the complex, interconnected nature of modern socio-technical systems, as seen across diverse sectors like healthcare, aviation, and maritime industries. Such evolution from predominantly qualitative methods to a balanced mix of qualitative and quantitative techniques, along with the integration of emerging technologies, highlights the ongoing refinement and expansion of the FRAM's capabilities.

INDUSTRIAL DOMAINS

From a domain application standpoint, healthcare stands out prominently, with a total of 64 documents, indicating its substantial prevalence and possibly underscoring its critical role and extensive operations within the dataset (Table 4.6). The high frequency suggests a significant emphasis on healthcare-related processes, reflecting the intricate and expansive nature of this sector.

TABLE 4.6
Occurrences of Domain Classifications

Domain	N. of Documents
Healthcare	64
Industrial Operations	29
Other	27
Maritime	24
Land Mobility	23
Construction	16
Aviation	14
Oil & Gas	14
Information Technology	8
Maintenance	6
Nuclear	5
Natural Hazards	4

Note Healthcare (64) is the most frequent, followed by Industrial Operations (29), and Maritime (24). Land Mobility (23). Construction (16), Aviation and Oil & Gas have all 14 occurrences. IT (8), Maintenance (6) concerns work not directly relatable to other domains, for example, Industrial Ops or Aviation. Nuclear (5) and Natural Hazards (4) are the less addressed domains. The Other (27) category highlights the distribution of focus across various further sectors.

Many studies emphasize the inherent complexity and nonlinear interactions in these kinds of socio-technical systems, starting with the review by McGill et al. (2022). This scoping review of FRAM's application in healthcare extensively analyzes how FRAM has been used to reveal complex interactions and variabilities within healthcare processes. By examining work as done versus work as imagined, the review highlights how the FRAM can map essential functions, identify performance variability, and uncover interdependencies that impact patient safety. The review provides clear examples of the FRAM's ability to enhance safety and operational efficiency, such as improving medication administration practices, streamlining transitions in patient care, and reducing the risks of adverse events like delayed sepsis detection or medication errors. These findings offer valuable practical recommendations for healthcare practitioners and policymakers aiming to optimize system performance and patient outcomes. O'Hara et al. (2020) utilized the FRAM to analyze transitional care, emphasizing the need to consider multiple stakeholder perspectives to improve patient outcomes post-discharge. This study highlighted the importance of supporting patient-facing functions in hospitals to ensure safe transitions. Similarly, Kaya et al. (2019) applied the FRAM to the drug administration process in neonatal intensive care units (NICUs), aiming to understand performance variability and its impact on system success or failure. This research underscored the necessity of anticipating how changes in conditions could affect both safety and efficiency in high-stakes environments like NICUs. Another significant contribution comes from Schutijser et al. (2019), who examined the double-check protocol for injectable medication administration. By contrasting the Work-As-Imagined with Work-As-Done, they identified discrepancies and suggested that nurses' adaptations based on real-time assessments could potentially enhance workflow and safety, though the variability's impact on patient harm remains uncertain. Salehi et al. (2021) extended the application of the FRAM to model the hospital-to-home transition processes for frail older adults. Their findings revealed critical challenges such as waitlists, the need for team-based care, and financial concerns, providing insights that could help improve transition processes and overall patient care. These studies demonstrate the FRAM's versatility and effectiveness in addressing various aspects of healthcare operations, from drug administration and transitional care to broader systemic challenges.

The domain of industrial operations accounts for 29 occurrences, marking it as a noteworthy area of focus. This domain's considerable presence signifies the broad spectrum of activities and processes involved in industrial operations, which encompass manufacturing, production, and logistics. The data suggests a substantial investment in monitoring and optimizing these operations to enhance efficiency and productivity, as well as operational safety.

Studies like "Functional Resonance Analysis Method and Human Performance Factors Identifying Critical Functions in Chemical Process Safety" by Menezes et al. (2021) delve into the integration of human factors and system functions to enhance safety outcomes in industrial operations. These works highlight the benefits of combining methodologies to better understand and manage risks in complex systems. The paper "A Simulation-Based Optimization Methodology for Facility Layout Design in Manufacturing" applies the FRAM by analyzing the variability in production processes and logistics systems (Zúñiga et al., 2020). By integrating the FRAM, the authors provide a structured approach to identifying potential inefficiencies and bottlenecks in facility layout designs, thus enhancing operational performance. This methodology, focused on simulation-based optimization, highlights the critical role of understanding functional interactions in complex manufacturing environments to achieve efficient and optimized facility layouts. In industrial operations linked to cobots and robots, Adriaensen et al. (2022) developed a framework to assess co-agency in human-robot interactions. They highlighted the importance of Distributed Situation Awareness and shared control in safe and efficient human–robot interaction. By combining the FRAM with Interdependence Analysis, they provided a comprehensive methodology for analyzing and improving the safety and efficiency of collaborative robot applications, further underscoring the value of systems thinking in this evolving field. De Nicola et al. (2023) explored the transition from Work-As-Imagined to Work-As-Done using semantics-driven approaches. This study addressed the misalignments between formal process descriptions and actual work practices, proposing a computational creativity approach to improve industrial resilience. The application of the FRAM in this context provided a structured method to identify and prioritize critical work analyses, thereby enhancing operational success.

The maritime sector appears with 24 occurrences, mainly about shipping and naval safety protocols and risk assessment.

Documents such as those by Salihoglu and Besikçi (2021) illustrate the FRAM's utility in qualitative risk analysis of maritime accidents, specifically analyzing the Prestige Oil Spill to identify underlying event variabilities and propose safety enhancements. This approach is critical for developing consistent opinions on system operations and enhancing accident risk analysis. Similarly, Chae et al. (2020) focus on the development of Maritime Autonomous Surface Ships (MASS), identifying technological improvements and human factors crucial for their successful implementation. They highlight the importance of robust communication systems, cybersecurity, and hazard analysis using methods like FRAM and System-Theoretic Process Analysis (STPA) to ensure the safe and efficient operation of MASS. Guo et al. (2023) integrate the FRAM with Dynamic Bayesian Networks to analyze the risk evolution in ship pilotage operations, identifying collision risk influencing factors and their dynamic interactions. This integrated model helps in understanding the spatiotemporal risk evolution and formulating targeted risk control strategies, emphasizing the sensitivity of collision risks to regional locations and human performance variability. In artisanal fishing, Saldanha et al. (2020) apply a Safety-II approach based on the FRAM to improve safety management in raft fishing. Their study focuses on fishermen's activities and strategies, uncovering the knowledge and expertise that inform safety-related decisions during fishing expeditions. This approach underscores the need for a broader, systemic perspective to enhance safety in such activities where humans are central via continuous improvement and support for sensemaking in dynamic and unpredictable environments. Lee et al. (2020) propose a framework for analyzing human collaboration in maritime safety using the FRAM. This framework considers both formal and informal human collaborations within and between ships, emphasizing the importance of human factors in accident analysis and safety management. By analyzing human collaborative relationships, the framework aims to enhance decision-making and adjust to variable and uncertain situations in maritime operations. Adhita et al. present a Safety-II case study of ship navigation, using the FRAM to analyze the everyday performance of ship officers (Adhita et al., 2023). Their study highlights the adaptability and flexibility of ship officers in contributing to system resilience, demonstrating the benefits of human functions in coping with the complexities of modern maritime transportation. These studies collectively underscore the significance of the FRAM in enhancing maritime safety, a sociotechnical system particularly prone to the consequences of systemic risks.

Land mobility, with 23 entries, examines the dynamics of land transport systems with a focus on safety, efficiency and sustainability. Many of these studies aim to tackle the challenges posed by the increasing urbanization and the demand for safe and efficient transportation solutions.

Method Geek

Combining the FRAM with the Internet of Everything (IoE) and big multimedia data approaches has proven to significantly enhance the resilience of urban transport systems (UTS) in smart cities. Bellini et al. (2021) illustrate how these data-driven techniques enable better monitoring, prediction, and decision-making in response to critical events like flash floods and traffic disruptions. These approaches highlight the potential of the FRAM to address complex urban mobility challenges by utilizing real-time data streams from IoT devices, public Wi-Fi, and social media, ultimately improving safety and operational efficiency in smart city infrastructures. These innovative approaches show the potential of the FRAM in addressing complex urban transportation challenges through data-driven methods. Cheberiachko and colleagues have contributed significantly to this domain by developing models to improve passenger road transportation safety (Bazaluk et al., 2022; Cheberiachko et al., 2023). Their research uses the FRAM to identify factors influencing transportation reliability and assess accident risks. They emphasize the importance of monitoring the psychophysiological state of drivers and establishing a strong relationship between transportation functions and safety criteria. Their studies propose enhanced control measures to reduce the likelihood of failures and to increase the reliability of transportation processes. Grabbe et al. (2022) further explore the application of the FRAM in road traffic scenarios, analyzing the performance variability mechanisms between human drivers and automation. Their research supports the development of adaptive automation strategies that promote human–machine collaboration, thereby enhancing traffic safety. Hlaing et al. applied the FRAM to analyze road accidents in Yangon, Myanmar, using a quantitative model to predict functional couplings and assess the impact of variability in road accident factors (Hlaing et al., 2021). Their findings highlight the significance of upstream function variability on downstream outcomes, providing insights into managing and reducing accident severity. Gill and colleagues' research focuses on measuring the variability of the pedestrian crossing function in urban road transport systems using the FRAM (Gill et al., 2022). Their methodology captures "functional vibrations" to predict undesired states based on normal operation, offering a proactive approach to safety management.

Construction, with 16 occurrences, reflects a non-negligible interest rate. This domain's inclusion highlights the importance of infrastructure development and the role of construction activities in shaping the built environment. The relatively lower frequency might indicate a more specialized or concentrated focus on particular aspects of construction.

Method Geek

To begin with, one cannot miss this paper by Tierra-Arévalo et al. (2023), in which a literature review on the application of the FRAM in the construction sector is carried out. The contributions of the FRAM serve as a solid basis for an in-depth and systematic analysis of daily performance in the construction sector. The construction industry has increasingly recognized the importance of integrating sophisticated safety protocols and leveraging technology to enhance site safety. Studies in this domain focus on identifying and mitigating hazards in dynamic environments, emphasizing safety training, real-time monitoring, and best practices to reduce accidents. This proactive approach aims to create safer work environments and improve project outcomes through effective risk management and safety planning. Li et al. integrated the FRAM with a multiplex network to model and analyze multiteam coordination task safety risks in the construction industry (Li & Wang, 2023). This study addressed the complex interactions and interdependencies among different teams, revealing how coordination performance influences task safety. The method's application in case studies demonstrated its effectiveness in detailing the impact of multiteam performance on safety, providing insights for developing targeted countermeasures to control risks within multiteam systems. Saldanha et al. introduced the ResiliFRAM method, which integrates ergonomic analysis with the FRAM to assess resilient performance in construction (M. Saldanha et al., 2022). Their research focused on how resilience abilities – monitoring, anticipating, responding, and learning – are interconnected to produce resilient outcomes. Applied to a case study in the construction sector, ResiliFRAM highlighted organizational practices that support resilient performance, offering a detailed framework for understanding and enhancing resilience in construction operations. Pardo-Ferreira et al. (2020) utilized the FRAM to understand everyday construction activities for building concrete structures. Their analysis, based on documentation, interviews, and on-site observations, identified several safety challenges, including the underutilization of health and safety plans, organizational pressures affecting safety, and the lack of leading indicators to capture normal work. The study emphasized the role of the FRAM in conducting in-depth and systematic analyses of daily performance, revealing critical issues that had been previously overlooked. These studies illustrate the growing recognition of FRAM's value in the construction industry, particularly for its ability to identify and manage risks proactively. By analyzing the variability in work processes and the interactions between different functions, the FRAM provides a comprehensive framework for improving safety protocols and ensuring better project outcomes. This evolution underscores the industry's shift towards more detailed system-based risk analyses with the aim of enhancing overall safety and resilience in construction projects.

Aviation accounts for 14 entries. The focus of the analyzed studies is often on understanding how variability in human performance can impact overall system safety and efficiency (Tengiz & Unal, 2023). This domain reflects the ongoing efforts to enhance safety protocols, mitigate risks, and optimize operational procedures in aviation, leveraging advanced analytical and probabilistic methods to anticipate and manage potential hazards.

Method Geek

The systematic review by Tian and Caponecchia presents how the FRAM is applied across various aviation functions, such as air traffic control, cockpit operations, and ground handling (Tian & Caponecchia, 2020). Kwasiborska and Kadziola (2023) applied the FRAM to explore ground handling processes at airports, highlighting interactions between technical procedures and human factors. The shift from purely retrospective to more balanced prospective and retrospective analyses is notable, reflecting a broader interest in using the FRAM for both understanding past incidents and improving future aviation safety protocols.

Guo et al. discuss the application of FRAM in analyzing and improving the resilience of airport operations, introducing a Security Resilience Management (SRM) model that assesses the loss of security resilience and helps managers assess and improve emergency response capabilities. This approach was applied in a case study of a corridor bridge safety incident, where the FRAM helped to identify key areas for improving airport safety management (J. Guo et al., 2022).

The representation of Oil and Gas in research is marked by 14 occurrences. Studies within this domain often concentrate on implementing advanced safety protocols, utilizing predictive maintenance technologies, and analyzing human factors that influence safety outcomes. This research is critical for preventing incidents, ensuring regulatory compliance, and enhancing the overall safety and efficiency of oil and gas operations.

Method Geek

Cantelmi et al. (2022) explored resilience in emergency management on oil and gas platforms during the COVID-19 pandemic. Their study used the FRAM to analyze the adaptive capacities of a Norwegian organization in managing COVID-19 outbreaks. This research highlighted the importance of organizational learning and adapting emergency plans to specific contexts, leading to the development of revised emergency management procedures and confirming the applicability of the FRAM in broader crisis management scenarios. Further, França et al. (2021) applied the FRAM to analyze human factors and non-technical skills in offshore drilling operations. By modelling the activities of drillers on offshore rigs, the study focused on the variability of human behaviour and the correlation between Work-As-Imagined and Work-As-Done. The FRAM model provided insights into the work demands and the impact of variability, both positive and negative, on task execution, highlighting the critical role of human factors and non-technical skills in maintaining productivity and safety in normal and emergency situations. Bjørnsen, Jensen, and Aven (2020) discussed the integration of qualitative risk assessments with FRAM to enhance system resilience, providing an example within the Oil and Gas sector. They proposed adding strength of knowledge considerations and qualitative sensitivity analysis to the FRAM to address risk and uncertainty, providing a more robust basis for assessing and strengthening system resilience. Henriqson et al. (2022) examined the challenges of digital transformation in oil and gas integrated operations through the FRAM modelling. Their study identified key challenges, including supporting coordination demands, ensuring adequate human supervision of automated functions, reducing system opacity, supporting adaptations, and enhancing operators' non-technical skills. The analysis underscored the reliance on human-machine interactions for resilient operations, with operators managing system variabilities to ensure safe and continuous operations.

Information technology (IT), as a domain, accounts for 8 items. The contribution listed in this category refers to cases that predominantly take advantage of IT developments, even though mixed with use cases in various fields.

Method Geek

The paper by Salehi et al. introduces a novel reinforcement learning (RL) approach to the traditional FRAM model to enhance its ability to assess complex systems by allowing an artificial intelligence (AI) agent to explore functional routes based on rewards (Salehi et al., 2022). This method integrates reinforcement learning into the FRAM to simulate agent-based decision-making, providing deeper insights into how different management structures and operational environments can affect system performance. The study demonstrates this approach using healthcare operations, where the AI agent explores the system's functional environment and selects routes based on accumulated rewards, allowing for the identification of functional paths that might enhance or degrade overall system performance. Similarly, Costantino et al. focus on applying the FRAM within a safety management framework in socio-technical systems (Costantino et al., 2020). They propose the H(CS)2I framework, leveraging FRAM-based modelling to identify discrepancies between Work-As-Imagined and Work-As-Done. Their framework incorporates mobile participatory crowd-sensing and ontology-based data collection to streamline the process of Work-As-Done knowledge elicitation. Both papers highlight the growing applications of the FRAM in complex systems analysis, offering advanced techniques like reinforcement learning and real-time data collection, enhancing the tool's ability to manage and predict operational risks effectively. These advancements in the FRAM present a critical approach for proactive and adaptive system management, particularly relevant in dynamic IT environments.

Maintenance research with the FRAM, although less frequent, explores maintenance practices with respect to the influence of human factors on their execution in terms of reduced downtime, failure prevention, and operational success.

Method Geek

For example, de Souza et al. investigated the use of a layered FRAM approach to analyze Work-As-Done in HVAC maintenance, addressing the complex interactions between human agents and equipment, revealing that the layered FRAM could simplify the analysis of functional interactions and reduce cognitive workload, thus promoting a more user-friendly method to handle complex maintenance tasks (de Souza, Rosa, Evangelista, et al., 2021). In a complementary study, the authors integrated information technology systems with the FRAM to improve the quality and productivity of work environments in building maintenance (de Souza, Rosa, Vidal, et al., 2021). Hosseinnia et al. focused on the risks associated with hydrocarbon release during maintenance operations in the oil and gas industry (Hosseinnia et al., 2019). They employed the FRAM to model and assess these risks, integrating a dynamic graph approach to evaluate the impact of performance variability and prioritize critical stages and interactions during maintenance work. Their findings highlight the need for a systematic approach to managing the dynamic interactions and risks inherent in chemical maintenance operations. A work by Costella et al. examined the relationship between safety and maintenance in industrial refrigeration systems using the FRAM (Costella et al., 2023). Their case study identified key areas for improvement in the management of occupational health and safety (OHS) measures and emphasized the importance of comprehensive safety approaches, especially when dealing with hazardous substances like ammonia.

In the nuclear domain (5 records), characterized by stringent safety and regulatory standards, the emphasis lies on risk management, safety protocols, and human factors. The research here focuses on enhancing the safety and reliability of nuclear operations through advanced safety systems, comprehensive training programmes, and robust management frameworks.

Method Geek

For instance, Laarni et al. (2020) employed the FRAM to evaluate operational procedures, offering detailed analyses of systemic interactions. Their study revealed that creating an initial FRAM model from the perspective of the nuclear process, followed by a detailed model from the control room operators' perspective, is beneficial. This method identified potential variability in functions, particularly in communication and collaboration. Similarly, Talarico et al. (2023) compared two methods for identifying variability in a FRAM model representative of operations in a nuclear facility. They concluded that an indirect method for acquiring variability information is adequate, as it showed comparable results to the prescribed FRAM.

The field of natural hazards management (4 records) underscores the FRAM's potential to improve risk management by providing a detailed understanding of the functional variability of natural phenomena and its impact on infrastructural system resilience. These insights are crucial for proactive disaster preparedness and for refining emergency response strategies in the face of increasing natural hazards.

Method Geek

The first study by Togawa et al. (2023) documents the application of the FRAM to the emergency response process during Typhoon Hagibis in Koriyama City, Japan. This case study emphasizes how the FRAM can map out the interactive relationships between key municipal functions, such as public communication, evacuation management, and rescue operations, during a natural disaster. By identifying how variability in function performance resonated across the system, the authors demonstrated the model's ability to evaluate and improve disaster response retrospectively. This analysis led to critical recommendations for minimizing variability in evacuation behaviour and improving public communication systems, thus enhancing the overall resilience of flood risk management at the community level. Similarly, Steen and Ferreira (2020) employ the FRAM to analyze flood risk management at the municipal level, with a particular focus on Norwegian municipalities. Their study identifies how key functions like contingency planning, resource allocation, and evacuation operations interact and adapt during flood events. Using the FRAM, they reveal how interdependencies and performance variability affect resilience capabilities, particularly the ability to anticipate, monitor, and respond to crises. The study highlights the importance of integrating resilience engineering into flood risk management to ensure that municipalities can adapt to complex and dynamic flooding scenarios.

CONCLUSIONS

As this initial chapter reviewing recent FRAM literature draws to a close, it is clear that our exploration is part of a much broader and evolving narrative. From its beginnings as a theoretical model to its current status as a cornerstone of safety science, the FRAM has greatly enhanced our understanding of complex socio-technical systems. Under the PRISMA methodology lens and by exploiting the analytical power of KNIME, we have carefully analyzed and gained valuable insights from an extensive set of documents about the FRAM. The FRAM has demonstrated extensive versatility and applicability across various high-risk domains. When comparing recent advancements (2019–2024) with earlier studies (pre-2019), a clear evolution in focus and application areas emerges, reflecting the dynamic nature of the FRAM in methodological terms. Additionally, the widespread usage of the FRAM confirmed its extensive applicability and transformative potential across diverse domains. The comparison of recent advancements with earlier literature highlights shifts in focus, increased integration of quantitative methods, and broader adoption of the FRAM in conjunction with other approaches to guarantee higher flexibility with complex scenarios. This review, building upon the foundational work that has traced the development and application of the FRAM, reinforces the method's significance in addressing current and emerging complexities in various high-risk sectors.

REFERENCES

Adhita, I., Fuchi, M., Konishi, T., & Fujimoto, S. (2023). Ship navigation from a Safety-II perspective: A case study of training-ship operation in coastal area. *Reliability Engineering & System Safety, 234.* https://doi.org/10.1016/j.ress.2023.109140

Adriaensen, A., Berx, N., Pintelon, P., Costantino, F., Di Gravio, G., Patriarca, R., 2022. Interdependence Analysis in collaborative robot applications from a joint cognitive functional perspective, *International Journal of Industrial Ergonomics, 90,* 103320, ISSN 0169-8141, https://doi.org/10.1016/j.ergon.2022.103320

Bazaluk, O., Koriashkina, L., Cheberiachko, S., Deryugin, O., Odnovol, M., Lozynskyi, V., & Nesterova, O. (2022). Methodology for assessing the risk of incidents during passenger road transportation using the functional resonance analysis method. *Heliyon, 8*(11). https://doi.org/10.1016/j.heliyon.2022.e11814

Bellini, E., Bellini, P., Cenni, D., Nesi, P., Pantaleo, G., Paoli, I., & Paolucci, M. (2021). An IOE and big multimedia data approach for urban transport system resilience management in smart cities. *Sensors (Switzerland), 21*(2), 1–35. https://doi.org/10.3390/s21020435

Bellini, E., Coconea, L., & Nesi, P. (2020). A functional resonance analysis method driven resilience quantification for socio-technical systems. *IEEE Systems Journal, 14*(1), 1234–1244. https://doi.org/10.1109/JSYST.2019.2905713

Berthold, M. R., Cebron, N., Dill, F., Gabriel, T. R., Kötter, T., Meinl, T., Ohl, P., Thiel, K., & Wiswedel, B. (2009). KNIME - The Konstanz information miner: Version 2.0 and beyond. *SIGKDD Explor. Newsl., 11*(1), 26–31. https://doi.org/10.1145/1656274.1656280

Bjørnsen, K., Jensen, A., & Aven, T. (2020). Using qualitative types of risk assessments in conjunction with FRAM to strengthen the resilience of systems. *Journal of Risk Research, 23*(2), 153–166. https://doi.org/10.1080/13669877.2018.1517382

Cantelmi, R., Steen, R., Di Gravio, G., & Patriarca, R. (2022). Resilience in emergency management: Learning from COVID-19 in oil and gas platforms. *International Journal of Disaster Risk Reduction, 76*, 103026. https://doi.org/10.1016/j.ijdrr.2022.103026

Chae, C., Kim, M., & Kim, H. (2020). A study on identification of development status of MASS technologies and directions of improvement. *Applied Sciences, 10*(13). https://doi.org/10.3390/app10134564

Cheberiachko, S., Yavorska, O., Deryugin, O., & Lantukh, D. (2023). Improving safety of passenger road transportation. *Transactions on Transport Sciences.* https://doi.org/10.5507/tots.2023.003

Costantino, F., Di Gravio, G., Falegnami, A., Patriarca, R., Tronci, M., De Nicola, A., Vicoli, G., & Villani, M. L. (2020). *Crowd sensitive indicators for proactive safety management: A theoretical framework.* 1453–1458. https://doi.org/10.3850/978-981-14-8593-0_3928-cd

Costella, M., Pelegrini, G., Bortolosso, H., Vicari, P., & Dalcanton, F. (2023). Exploring the relationships between safety and maintenance in the cold generation process: Insights from the functional resonance analysis method. *International Journal of Occupational Safety and Ergonomics, 29*(1), 216–223. https://doi.org/10.1080/10803548.2022.2038960

De Nicola, A., Villani, M., Sujan, M., Watt, J., Costantino, F., Falegnami, A., & Patriarca, R. (2023). Development and measurement of a resilience indicator for cyber-socio-technical systems: The allostatic load. *Journal of Industrial Information Integration, 35.* https://doi.org/10.1016/j.jii.2023.100489

de Souza, I., Rosa, A., Evangelista, A., Tam, V., & Haddad, A. (2021). Modelling the work-as-done in the building maintenance using a layered FRAM: A case study on HVAC maintenance. *Journal of Cleaner Production, 320.* https://doi.org/10.1016/j.jclepro.2021.128895

de Souza, I., Rosa, A., Vidal, M., Najjar, M., Hammad, A., & Haddad, A. (2021). Information technologies in complex socio-technical systems based on functional variability: A case study on HVAC maintenance work orders. *Applied Sciences, 11*(3). https://doi.org/10.3390/app11031049

Delikhoon, M., Habibi, E., Zarei, E., Banda, O. A. V., & Faridan, M. (2024). Towards decision-making support for complex socio-technical system safety assessment: A hybrid model combining FRAM and dynamic Bayesian networks. *Process Safety and Environmental Protection, 187*, 776–791.

Fargnoli, M., & Murgianu, L. (2023). A resilience engineering approach for the risk assessment of IT services. *Applied Sciences, 13*(20). https://doi.org/10.3390/app132011132

França, J., Hollnagel, E., dos Santos, I., & Haddad, A. (2021). Analysing human factors and non-technical skills in offshore drilling operations using FRAM (functional resonance analysis method). *Cognition Technology & Work*, *23*(3), 553–566. https://doi.org/10.1007/s10111-020-00638-9

Garito, M. A., Caforio, A., Falegnami, A., Tomassi, A., & Romano, E. (2023). Shape the EU future citizen. Environmental education on the European green deal. *Energy Reports*, *9*, 340–354. https://doi.org/10.1016/j.egyr.2023.06.001

Gill, A., Smoczynski, P., & Lawniczak, D. (2022). Measuring the variability of the pedestrian crossing function in the socio-technical system of urban road transport. *Scientific Journal of Silesian University of Technology-Series Transport*, *117*, 57–68. https://doi.org/10.20858/sjsutst.2022.117.4

Grabbe, N., Arifagic, A., & Bengler, K. (2022). Assessing the reliability and validity of an FRAM model: The case of driving in an overtaking scenario. *Cognition Technology & Work*, *24*(3), 483–508. https://doi.org/10.1007/s10111-022-00701-7

Guo, J., Li, H., & Yang, C. (2022). A semi-quantitative method of safety resilience assessment for airport operation system based on FRAM model. *2022 8th International Symposium on System Security, Safety, and Reliability (ISSSR)*, 6–13. https://doi.org/10.1109/ISSSR56778.2022.00010

Guo, Y., Jin, Y., Hu, S., Yang, Z., Xi, Y., & Han, B. (2023). Risk evolution analysis of ship pilotage operation by an integrated model of FRAM and DBN. *Reliability Engineering & System Safety*, *229*. https://doi.org/10.1016/j.ress.2022.108850

Guo, Z., She, J., Li, Z., Du, J., & Ye, S. (2024). Integrating FRAM and BN for enhanced resilience evaluation in construction emergency response: A scaffold collapse case study. *Heliyon*, *10*(3). https://doi.org/10.1016/j.heliyon.2024.e25342

Henriqson, E., Rodrigues, F. S., Werle, N. J. B., & Lando, F., Trancoso, R. D. S., & Fogaça, L. B. (2022). Resilience and digital transformation challenges in oil and gas integrated operations. *Resilience in a digital: Global challenges in organisations and society* (pp. 19–38) https://doi.org/10.1007/978-3-030-85954-1_3

Hirose, T., & Sawaragi, T. (2020). Extended FRAM model based on cellular automaton to clarify complexity of socio-technical systems and improve their safety. *Safety Science*, *123*. https://doi.org/10.1016/j.ssci.2019.104556

Hlaing, K., Aung, N., Hlaing, S., & Ochimizu, K. (2021). *Performance variability analysis on road accident in Yangon* (J. Filipe, M. Smialek, A. Brodsky, & S. Hammoudi, Eds; pp. 329–336). https://doi.org/10.5220/0010451803290336

Hosseinnia, B., Khakzad, N., Patriarca, R., & Paltrinieri, N. (2019). *Modeling risk influencing factors of hydrocarbon release accidents in maintenance operations using FRAM* (pp. 290–294). Scopus. https://doi.org/10.1109/ICSRS48664.2019.8987694

Ju, W., Li, J., Yu, W., & Zhang, R. (2016). iGraph: An incremental data processing system for dynamic graph. *Frontiers of Computer Science*, *10*(3), 462–476. https://doi.org/10.1007/s11704-016-5485-7

Kaya, G., Ovali, H., & Ozturk, F. (2019). Using the functional resonance analysis method on the drug administration process to assess performance variability. *Safety Science*, *118*, 835–840. https://doi.org/10.1016/j.ssci.2019.06.020

Kwasiborska, A., & Kadziola, K. (2023). Application of causal analysis of disruptions and the functional resonance analysis method (FRAM) in analyzing the risk of the baggage process. *Scientific Journal of Silesian University of Technology-Series Transport*, *119*, 63–81. https://doi.org/10.20858/sjsutst.2023.119.4

Laarni, J., Tomminen, J., Liinasuo, M., Pakarinen, S., Lukander, K., Harris, D., & Wen-Chin Li. (2020). *Promoting operational readiness through procedures in nuclear domain.* Springer International Publishing; Inspec. https://doi.org/10.1007/978-3-030-49183-3_4

Lee, J., Yoon, W., & Chung, H. (2020). Formal or informal human collaboration approach to maritime safety using FRAM. *Cognition Technology & Work*, *22*(4), 861–875. https://doi.org/10.1007/s10111-019-00606-y

Li, J., & Wang, H. (2023). Modeling and analyzing multiteam coordination task safety risks in socio-technical systems based on FRAM and multiplex network: Application in the construction industry. *Reliability Engineering & System Safety, 229*. https://doi.org/10.1016/j.ress.2022.108836

McGill, A., Smith, D., McCloskey, R., Morris, P., Goudreau, A., & Veitch, B. (2022). The functional resonance analysis method as a health care research methodology: A scoping review. *JBI Evidence Synthesis, 20*(4), 1074–1097. https://doi.org/10.11124/JBIES-21-00099

Menezes, M., Haddad, A., & Nascimento, M. (2021). Functional resonance analysis method and human performance factors identifying critical functions in chemical process safety. *IEEE Access, 9*, 168368–168382. https://doi.org/10.1109/ACCESS.2021.3135747

Moher, D., Liberati, A., Tetzlaff, J., & Altman, D. G. (2009). Preferred reporting items for systematic reviews and meta-analyses: The PRISMA statement. *PLoS Medicine, 6*(7), e1000097. https://doi.org/10.1371/journal.pmed.1000097

O'Hara, J., Baxter, R., & Hardicre, N. (2020). "Handing over to the patient": A FRAM analysis of transitional care combining multiple stakeholder perspectives. *Applied Ergonomics, 85*. https://doi.org/10.1016/j.apergo.2020.103060

Pardo-Ferreira, M., Rubio-Romero, J., Gibb, A., & Calero-Castro, S. (2020). Using functional resonance analysis method to understand construction activities for concrete structures. *Safety Science, 128*. https://doi.org/10.1016/j.ssci.2020.104771

Pardo-Ferreira, M. C., Martínez-Rojas, M., & … (2019). Evolution of the Functional Resonance Analysis Method (FRAM) through the combination with other methods. *Direccion y Organizacion* https://www.revistadyo.es/DyO/index.php/dyo/article/view/550/572

Patriarca, R., Di Gravio, G., Woltjer, R., Costantino, F., Praetorius, G., Ferreira, P., & Hollnagel, E. (2020). Framing the FRAM: A literature review on the functional resonance analysis method. *Safety Science, 129*. https://doi.org/10.1016/j.ssci.2020.104827

Saldanha, M., Araújo, L., Arcuri, R., Vidal, M., de Carvalho, P., & de Carvalho, R. (2022). Identifying routes and organizational practices for resilient performance: A study in the construction industry. *Cognition Technology & Work, 24*(3), 521–535. https://doi.org/10.1007/s10111-022-00703-5

Saldanha, M., de Carvalho, R., Arcuri, R., Amorim, A., Vidal, M., & de Carvalho, P. (2020). Understanding and improving safety in artisanal fishing: A safety-II approach in raft fishing. *Safety Science, 122*. https://doi.org/10.1016/j.ssci.2019.104522

Salehi, V., Hanson, N., Smith, D., McCloskey, R., Jarrett, P., & Veitch, B. (2021). Modeling and analyzing hospital to home transition processes of frail older adults using the functional resonance analysis method (FRAM). *Applied Ergonomics, 93*. https://doi.org/10.1016/j.apergo.2021.103392

Salehi, V., Tran, T., Veitch, B., & Smith, D. (2022). A reinforcement learning development of the FRAM for functional reward-based assessments of complex systems performance. *International Journal of Industrial Ergonomics, 88*. https://doi.org/10.1016/j.ergon.2022.103271

Salehi, V., Veitch, B., & Smith, D. (2020). Modeling complex socio-technical systems using the FRAM: A literature review. *Human Factors and Ergonomics In Manufacturing, 31*(1), 118–142. https://doi.org/10.1002/hfm.20874

Salihoglu, E., & Besikçi, E. (2021). The use of functional resonance analysis method (FRAM) in a maritime accident: A case study of Prestige. *Ocean Engineering, 219*. https://doi.org/10.1016/j.oceaneng.2020.108223

Schutijser, B., Jongerden, I., Klopotowska, J., Portegijs, S., de Bruijne, M., & Wagner, C. (2019). Double checking injectable medication administration: Does the protocol fit clinical practice? *Safety Science, 118*, 853–860. https://doi.org/10.1016/j.ssci.2019.06.026

Slim, H., & Nadeau, S. (2020). A mixed rough sets/fuzzy logic approach for modelling systemic performance variability with FRAM. *Sustainability*, *12*(5). https://doi.org/10.3390/su12051918

Stancin, I., & Jovic, A. (2019). *An overview and comparison of free Python libraries for data mining and big data analysis* (pp. 977–982). https://doi.org/10.23919/MIPRO.2019.8757088

Steen, R., & Ferreira, P. (2020). Resilient flood-risk management at the municipal level through the lens of the functional resonance analysis model. *Reliability Engineering and System Safety*, *204*. https://doi.org/10.1016/j.ress.2020.107150

Talarico, M., Melo, P., & Gomes, I. (2023). Comparison of two methods for obtaining the variabilities in a FRAM model of a combined nuclear facility project. *Nuclear Technology*, *209*(5), 745–764. https://doi.org/10.1080/00295450.2022.2155021

Tengiz, E., & Unal, G. (2023). A fuzzy logic evolution of the functional resonance analysis method (FRAM) to assess risk in ground operation. *Aircraft Engineering and Aerospace Technology*, *95*(10), 1614–1623. https://doi.org/10.1108/AEAT-01-2023-0007

Tian, W., & Caponecchia, C. (2020). Using the functional resonance analysis method (FRAM) in aviation safety: A systematic review. *Journal of Advanced Transportation*, *2020*. https://doi.org/10.1155/2020/8898903

Tierra-Arévalo, J. M., Del Carmen Pardo-Ferreira, M., Herrera-Pérez, V., Rubio-Romero, J. C., García Márquez, F. P., Segovia Ramíez, I., Bernalte Sánchez, P. J., & Muñoz Del Río, A. (2023). *FRAM in the construction sector* (pp. 473–477). Springer Nature; Inspec. https://doi.org/10.1007/978-3-031-27915-7_83

Tierra-Arévalo, J. M., Pardo-Ferreira, M. D. C., Herrera-Pérez, V., & Rubio-Romero, J. C. (2024). Review of the functional resonance analysis method and other tools for resilience analysis in the construction sector. *Dirección y Organización*, 70–80. https://doi.org/10.37610/1wcvsd31

Togawa, T., Morita, H., & Tsuji, T. (2023). Applying the functional resonance analysis method (FRAM) to flood risk management at a community level: Koriyama city's emergency-response process during Typhoon Hagibis. *Progress in Disaster Science*, *19*, 100291. https://doi.org/10.1016/j.pdisas.2023.100291

Tomassi, A., Caforio, A., Romano, E., Lamponi, E., & Pollini, A. (2024). The development of a competence framework for environmental education complying with the European qualifications framework and the European green deal. *Journal of Environmental Education*, *55*(2), 153–179. https://doi.org/10.1080/00958964.2023.2259846

Tomassi, A., Falegnami, A., & Romano, E. (2024). Mapping automatic social media information disorder. The role of bots and AI in spreading misleading information in society. *PLOS ONE*, *19*(5), e0303183. https://doi.org/10.1371/journal.pone.0303183

Van Eck, N. J., & Waltman, L. (2011). Text mining and visualization using VOSviewer. *arXiv Preprint arXiv:1109.2058.*

Zarei, E., Khan, F., & Abbassi, R. (2022). A dynamic human-factor risk model to analyze safety in sociotechnical systems. *Process Safety and Environmental Protection*, *164*, 479–498. https://doi.org/10.1016/j.psep.2022.06.040

Zinetullina, A. (2020). *Quantitative Dynamic Resilience Assessment of Chemical Process Units Using Dynamic Bayesian Network.* https://nur.nu.edu.kz/handle/123456789/4661

Zúñiga, E. R., Moris, M. U., Syberfeldt, A., Fathi, M., & Rubio-Romero, J. C. (2020). A simulation-based optimization methodology for facility layout design in manufacturing. *IEEE Access*, *8*, 163818–163828. https://doi.org/10.1109/ACCESS.2020.3021753

Section I

Step 1
To Identify and Describe the System's Functions

5 The Use and Implications of Small FRAM Models
A Case from Healthcare

Jeanette Hounsgaard

CONTEXT: HOSPITAL MATTRESS MANAGEMENT

As explained in previous chapters, the FRAM (Hollnagel, 2012, 2021) is a method to analyze how the daily work is done. By the FRAM, analysts can model and visualize the daily activities, their dependencies, and relations through a model. This model can then be used for specific types of analysis, whether to determine how something went wrong, to look for possible bottlenecks or risks, to check the feasibility of proposed interventions, or simply to understand how the daily work takes place.

During more than 12 years, I have probably contributed to and completed more than 250 FRAM analyses. Among FRAM users worldwide, I have been told that I use the method in a simple way and create valuable and useful information about how the daily work is done through small FRAM models.

Let me describe how I define a small FRAM model:

- The model normally includes 10 to 20 functions.
- In numbers, the model has fewer foreground functions than background functions.
- Only the foreground functions in focus have information on aspects other than input/output.
- You can talk through the FRAM model.
- There are a few instantiations.
- You can talk through the instantiations.
- You can easily explain how variability can emerge into something unwanted and unexpected.
- The model is easy to present to the frontline staff.
- Even though staff do not know anything about the method FRAM, they immediately recognize their daily job in the FRAM model – they are not lost – and the most important thing is that they get the insight into how their job relates to the job of their colleagues.

DOI: 10.1201/9781003518167-7

- The staff also realizes how the adjustments they make affect the job of their colleagues.
- The model gives sufficient insight for the staff to go into a dialogue with management about potential improvements and ensures a bottom-up process, the motivation and the commitment of staff to change.

In general, I find a FRAM analysis useful whenever you need to know what is going on before you intervene in the system.

For more than 26 years in my early work career, I used root cause analysis (RCA) to analyze unwanted outcomes in a complex system. The RCA is a strong tool, but has limitations when used to intervene in complex systems. Today, I know that many of the interventions in my early career were not grounded in insight and understanding of how the system worked. They were based only on how we *thought* the system worked. Now I believe that this can explain why the same unwanted outcome appeared again.

When compared to linear approaches like the RCA, I find that an FRAM model shows a higher number of possible interventions, grounded in how the system works. At the same time, the FRAM model can show the consequences of a change, especially how the change affects other activities in the system. In other words, the FRAM can help you avoid the idea that an intervention just is a change and not an improvement (thus useless in my eyes). When you do not know what is going on, the danger is that you intervene in a system that works and implement a change that makes the system work in a less safe, less efficient, and ultimately dangerous way.

Through a case, I want to give you knowledge about how the FRAM was used to unveil why monitoring data looked as it did and how the FRAM analysis gave possible answers to the question, what to do about it if not acceptable?

THE CASE OF OVERCONSUMING PRESSURE-RELIEVING MATTRESSES

The FRAM analysis was completed in 2023 at a hospital in the Region of Southern Denmark. A bed ward was overconsuming pressure-relieving mattresses compared to similar wards at other hospitals in the region, and the management wanted to know why.

The management was concerned because a pressure-relieving mattress is not always the best choice of mattress for the recovery of the patient; the mattress is also the most expensive type of mattress to buy and needs to be powered by electricity. Several adverse events had been reported to management during the years when the mattress had not been powered, resulting in the patient suffering from a pressure ulcer.

To reduce the consumption, management had decided to intervene by introducing a new type of mattress, solid and of a higher quality. However, after several months, the consumption of pressure-relieving mattresses had not decreased as expected.

The management hoped that a FRAM analysis could answer why the intervention failed and give new perspectives on the challenge with the overconsumption of pressure-relieving mattresses. Since the FRAM has been used in the Region of Southern Denmark since March 2012, the FRAM already belongs to their toolbox whenever management faces complexity. In this case, the management believed that introducing a solid mattress of a higher quality would make the staff choose this mattress instead of the pressure-relieving mattress.

DATA COLLECTION

Once the purpose of the analysis was stated and the respective starting point was described and agreed upon with management, I started familiarizing myself with the area to be analyzed by talking to the local management and reviewing the available documentation to prepare for collecting data in Step 1. This documentation normally includes procedures and instructions. If available, these latter are of interest because they constitute a basis for preparing a FRAM model of the planned activities. This model is usually able to capture a variety of work called Work-As-Imagined. If they are not available, I interview the local management on how they *think* (imagine) the job is done.

Another set of interviews is then necessary to capture how the staff doing the job actually do the job. Even though the interviewing technique is not included in this chapter, I would like to share my experience with collecting data through interviews for the building of the FRAM model, Work-As-Done. The source of this data is the staff doing the job. The data collection is conducted through interviews (approx. 80%) and observations (approx. 20%) about how the work is done.

The interviews are semi-structured according to the 6 aspects: Input (I), Output (O), Precondition (P), Resource (R), Time (T), and Control (C). The semi-structured perspective strengthens the dialogue between me and the staff member and allows me to ask for details of the work in a natural order.

Before I start asking for the six aspects, I ask the person to briefly talk about their job regarding the function in focus.

The interviewed persons are both experienced and inexperienced, as well as newly employed. The experienced know the adjustments and trade-offs (usually referred to as "work-a-rounds" and "short-cuts") that are sometimes needed to make the routines work. The inexperienced and newly employed do not, but they often have questions concerning the routines and provide valuable information. Wonders that may have a background in the experience they have from another workplace or from their education. Common to all the interviewed persons is that they do the work that is in focus.

Depending on the purpose of the FRAM analysis, the interviewed person can represent various professions, responsibilities, authorities, and assignments. They have one thing in common – they do the job.

If I use the FRAM to illuminate a serious or fatal incident, I would rather not interview persons who have been involved in the incident, since it is not the exact incident that is the focus but the routines around the incident. These routines can illustrate why something usually goes well but can sometimes go wrong.

I prefer to talk to one person at a time for the person to speak freely about adjustments. The duration of an interview is typically 30 minutes. I interview about one function at a time, but several functions can be included if the interviewed person is doing the job.

The interviews are visually facilitated to support reflection by the person being interviewed about their daily work. I place myself next to the person being interviewed with a large piece of paper on which a FRAM hexagon with the function in focus is inscribed. The paper is placed so that both I and the interviewed person can see it throughout the interview. I apply notes during the interview, visible to the interviewed person. The interviewed person can therefore correct the notes along the way and make additional comments. In this way, the collected data are validated at the same time.

Domain Expert

Table 5.1 shows the notes from an interview with a social and health assistant with 3 years of experience in the job. The interview was completed within 30 minutes. The notes are the basis of preparing a function in an FRAM model.

TABLE 5.1
Example of Notes from a FRAM Interview

Aspect	**Function**: To make the outpatient ready for ECT (electroconvulsive therapy) The function consists of several activities, for example, to admit a patient, to complete the patient record, to call First Call Physician, and to order an IV line
Input (I)	I1: A list on paper showing the patients expected to be treated on that date is handed over at 8.30 am I2: The order of the patients is fixed – first time patients, diabetics, the state of the patient I3: All patients have arrived (have been invited previously by the secretary)
Output (O)	O1: Ready for treatment O2: A call to the First Call Physician – ready to begin the treatment
Preconditions (P)	P1: The patient's vital parameters have been measured (Blood Pressure, Saturation, Respiratory Rate) P2: The patient has fasted P3: Observation of the patient P4: The ECT data if medication has been given, has been checked P5: The outpatient's contact person is there
Resources (R)	R1: The ECT Team R2: The outpatient's contact person R3: Patient records
Control (C)	C1: The printed list of patients C2: The list is shown on the CETREA screen C3: The monitoring and guiding of less experienced staff
Time (T)	T1: Other patients who want to talk T2: Alarms, where the staff must run to help

During the interviews, the ETTO rules (Hollnagel, 2009) helped me identify possible adjustments and trade-offs and thus details of the analysis. There are 17 ETTO rules, but especially 7 of them I have met often, to explain that an activity has been performed differently than planned:

Rule no. 1: *"It looks right"*
Rule no. 5: *"It will be checked later by another"*
Rule no. 6: *"It has been checked earlier by another"*
Rule no. 7: *"It is much faster in this way"*
Rule no. 11: *"We have always done it this way"*
Rule no. 13: *"It normally works"*
Rule no. 15: *"It must be ready in time"*

I use a very simple interview guide when I ask about the six aspects. It consists of 1–3 main questions for each aspect, and no more, because the next question lies in the answer from the interviewed person (Table 5.2).

When asking these questions, I have some advice for you:

- **Input (I)** is the start signal of a function. When I interview, I am persistent in ensuring accurate information about what starts the function. In an example from the function <To do a Ward round>, the first answer was that the ward round begins when the physician is ready. In-depth questions showed that the function only started when both the physician and the nurses were ready and had found each other.

TABLE 5.2

The Main Questions to Guide the Interview Process, as Linked to the Six FRAM Aspects

Aspect	Main Question
Input (I)	What makes you start the function?
	How do you know you need to get started with the function?
Output (O)	What is the result of the function?
	Who needs the result?
Resources (R)	What do you normally use when performing the function for the result to come out as expected?
Time (T)	Is there anything that disturbs your work?
	Should the function be performed within a specified period?
	Is it important that the function is performed before other functions?
Control (C)	How do you know what to do?
	Why do you do it in this way?
Preconditions (P)	What should be in place, completed before you start for the function to have the expected result?

If there are more inputs to a function, I will always ask in more detail which Input starts the function or whether it is random. It can be risky if a function can be started by several Inputs because we do not know which Input started it and when.

This will also apply if an Input is missing; that is, the person doing the job must remember to start the function. It is risky to have functions that depend on whether the staff remembers to start them. The consequence may be that the function is not started or started too late.

- **Output (O)** is what comes out of the function and which downstream functions need in one of the other aspects – Input, Preconditions, Resources, Time, or Control. I ask what the result of the function is – there may be several – and how the interviewed person knows that the work has been completed. If there are many Outputs, we consider going into depth with the function.

  If quantitative data exists for the Output of one or more functions and they are relevant to the purpose, I supplement the analysis with this data. The interaction with quantitative data is not described in this chapter.

- **Precondition (P)** is what must be in place, that is, completed, before a function is started to achieve an acceptable result. One example is that before starting an operation, the cleaning must be completed, and the patient must have consented. If there are multiple Preconditions for a function, I will ask if the interviewed person sometimes starts the function even if not all the Preconditions are in place. Sometimes this answer results in variability of the Output.

- **Resource (R)** is what the interviewed person needs or consumes while doing the job. If many different Resources are needed, it will also trigger me to ask more detailed questions, including the consequences of whether those resources are not present, they come too late or arrive in an imprecise format.

- In the **Time (T)** aspect, I focus on whether there is something that interrupts or disturbs the interviewed person while the job is being done. Moreover, what time pressure means for the function: whether the function is not performed at all or whether it is performed in a different way.

- In the **Control (C)** aspect, I ask how the person knows that the job should be done in the way the person has explained and demonstrated. Here the answer will often be *"We usually do it in this way," "The others also do it this way," "We have agreed to do it in this way," "We follow a checklist," "I have learned from a more experienced colleague," and "I have been supervised by a more experienced colleague."* The answer is often reflecting the culture and not the written procedures and instructions.

Moreover, it is crucial to listen without commenting and without making suggestions in this aspect. I never ask for procedures or instructions, but make a note if the interviewed person mentions it themselves. At times, I may have doubts as to which aspect the information it belongs to. However, if it cannot be clarified through the interview of the person, the most important thing is that the information is included, in whatever aspect.

To reduce the distance between what the interviewed person thinks they do and what they do, the interview is conducted where the job is done. If this is not possible, the interview can be completed by the person showing me the workplace. I see the workplace with new eyes and may see something that the interviewed person no longer sees. At the same time, I have experienced that the interviewed person remembers further details of the workflow.

I never record the interviews, as it affects the safe and secure environment between me and the interviewee.

Once information about how the job is done is collected, I build the FRAM model Work-As-Done.

The model is qualified by the interviewed person and subsequently by the colleagues who perform the same function. My experience is that the interviewed person and the colleagues confirm its representativeness when they state that the model gives a good picture of how they do the job: *"That's how it is!"*

METHODOLOGY AND RESULTS

This section describes the steps used to develop the FRAM analysis. Since the modelling of this use case is completely based on the traditional FRAM building steps (explained earlier in this book), this section presents the methodological contribution jointly with the empirical findings.

In particular, after posing particular emphasis on framing the purpose of the analysis (Step 0) and providing empirical evidence on how I iterated Step 1 to identify the system functions, a short summary of the intermediate building steps (Steps 2 and 3) is derived, reserving instead a longer discussion to Step 4 about the management of the variability.

STEP 0: RECOGNIZE THE PURPOSE OF THE ANALYSIS

The start of the FRAM analysis was to agree with management on its purpose. What should the FRAM analysis answer? Normally, the purpose of a FRAM model is aligned with management at a meeting prior to the start of the FRAM analysis. At the same meeting, I'm used to balancing expectations with the result of an FRAM analysis and to ensure that the management is aware of and accepts that they must go into a dialogue with the staff to identify possible interventions.

This step should be in place prior to the start of the FRAM analysis because its purpose determines the functions that are necessary and sufficient to answer the question. This action thereby limits the number of functions in the model. At the same time, precise wording in the purpose determines the degree of detail of the individual functions in the model, keeping in mind that it is not always necessary to describe all 6 aspects of each function to fulfil the purpose.

To ensure an overview of the work in focus, the method book of FRAM (Hollnagel, 2012) recommends staying in width before depth when building the FRAM model. Step 0 makes it possible to do so, especially when each function in the FRAM model includes many activities.

The purpose of the FRAM in the proposed use case was to identify the possible reasons for the overconsumption of pressure-relieving mattresses in the ward. To make it more precise, the purpose was to understand in depth the function <To select type of mattress>. Consequently, the starting point of the FRAM model is this function and its 6 aspects. The entire FRAM model develops from here. For example, the FRAM analysis showed that a planned resource (R) for selecting the type of mattress was "training of staff." It was not the purpose of this FRAM analysis to go deeper into the training, but the Output of the training is needed when selecting a mattress. The function <To train new employees in type and use of mattresses> is therefore a background function. As such, interviews are not further investigating other aspects of this latter function.

The purpose did not focus on why the previous intervention had failed, but on the daily routines of selecting the type of mattress and how the staff is doing this every day. Experience has shown that the insight and understanding of how the daily routines are executed can explain why a previous intervention failed.

Step 1: Identify and Describe System Functions

During the building of the FRAM model, I often ask: *Is this important to fulfill the purpose?* If not, the function will only be described with an Input or an Output (background functions) or an Output and Input. In such a case, information about the other four aspects (i.e., Precondition, Resource, Time, and Control) will not be collected during the interviews.

A procedure for selecting a type of mattress was available and as such was used to develop the model of the planned work (Work-As-Imagined). The resulting FRAM model (depicted in Figure 5.1) was used in the preparation for the interviews of the staff doing the job.

As mentioned earlier, the starting point of the FRAM analysis, that is, the function <To select type of mattress> is reported as a hexagon on the left side of Figure 5.1. When I build a model, I always colour the hexagons to distinguish the operator doing the work, or the patient's activities. A person on a higher hierarchical level can also perform specific functions (e.g., preparing procedures, defining criteria, principles, or policies, or training staff), in this case, deserving a different shade of grey.

It is possible mentally "to go through" or "to think through" a small FRAM model, as described in this chapter. However, the same exercise becomes harder for larger models. If I have to work with a larger FRAM model, I split the model into smaller

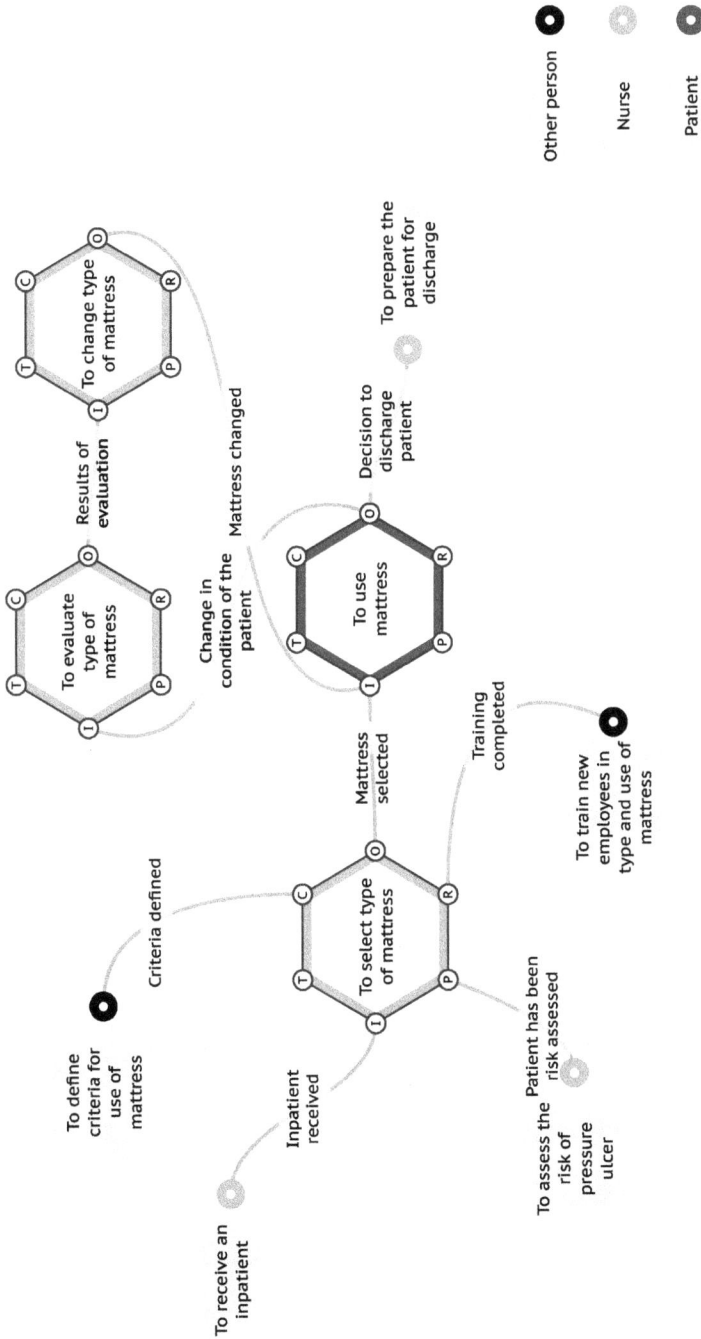

FIGURE 5.1 Work-As-Imagined FRAM model for mattress management.

models but keep the larger model to give an overview and show how the smaller FRAM models fit in like puzzle pieces. Since the Work-As-Imagined model has 10 functions, it qualifies as a small model.

___

Domain Expert

___

Let me talk to you about the model itself from left to right. Obviously starting from <To select a type of mattress>, it is clear that its Input is the Output coming from the function <To receive a patient>, i.e., 'Inpatient received'. The aspects Input, Precondition, Resource, Time, and Control do not need to be described for this function because it is not needed to fulfil the purpose as described in Step 0.

The Precondition for selecting a mattress is related to the function <To assess the risk of pressure ulcer>, and the Resource is that staff have been trained in selecting the professionally correct mattress, as for the function <To train employees in type and use of mattresses>. A Control of the function comes from <To define criteria for use of mattress>. The time aspect was not mentioned in the procedure of selecting a type of mattress, and as such not reported.

All the 4 mentioned functions are background functions to the function <To select type of mattress>. The Output from the functions is needed to complete the model, but to fulfil the purpose as described in Step 0, you do not need to go into details with the other 5 aspects of these functions.

Depending on the result of the risk assessment, the knowledge of staff through the training and the criteria for use of mattresses, a type of mattress is selected by the nurse receiving the patient.

Patients who are transferred to the ward from intensive care are always lying on a pressure-relieving mattress due to their vulnerable condition. This is indicated by the background function <To receive patients from intensive care>.

At this stage, the patient uses the mattress.

When there is a change in the condition of the patient, the nurse must evaluate the type of mattress in use, that is, <To evaluate type of mattress>, and the nurse must ask themselves and answer the question: Is the mattress in use still professionally correct to support the rehabilitation of the patient?

This evaluation can result in using a new type of mattress, and the mattress in use must be changed, that is, <To change the type of mattress>. The latter function loops back to the function <To use the mattresses>. The patient is using the mattress until the decision to discharge the patient has been made. The Output from the function <To use the mattress> is connected to the function <To prepare the patient for discharge> as an Input.

Even through these only a few exemplary aspects, it appears evident how the FRAM model of Work-As-Imagined (Figure 5.1) gave the necessary and sufficient overview for starting to interview the staff doing the job.

___

To complement the Work-as-Imagined with the Work-As-Done, for this case, I interviewed 2 nurses and 1 doctor, in individual sessions lasting 30 minutes each, as sketched in Figure 5.2. The result surprised management and reflected fully the wording of Peter Drucker: *Culture eats strategies for breakfast!*

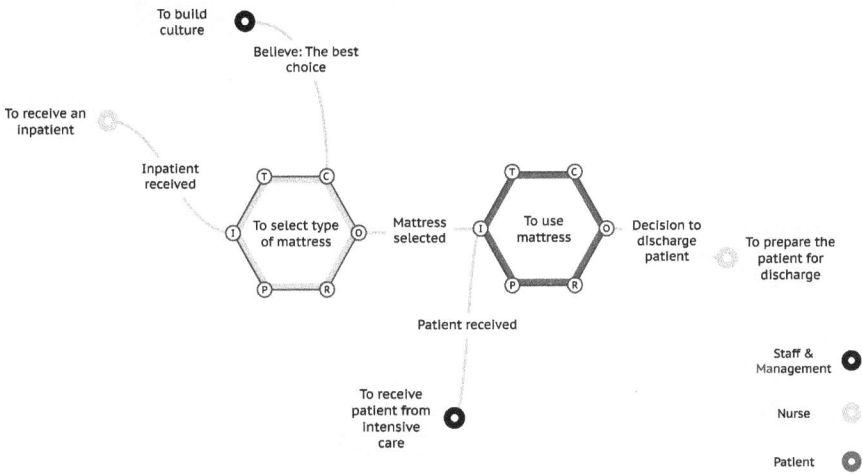

FIGURE 5.2 Work-As-Done FRAM model for mattress management.

Domain Expert

Let me once again talk to you through the model for how the work is done.

The interviewed nurse told me: *"When I receive a patient, I choose a pressure-relieving mattress. No matter what. It is the best mattress, and therefore I do not need to assess the risk of pressure ulcers. I can't choose a better one."* This statement is covering both Input (*When I receive a patient,....*) and Output (..., *I choose a pressure-relieving mattress. No matter what.*) of the function <To select type of mattress>. The Input to the function must be an Output from an upstream function <To receive an inpatient>. This function is included in the FRAM model for Work-As-Done. The function is not in focus in relation to the purpose of the FRAM analysis and is only described with the Output aspect, and therefore a background function. The statement also showed that the risk of pressure ulcers was not assessed. The planned function <To assess the risk of pressure ulcer> is not activated and therefore not shown in the FRAM model for Work-As-Done.

When I asked about the training in selecting a type of mattress, the nurse answered: *"Yes, I had brief training in the different mattresses and the criteria for using them. But it is not so important, because the pressure-relieving mattress is the best choice."* This statement is related to the Resource and the Control of the function <To select type of mattress>. In the FRAM model, the Output from the function <To train new employees in type and use of mattresses> is not used by the nurse and therefore not shown in the FRAM model for the Work-As-Done. The nurse does not need to use the knowledge gained through the training because the nurse had already chosen the best mattress for the patient – the pressure-relieving mattress. The planned control of the function to select a type of mattress is the function <To define criteria for use of mattress>. The nurse knows that the criteria are there, but by choosing the pressure-relieving mattress, she has already made the best choice. The function <To define criteria for use of

mattress> is therefore not controlling the function <To select type of mattress> and is not shown in the FRAM model for Work-As-Done.

I asked the nurse if the job of selecting the type of mattress was done under time pressure or disturbances/disruptions by other jobs or colleagues. The answer was: *No!* The FRAM model in Figure 5.2 visualizes this behaviour by showing no connections to the time aspect.

The nurse confirmed that patients from intensive care were transferred to the ward on a pressure-relieving mattress. Since the nurse believed this was the best choice, the type of mattress was not changed. In the FRAM model, this behaviour is visualized by connecting the Output from the function <To receive patient from intensive care> to the Input of the function <To use mattress>.

I got the same answer when asking about evaluating the type of mattress by a change in the condition of the patient: "*The pressure relieving mattress is the best choice.*" The consequence of this statement is that the planned functions <To evaluate type of mattress> and <To change type of mattress> are not activated and therefore not shown in the FRAM model for Work-As-Done.

The belief that the pressure-relieving mattress is the best choice is shown in the FRAM model for Work-As-Done as a Control of the function <To select the type of mattress>.

The other nurse gave similar answers.

One of the nurses mentioned that they did not have to turn the patient so often when they were lying on a pressure-relieving mattress. It saved time for other things to do. In addition, not changing the mattress saved time because no cleaning or logistics of mattresses were needed.

The interview of a doctor revealed that a pressure-relieving mattress does not always represent the most professional, correct choice. At a certain point in the recovery of the patient, the patient must be able to turn themself; something we all do during sleep.

The doctor also informed me that the criteria for selecting a type of mattress were based on evidence.

The FRAM model for Work-As-Done (Figure 5.2) showed that crucial functions to achieve an acceptable and satisfactory use of pressure-relieving mattresses were bypassed/not activated and how strong culture was controlling how the job is done.

The belief of the nurses that a pressure-relieving mattress always, no matter what, was the best type of mattress for the patient was so strong that it overruled procedures of risk assessment and evaluation, training, and defined criteria for selecting the correct mattress.

Building the FRAM model is the last thing you do in Step 1.

STEP 2: IDENTIFY VARIABILITY

Step 2 about variability was very easy to complete. The variability from the function <To select a type of mattress> is very low, since normally the only selected mattress was the pressure-relieving mattress and no other types of mattresses.

STEP 3: AGGREGATE VARIABILITY, ACTUAL AND POTENTIAL

Step 3 is where you walk through the FRAM model. In larger FRAM models, there are often different ways through the model, the so-called instantiations. In this FRAM model, there is only one way through and, as such, one single instantiation.

STEP 4: ASSESS THE CONSEQUENCES OF THE ANALYSIS

In Step 4, management and staff together discuss the result of the FRAM analysis, identify, and agree on possible interventions. The FRAM models "Work-As-Imagined" and "Work-As-Done" are a common starting point for their discussions.

In the case at hand, a common meeting was set up. I presented the result of the FRAM analysis for the management and the staff (i.e., doctors and nurses) both in terms of Work-As-Imagined (Figure 5.1) and Work-As-Done (Figure 5.2). The presentation was supported by showing or talking through the two FRAM models.

During my presentation, I pointed out several issues to support the understanding of the FRAM model:

- A risk assessment of pressure ulcers is indicated as a Precondition for choosing a mattress. The nurses expressed that this was not carried out. There was no need to prepare the risk assessment of pressure ulcers because they had already made the best choice. They believed that *"the pressure-relieving mattress, no matter what, was the best choice for the patient."* The informal belief weighs more heavily than the formal risk assessment.
- The strongest Control when choosing a mattress was not the clinical criteria but the belief that *"the pressure-relieving mattress is the best choice for the patient."*
- The evaluation (re-assessment) of the initial selection of mattress is not carried out when the patient's condition changes because the patient is already lying on *"the best mattress."*
- Additionally, the selection of the pressure-relieving mattress saved time in many ways: no time-consuming risk assessment and evaluations, less turning of the patient, no change of mattress and therefore less cleaning and logistics.
- In the intensive care unit, patients are always placed on a pressure-relieving mattress, as the patient is in a critical condition, that is, when the patient is transferred to the ward, they will be on a pressure-relieving mattress. After being received in the ward, the choice of a pressure-relieving mattress is not re-assessed, because *"the patient has already the best type of mattress."*
- Staff are taught how to select and use the mattresses that are professionally correct. The ward is responsible for the training, and there is no standardized training material, but the training is overruled by the belief that the pressure-relieving mattress is the best choice.
- The high-quality solid mattresses had not been included as a possibility for selection and had not been introduced to the nurses.

The visualization of the functions in the two models and how the functions depend on and interact with each other, and the consequences of the strong belief among the nurses that a pressure-relieving mattress is the best choice, was a very strong basis for a discussion between management, doctors, and nurses. They all understood that they believe in the pressure-relieving mattress as "*the best choice for the patient*," overruled procedures, training, and defined criteria for selecting the professional correct mattress.

No one blamed each other. At the common meeting management, doctors and nurses agreed that change was necessary and that they had to implement several actions to support the change in the culture of "best choice." During the dialogue, they agreed on the following interventions:

- Include high-quality, solid mattresses in the criteria for selecting a professionally correct mattress to support the recovery of the patient.
- Retrain the nurses in the selection of the type of mattress with a focus on how the type of mattress supports the rehabilitation of the patient, the criteria, and the risk of pressure ulcers. Both doctors, nurses and management wanted the nurses to continue selecting the mattress when receiving the patient.
- Prepare standardized training material for the training of new employees in the selection of mattresses.
- Centralize the training to dampen the variability of the result of the training.
- Include the evaluation of the type of mattress in the ward rounds so that both the doctors and the nurses are involved in the decision to change the type of mattress during hospitalization.
- Look into the cleaning and logistics issues concerning the change of a mattress.
- Look into how the changes affect the nurses' time to do the job.

Management and staff believed that these changes would improve reliability, quality, and patient safety (all patients get the professionally best mattress in relation to recovery) and would reduce the consumption of pressure-relieving mattresses and improve economics at the same time.

Since the FRAM is not an implementation tool, the list of possible interventions was used in the improvement work and typically structured according to the PDSA Cycle (Plan-Do-Study-Act) and the consumption of pressure relieving was monitored to see if the changes have an effect on the consumption.

Some of the interventions were easy to implement, some others needed more time and resources.

My experience is that the FRAM analysis often results in numerous interventions affecting all 5 aspects of a function, for example, in the conditions for doing the work. It is worth remembering that further training efforts mainly affect the resource of the function and sometimes the control aspect if in-situ training is offered, and that a new procedure only affects the control aspect, but not always. Therefore, it is important to bring all 5 aspects into the game when interventions are discussed and decided.

DISCUSSION

The case of the consumption of pressure-relieving mattresses shows that the FRAM provides a broader perspective on the problem and strengthens the basis for gaining common insight and understanding, both for management and staff. The visualization of the FRAM model for Work-As-Done is a strong basis for discussing possible interventions for improvement. In the case, the differences between the two FRAM models for "Work-As Imagined" and "Work-As-Done" were a surprise for both the staff and management. Conversely, the same differences made them discuss in a constructive way during the joint meeting.

The next step of using the FRAM in the case is to follow up on the implementation of the interventions approximately after one year. Another possibility is to describe the other aspect of the background functions to gain extra knowledge on diverse aspects (e.g., about how the training was done before, with the intention to identify the right focus for centralized and standardized training).

Compared to many of the other FRAM models I have completed during the last 12 years, this FRAM analysis offers empirical evidence on how strong culture is in controlling how the work is done, overruling procedures, training, and requirements in the form of criteria.

It also showed how people doing the job can simplify the planned work, believing they are doing the best for the customer (the patient); the system productivity, that is, no time-consuming risk assessment and evaluations, less turning of the patient, less cleaning, and less logistics of mattresses; and by-passing activities when they are not meaningful to sustain the goal linked to the recovery of the patient.

CONCLUSIONS: MY PERSONAL COMMENTS

In general, interventions should aim at supporting the day-to-day work that is already working and to dampen unwanted variability, as well as reducing dependence on human memory. The implementation of change can be supported by monitoring (quantitative data) and using the FRAM to follow up on the implementation of the change, for example, after 60, 90 days, and 365 days. New models are being prepared for each follow-up to see if the intervention is reflected in daily practice or whether other actions need to be taken.

The critics of using the FRAM often say that it is very time-consuming. My experience is that smaller, focused FRAM models typically take 12 hours to be developed: 1–2 meetings with management, 5 interviews of 30 minutes each, preparation and model building, and 1 joint meeting between staff and management to discuss the result and possible interventions.

Except for the time management invested in the project when joining the meetings, the FRAM analysis only takes the staff out of production for 2.5 hours. The majority of the time is used by the analyst to prepare, collect data, and build the model.

In the Region of Southern Denmark, the management is taught: "*Know your business. Do your business. Improve your business!*" The FRAM is a powerful tool for at least two of these actions, that is, whenever you want to know what is going on in your business and want to identify possible improvements.

REFERENCES

Hollnagel E. (2009), *The ETTO Principle – Efficiency-Thoroughness-Trade-Off: Why Things That Go Right Sometimes Go Wrong*. Ashgate Publishing.

Hollnagel E. (2012), *FRAM: The Functional Resonance Analysis Method, Modelling Complex Socio-technical Systems*, Ashgate Publishing.

Hollnagel E. (2021), *Synesis, The Unification of Productivity, Quality, Safety and Reliability*, Routledge.

6 Function Identification and FRAM Modelling of an Aviation Near-Miss Using Convergent Mixed Methods

Riana Steen and James Norman

CONTEXT: AVIATION OPERATIONS AS A SYSTEM

From 2009 to 2024—15 years—there was not a *single* mass casualty accident in US aviation and nearly the same record in Europe.[1] It was as if Hollnagel anticipated this condition and preemptively devised a theory that gave normal work, incident, and accident analysis a similar relational viewpoint. This is doubly profound, given that global aviation operates approximately 100,000 flights per day, transporting 11 million passengers, while maintaining this level of safety (IATA, 2024). Therefore, aviation provides ample opportunity to apply the concepts related to Safety-II, including the ETTO principle, resilience, and the focus of this book, the FRAM for modelling socio-technical systems.

In this chapter, we offer the reader what we aim to be both interesting and useful by way of a case study of a ground-based near miss at JFK airport in New York, United States. This non-injurious and non-impact event triggered a full investigation by the National Transportation Safety Board (NTSB), resulting in 613 pages of textual report data, as well as audio files from air traffic transmissions, ADS-B trajectory files, and flight data recorder 'black box' files.

Although the situation of a near miss in aviation as well as other domains is an age-old phenomenon, this study in part seeks to address what we see from this event: an increasing, if not overwhelming, amount of data available to safety investigators today (Thoroman et al., 2018). Historically, only a substantial mishap or fatal event would trigger a full government investigation. The National Transportation Safety Board (NTSB) defines their investigation trigger as a serious injury or death or substantial damage to the aircraft (NTSB, 2024). Similarly, ICAO Annex 13 states that an investigation is mandatory after a fatality, aircraft damage, or the aircraft goes missing (ICAO, 2020).

DOI: 10.1201/9781003518167-8

However, today, the NTSB has broadened its strict interpretation of the 'accident' mandate and may conduct full investigations at its discretion. Due to the dearth of commercial aviation accidents, investigators' skills may become diminished over time. Perhaps recognizing that investigative skills are perishable, the NTSB has begun to provide full investigation resources and publish reports even when no collision occurred or persons were injured. This has provided researchers with new sources of material to identify factors, relationships, and interdependencies with respect to aviation safety. However, at the same time, the increasing amount of 'big data' published after safety investigations can approach the cognitive limits of sense-making by researchers alone. For this reason, we introduce elements of machine learning into the process. Our goal is to provide a robust FRAM using an enhanced methodology that allows for a nuanced understanding of the near miss, leading to wisdom elicitation for the industry.

Although a combination of quantitative and qualitative methods (i.e., mixed methods) exists for exploration of resonance and interdependencies (Patriarca et al., 2017), a mixed methods approach is reflected only limitedly in the available literature, specifically concerning the identification of FRAM functions. At the same time, an increasing amount of data is available for the study of near misses, yet pure qualitative researchers may struggle to adequately absorb such volume. To address this challenge, cybernetic theory, evolved as the Viable System Model (VSM) (Beer, 1984), is useful due to its emphasis on regulatory, autonomous, and adaptive mechanisms, which align with the approach of the FRAM to understanding and managing dynamic interactions and variabilities in aviation safety. This offers a systems approach towards organizing such data, because machine learning can aid in identifying themes, yet it has not been leveraged with the goal of a unified FRAM model.

Domain Expert

For this study, we follow the use by Madsen et al. (2016) of the ICAO incident framework to define *near miss* as 'outcomes that could have been worse, but for the intervention of good fortune'. It is important to note that the ICAO does not define near miss, rather using the term AIRPROX with respect to the distance between aircraft as compromising safety (ICAO, 2023). We chose near-miss rather than incident, following Madsen et al. (2016) methodology that near-misses are a) a subset of incidents and b) near-misses have a more prominent connotation in aviation.

In aviation, accidents, incidents, and near misses often occur due to the interactions and interdependencies among multiple components within a system. Oftentimes, these interactions are not anticipated through design. Therefore, understanding how different elements influence each other and contribute to the dynamics is crucial (Milch & Laumann, 2016).

For example, the Air France 447 accident in 2009 revealed complex interrelationships between pitot tube design, autopilot logic, Airbus stall awareness, and pilot training (Palmer, 2013). The Qantas 32 engine failure in 2010 presented hundreds of computer messages to the crew, causing the captain to later say 'I lost all my free mental space... My mental model of the aircraft had failed' (Shorrock, 2019).

The objective of this study is to provide an integrated approach to identify, describe, and analyze system functions and the emergence of functional resonance, thus offering a foundation for understanding systemic behaviours and vulnerabilities and identifying functions. This objective emphasizes the need for advanced analytical methods capable of dissecting the nuanced interplay of human, technological, and organizational elements that characterize these systems. Addressing the intricate dynamics of socio-technical systems, the troika of FRAM, Natural Language Processing (NLP), and the VSM brings a new approach in methodological sophistication.

Furthermore, our contribution to Safety-II and safety science is threefold with respect to the theoretical, methodological, and practical aspects of the FRAM. First, the theoretical framework is triangulated by cybernetics, safety management, and artificial intelligence. These are then converged to construct a FRAM model of an aviation near miss. Second, we demonstrate a novel methodological approach by conducting a concurrent research method in a case study. Third, we offer practitioners a practical and pragmatic approach to determining the functions and scope of their FRAM construction, guided by the VSM framework.

System Theory and the FRAM

Over the past two decades, academic literature has contributed significantly to the analysis and understanding of sociotechnical systems through the development of concepts, methods, and tools. Among these developments, the application of systems theory stands out as a key contribution. One of the foundational definitions of system theory goes back to the 1950s, attributed to the pioneering work of Ludwig von Bertalanffy. He conceptualized system theory as a logic-mathematical field, fundamentally formal in its structure, yet broadly applicable across various scientific disciplines concerned with systems (Von Bertalanffy, 1950). Over the decades, scholars have attempted to refine and conceptualize this theory further. Despite the absence of a universally acknowledged explanation or application of system theory (Arnold & Wade, 2015), the emphasis shifts from isolated components to the intricate web of interrelations and dependencies that characterize a system (Espinosa & Walker, 2011; Hollnagel, 2016; Leveson, 2011).

The primary sources that guide FRAM practitioners describe a 'decomposition method', reducing complex systems into tractable components (Hollnagel, 2017). Although intuitive, this also presents challenges. As discussed by Li et al. (2019), the identification of FRAM functions varies by subjective judgement, leading to potential inconsistencies when performing a FRAM analysis. Furthermore, there is significant debate in the Safety-II and resilience engineering domains regarding a reductionist approach (Dekker et al., 2008; Dekker et al., 2011), as well as the possibility of inducing a reductive bias during incident investigation (Pruchnicki, 2018). The method described for this chapter considers all elements of the system simultaneously, rather than in isolation. We believe that this aligns well with systems theory and has the potential to move its theoretical application beyond accident investigation and into incident investigation (Muecklich et al., 2023).

Additionally, current FRAM construction may benefit from a more holistic view of system design. We believe that this will address the need for models that not only map detailed interactions but also integrate a broader system perspective. This balance is important for making FRAM both practical and insightful as a tool for practitioners. In this regard, we propose the application of the VSM.

The Viable System Model (VSM)

Stafford Beer developed the VSM, derived from the principles of how an organism functions to ensure its viability and adaptability (Beer, 1984). Beer recognized that the viability of any organism, or by extension any organization, depends on its ability to adapt and self-organize, regarding the dynamicity of its operating environment. In the VSM, the system is comprised of components, each with defined functions that are essential for the organization's cohesion and efficiency. This model delineates the functions of these components, capturing their complexity through recursive structure and communication patterns. Beer's taxonomy of organizational functions, designated Systems 1 to 5, each with its own specificity.

VSM comprises the system's components:

- System 1 is dedicated to performing the primary tasks the system is designed to perform.
- System 2 ensures coordination, preventing conflicts among primary activities.
- System 3 supervises technical controls, manages resources, and ensures subsystem integration.
- System 3* audits internal and external behaviours to enforce compliance with System 3's objectives and System 2's regulations.
- System 4 scans external environments, anticipating changes that could affect system performance.
- System 5 governs policy and strategic decision-making, setting overarching objectives by drawing on insights from System 4, even in highly uncertain situations.

The VSM principles underpin the essential interconnectivity of subsystems (Systems 1–5). To ensure organizational cohesion, these subsystems are to iteratively cooperate at all levels within a recursive structure of autonomous units.

The recursions of the viable system can be extended upward and downward. In the FRAM terminology, it is related to the 'foreground and background' functions, which contribute to performance variability (Hollnagel, 2012). However, within current research, the practice of defining functions in the development of FRAM predominantly emphasizes single-level analyses. To broaden this approach, Patriarca et al. (2017) proposed a framework that integrates abstraction hierarchy with FRAM. However, while the abstraction hierarchy concentrates on multi-level agents, it may not fully capture the complexities of the entire system and its dynamic operations.

Through the lens of VSM, subsystems are not exclusively tied to separate individuals, agents, or departments; especially in smaller organizations, a single person or unit may fulfil multiple subsystem roles based on their diverse activities (Hoverstadt, 2020, p.90). By analyzing the structures, components, and relationships between these subsystems, for example, key processes, communication, and information flows, the VSM provides a comprehensive overview of an organization's network model (Steen et al., 2024a, b). This structure is essential in aviation, where the autonomy of each division improves the overall responsiveness and resilience of the system, allowing it to adapt to operational variances and emerging situations effectively. Put another way, aviation components often operate in tightly coupled, decentralized settings, such as pilots on the flight deck, maintenance technicians in a hangar, or air traffic controllers in a tower.

In addition to its application in the design of systems (Ruiz-Martin et al., 2017), the VSM also offers a framework for the identification of pathologies and the performance of diagnostic analysis, encompassing structural, functional, and communicational aspects (Schwaninger & Scheef, 2016). In VSM, pathology refers to the inherent dysfunctions or maladaptations within a system, while diagnosis is the process of identifying and understanding these issues. Structural pathologies in complex systems stem from intricate interconnections between subsystems, complicating decision-making and responsibility allocation due to the vast number of network participants. Functional problems occur when certain organizational subsystems fail to meet their goals, leading to inefficiencies such as delays and errors. Communicational diagnostics affect decision-making and information sharing, affecting coordination activities between subsystems (Steen et al., 2024a, b).

Additionally, the principle of recursion characterizes an organization as a viable system nested within a broader set of viable systems. Each system, at every organizational layer, from ground operations to air traffic control, mirrors the structure and functionalities necessary for viability, creating a fractal pattern throughout the aviation organization. This ensures uniformity in operation from individual flight crews to the entire airline, enhancing stability and adaptability. Furthermore, integration of key subsystems, such as flight operations, maintenance, crew scheduling, and air traffic control, is essential for the functionality of sociotechnical systems such as aviation. These subsystems are vital as they work to create safety and efficiency, supporting sustainable flight operations through their interdependence and connectivity.

Applying the VSM to analyze aviation near-misses and their systemic causes facilitates the identification of subsystems and the relationships among them. The following sections introduce the case study of the JFK near-miss, illustrating the application of VSM and NLP in developing a FRAM-based analysis.

DATA COLLECTION

The last time you were on an aeroplane, did you happen to look out the window and wonder what all the signs with letters, numbers, arrows and symbols meant? Despite the notion that flying is the most challenging part of aviation, it is often ground operations. Every day, pilots and controllers participate in a delicate choreography to move aircraft where they need to go on the ground, which can be complicated by complex airport layouts, low visibility, poor weather, foreign accents, or even controller cadence. In fact, the greatest disaster in the history of aviation occurred on the ground, not in the air. On March 27, 1977, multiple aircraft were diverted to the island of Tenerife in the Canary Islands. Due to low visibility and confusion on the air traffic frequency, two 747s collided, killing 583 passengers.

Over the years, the hazards of airport operations have been mitigated by both technology and training.

Domain Expert

For example, there is ASDE-X, a ground-based algorithmic system that predicts aircraft and vehicle movements (NTSB, 2023a, 2023b, 2023c). Airport vehicles and aeroplanes are equipped with transponders that allow others to see them should visibility degrade. Pilots must read back all directions given by air traffic controllers in their entirety, as a closed-loop assurance system. Airport markings and signage have been standardized around the world by the International Civil Aviation Organization (ICAO), ensuring pilots have a predictable and consistent environment wherever they go. However, despite these layers of safety, failures occur.

In 2023, at JFK Airport, an American Airlines 777 mistakenly crossed runway 4L while a Delta 737 was taking off. The ASDE-X alert system triggered, and Delta aborted its takeoff, avoiding collision by 316 metres. The American crew continued to London, while Delta's flight was cancelled.

On January 13th, 2023, at John F. Kennedy International Airport in New York, United States, a London-bound American Airlines 777 received direction from air traffic control to taxi from the terminal to the runway, a distance of about 4 kilometres. While taxiing, the three pilots in the aeroplane were busy obtaining performance data for take-off, as well as being notified by their company of turbulence along their route. Furthermore, during this time, air traffic control were changing the JFK runway configuration, so that the new runway was not the original one that the pilots had been planning. As an added workload, American had recently implemented changes to procedures, which contributed to the divided attention. Workload has been shown to have insidious qualities among flight crews that are often dealt with using resilient strategies (Steen et al., 2024a, b).

As the 777 was moving southeast bound on taxiway B (Figure 6.1, dotted line initial and end points are marked on the left of the figure), it was told to turn right on taxiway K and cross runway 31 Left. Instead, it went straight ahead, crossing runway 4 Left (Figure 6.1, solid line: initial and end point are marked on the right of the figure). At the same time, a Delta Air Lines 737 was taking off

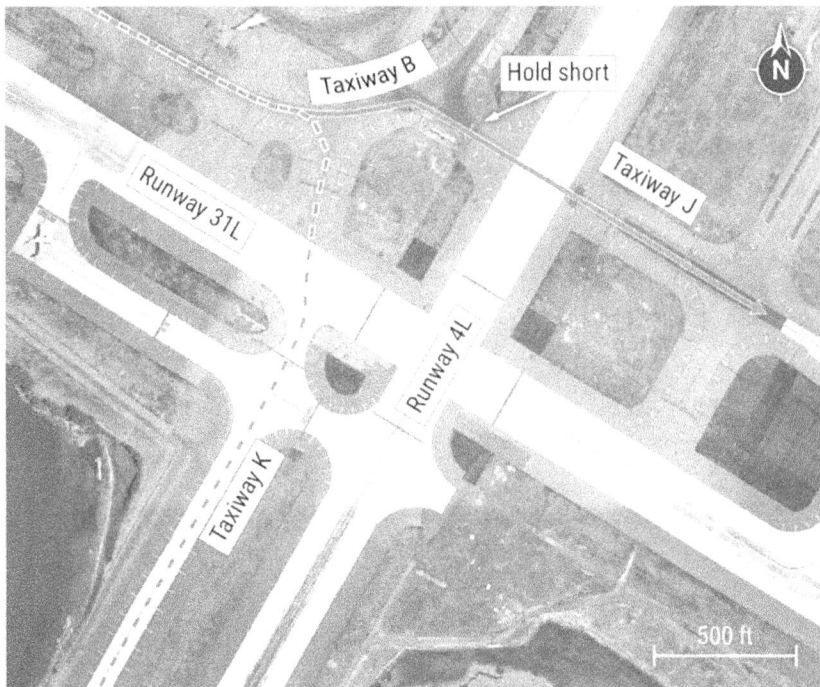

FIGURE 6.1 Satellite image of the incursion area. AAL106's assigned (dashed) and actual (solid) taxi routes overlaid.

from runway 4 Left. As the Delta aircraft was accelerating, the ASDE-X system issued an aural 'runway occupied alert' to the air traffic controller, who immediately instructed Delta to cancel its takeoff clearance. Delta immediately reacted, and the takeoff was discontinued.

After the event, the Delta aircraft returned to the gate, and the flight was cancelled. The American crew cleared the runway and received an official FAA notification of a possible pilot error. They then consulted with their company and pilot union and elected to continue to London.

Subsequent analysis found that the Delta 737 and American 777 came within 316 metres of collision (NTSB, 2023a, 2023b, 2023c). If this event had been catastrophic, it would have been the seventh-deadliest accident in aviation history.

This study used secondary sources, which consisted of information previously collected and repurposed for new research, sourced from existing publications, databases, or archives (Hox & Boeije, 2005). Using secondary data in this context to identify the FRAM function anchors this study, providing a reliable reference sourced from established databases such as the NTSB. This approach facilitates the confirmation of the findings by future researchers. We analyzed the following six documents, as shown in Table 6.1, which together comprised 467 pages.

TABLE 6.1
Data Sources Details

Document	NTSB Group	Title	Page Count	Word Count	Citation
D1	Air Traffic Control	Factual Report	20	3,503	NTSB (2023a)
D2	Air Traffic Control	ATC Interview Transcript	204	40,985	NTSB (2023b)
D3	Operational Factors & Human Performance	Factual Report	41	11,061	NTSB (2023c)
D4	Operational Factors & Human Performance	AAL Pilot Interview Transcripts	195	42,190	NTSB (2023d)
D5	Operational Factors & Human Performance	Delta Air Lines Pilots' Statements	3	839	NTSB (2023e)
D6	Operational Factors & Human Performance	AAL Pilots' Statements	4	1,423	NTSB (2023f)

METHODOLOGY

The researchers adopted a convergent mixed methods approach (Creswell & Creswell, 2017, p. 217). This approach involves collecting and analyzing qualitative and quantitative data simultaneously and separately, and then comparing the results to see if the findings confirm or disconfirm each other, thus improving the quality of the data analysis. The findings of both strands are compared, contrasted, and combined to draw conclusions. This approach allowed us to thoroughly investigate the complexities involved through four phases after the first data collection actions:

- *Phase 1: Applied thematic analysis and coding*
- *Phase 2: Data partitioning using VSM*
- *Phase 3: Data analysis using NLP*
- *Phase 4: FRAM model development and analysis*

The focus of this chapter centres on FRAM function identification, which comprises Phases 1–3, while we also formally derive a final FRAM model in Phase 4 and elaborate on its implications.

PHASE 1: APPLIED THEMATIC ANALYSIS AND CODING

We conducted our research using applied thematic analysis (ATA), a method known for its descriptive and exploratory orientation. ATA, a specific type of qualitative data inductive analysis, involved multiple analytic techniques, as outlined by Guest et al. (2012).

In this ATA application, the researcher-led analysis examined the six official investigation documents, reported in the data collection section. The analysis used a multilayered and iterative technique at three levels, according to different levels of abstraction.

Method Geek

The three levels of abstraction utilized a cumulative approach, starting with a descriptive level. This revealed 48 first-order codes. The second level then moved to an interpretive level, which sought to identify recurring themes relevant to the research purpose and to ensure they effectively captured commonalities. For example, codes such as 'distraction', 'shift in focus', and 'operational condition' were grouped under the code 'situational awareness'. The third level then combined these insights with NLP, allowing us to capture the nuances within the textual data. The resulting process was then applied in FRAM modelling, yielding the functions presented in the next sections.

PHASE 2: DATA PARTITIONING INTO VSM SYSTEMS

After first-order coding, we carried out a second-order analysis and partitioned them into the five VSM systems to support multi-level recursive analyses. A simplified version of the VSM for the case study is presented in Figure 6.2. For each system (i.e., S1-S5), we identified themes based on the ATA thematic coding process.

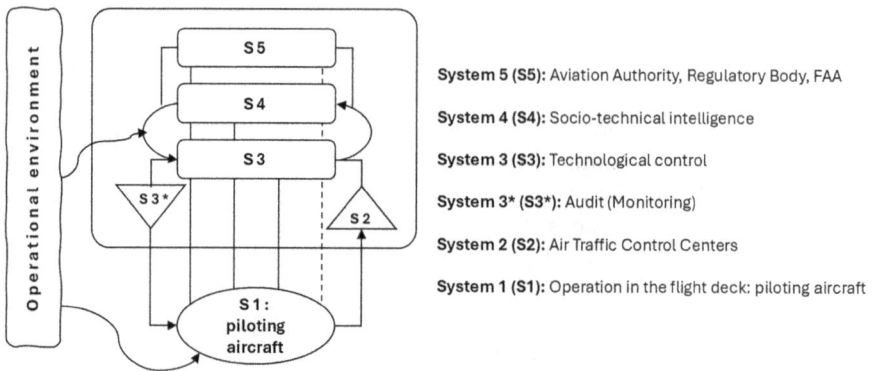

System 5 (S5): Aviation Authority, Regulatory Body, FAA

System 4 (S4): Socio-technical intelligence

System 3 (S3): Technological control

System 3* (S3*): Audit (Monitoring)

System 2 (S2): Air Traffic Control Centers

System 1 (S1): Operation in the flight deck: piloting aircraft

FIGURE 6.2 Simplified illustration of VSM components for the analysis of the JFK near miss.

For this study, partitioning the VSM in the manner of Figure 6.2 allows us to focus specifically on entities at the airport level: pilots, airlines, and the regulator. This analysis yielded 28 second-order codes, which were systematically categorized and stored in separate spreadsheets corresponding to Beer's systems 1 through 5.

Due to the preexisting theoretical framework of VSM, a deductive analysis was employed to partition textual data into VSM systems (Clarke & Braun, 2017; Fereday & Muir-Cochrane, 2006). This method allowed researchers to label qualitative data when categories were predefined. Deductive analysis contrasts with inductive analysis, where the researcher allows themes to emerge without predefined criteria, as in a grounded-theory approach. Deductive approaches in safety science have been employed for adaptive capacity modelling (Taarup-Esbensen, 2023), safety climate-performance modelling (Syed-Yahya et al., 2022), and fault tree analysis (FTA) (see Diop et al., 2022), among others.

The partitioning process involved an analysis of the operational context in terms of its environment. In the VSM view, the term *environment* refers to the external surroundings or context in which a system operates (Preece et al., 2015). Within the context of aviation flight operations, it encompasses various elements such as weather conditions, air traffic flow, runway conditions, and regulatory requirements, among others. Understanding and managing the dynamic interaction between the aviation system and its environment is essential to ensure safe and efficient operations.

Phase 3: Data Analysis Using NLP

To supplement the thematic coding, the researchers employed NLP for the analysis of the 467 pages of NTSB documents. The goal of this phase was to take advantage of computing power in identifying topics and themes in the data that may otherwise have remained opaque with a human-only derived analysis. A dedicated NLP pipeline was created using Python 3 programming in a Jupyter 6.5.2 notebook.

Based on the results of data partitioning as described in Section 4.4, a spreadsheet file was created including five sheets, each representing one of Beer's VSM Systems

(Systems 1–5). To ensure that each sheet was distinct enough to be considered a separate Beer system, a similarity matrix was computed using cosine similarity, as recommended in the academic literature (see, e.g., Huang, 2008).

Method Geek

A practical threshold often used is 0.5, with a cosine similarity below this value to be considered sufficiently dissimilar and treated as separate constructs (Crocetti, 2015). This threshold was met sufficiently, with a mean similarity of 0.22 and the highest value of 0.36.

A standard NLP pipeline was then developed to preprocess the textual data, including removal of standard English stopwords, lemmatizing, lowercasing, removing punctuation, and tokenizing. After inspecting the initial output, additional words deemed irrelevant by an aviation subject matter expert were manually removed by coding additional stop words into the Python script, such as 'go', 'lot', and 'try'. This process helped remove excess terms during analysis that did not add thematic value.

The preprocessed text from each sheet was combined into a corpus, and a Term Frequency-Inverse Document Frequency (TF-IDF) vectorizer was employed to extract the most significant terms for each system. TF-IDF identifies important terms by balancing their frequency in the document against their frequency across all documents or, in this case, VSM systems.

Method Geek

TF-IDF is calculated using the following equation:

$$\text{TF} - \text{IDF}\left(t,d,D\right) = \text{TF}\left(t,d\right) \times \text{IDF}\left(t,D\right)$$

where TF(t, d) represents the term frequency of term t in document d, and IDF (t, D) represents the inverse document frequency of term t across the corpus D.

For example, the term 'runway' received the highest TF-IDF score for System 1 (pilots) according to this calculation. This was calculated by a Term Frequency (TF) of

$$\text{TF}\left(\text{runway},d\right) = \frac{f\,\text{runway},d}{N_d}$$

where $f_{\text{runway},d}$ is the frequency of 'runway' in sheet d (System 1), and N_d is the total number of terms in the sheet.

The inversion document frequency (IDF) is calculated as:

$$\text{IDF}\left(\text{runway},D\right) = \log\left(\frac{N}{1+n_t}\right)$$

where N is the total number of sheets, and n_t is the number of documents containing the term 'runway'. Multiplying the TF and IDF scores resulted in a value of 0.055.

It should be noted that the Python library 'TfidfVectorizer' normalizes and adjusts its calculation in addition to the equations presented here. The highest value TF-IDF terms for each system were then identified and analyzed to identify the primary themes and concepts associated with each VSM system.

To uncover latent themes within each system, Latent Dirichlet Allocation (LDA) was applied.

LDA is a generative probabilistic model that assumes that documents are mixtures of topics and topics are mixtures of words (Blei et al., 2003).

Method Geek

The likely result of the word w appearing on sheet d is given by the equation:

$$P(w|d) = \sum \frac{K}{k=1} P(w|z=k)P(z=k|d)$$

where K is the number of topics, $P(w \mid z = k)$ is the probability of word w given topic k, and $P(z = k \mid d)$ is the probability of topic k given sheet d. Using a count vectorizer, topics within each system were transformed into a term frequency matrix, which served as input for the LDA model.

To summarize this step, TF-IDF and LDA were used together to reveal prominent terms and underlying themes present in the five systems delineated by Beer. This quantitative portion was then used to inform and strengthen the findings of the qualitative analysis led by researchers only. The text output of this step is shown in Table 6.2. Quantitative output is available on the associated GitHub page for the project.[2]

PHASE 4: FRAM MODEL DEVELOPMENT AND ANALYSIS

Based on this knowledge, the FRAM building process started, taking advantage of FMV (Hollnagel and Hill, 2020). To navigate the complexity of the modelling, we used a bespoke approach to FRAM construction, guided by the VSM in its five systems. Partitioning each VSM system allowed the researchers to identify how the data for the near miss event fit into Beer's existing framework, while adapting these systems into a FRAM model allowed the researchers to see interrelationships across systems and resultant synergies. This could be considered an emergent approach, where the notion of the system is greater than the sum of its constituent parts. Furthermore, the integration of theoretical frameworks brings about synergies and tensions, providing a holistic understanding of the systemic relationships and dependencies between system components within the FRAM framework (Preece et al., 2013).

TABLE 6.2
TF-IDF and LDA Prominent Terms

Score	System 1 Pilots		System 2 ATC		System 3 Tech. Control		System 4 Socio-Technical Intel.		System 5 Policy	
	TF-IDF	LDA	TF-IDF	LDA	TF-IDF	LDA	TF-IDF	LDA	TF-IDF	LDA
1	runway	taxi out	runway	runway change	light	ASDE	departure	scan	change	procedure
2	taxi	runway	change	yell at Delta	red	red	leave	incident	procedure	intersection
3	think	distraction	pilot	pilot deviation	ASDE	light	plan	threat brief	new	runway change
4	know	workload	deviation	runway change	hotspot	diagram	scan	runway change	message	new pilots
5	delta	communication	time	cancel clearance	cross	alert	brief	Swiss cheese	training	equipment

Step 1: Identify and Describe System's Functions

The initial step in FRAM development involves identifying and describing functions that can be categorized as background or foreground functions. In the context of aircraft operations, these functions have aspects that may encompass input (e.g., Flight plan available), output (e.g., Clearance received), precondition (e.g., Weather conditions checked), resource (e.g., Navigation system operational), control (e.g., Air traffic control instructions), and time (e.g., Departure window).

The FRAM model allows us to identify the specific relationships and resonance between an activity (function), that is, 'Maintaining adequate aircraft separation, and regulatory decisions about runway complexity such as a hotspot. It may not be evident that there are latent factors that influence pilots' ability to maintain situational awareness during taxi, and the designation by the regulator does not always exist at complex geographical points, as in this case study. Identifying functions separately in this manner helps us to understand that, in this example, pilots and regulators are indeed separate systems.

Additionally, the integration of the VSM facilitates the distinction between activities carried out on the flight deck (System 1) by the flight crew and those that require prioritization by the metasystem (Systems 2–5), including regulatory bodies such as the FAA, technical support teams, and air traffic control (ATC).

An illustrative example that emerged in our case involved ensuring adequate runway illumination by the Port Authority at the airport and implementing safety policies for hotspot identification (System 5 in VSM terminology). These relationships contribute to what we consider to be system-wide coherence, ensuring the safety and efficiency of operations. Communication between pilots (System 1) and ATC (System 2) exemplifies the need for clear and effective coordination between subsystems to achieve the overall objectives of the system.

Step 2: Characterize Variability

We characterized the actual and potential variability within the operations related to the near-miss incident. Based on official documents, we gained insight into how factors in the operational context affected pilot decision-making and performance, with the goal of understanding and predicting their impacts on safety and efficiency. The official documentation was supplemented by data extracted from interviews with air traffic controllers on duty during the incident, who were tasked with overseeing the overall operation or environment. Controllers spoke at length about the nuanced variability they must negotiate on a minute-by-minute basis. When not dealing with emergent variability, controllers often plan for future variability, such as a go-around clearance or a lost communications scenario.

From the VSM perspective, which focuses on the environment as one of the main components of modelling, we were able to particularly focus on studying the dynamic relationship between the coordination of a runway change by ATC and the adjustment of behaviour by the pilots. This resulted in variability in decision-making and attention capacities by pilots on the American 777 flight deck. Variability also propagated anticipatory adjustments necessary once the pilots received an update of the weather.

Method Geek

The application of VSM further strengthened our ability to uncover sources of what Beer called 'pathological diagnostics' (Beer, 1984), or dysfunctions in the system (Pfiffner, 2022a, 2022b). In the view of VSM, the term 'diagnosis' is used to identify disruptions or deviations from normal functioning within a system, while 'pathologies' describe the underlying factors causing these disturbances. In FRAM terminology, we would translate this as functional variability.

Step 3: Aggregate Variability

Variabilities in individual functions, such as communication breakdowns, monitoring inconsistencies, and procedural deviations, were identified in Step 2. Each of these variabilities represents an upstream function that could potentially influence downstream operations. The core of Step 3 is to analyze how these individual variabilities interact, leading to functional resonance in the entire system under study. For example, a communication breakdown between the flight deck and ground control could combine with a monitoring system, significantly complicating the situation by creating delays or incorrect actions. Step 3 investigates how these compounded variabilities affect critical functions such as aircraft handling and ground movement. The incident at JFK revealed that compounded variabilities led to a sequence of misperception and awareness of asymmetric severity. While the air traffic controller was faced with a highly salient ASDE-X alert and called for the abort of the Delta aircraft, the American crew was not aware of such severity and continued its flight.

From the VSM perspective, this means that the monitoring loop goes from System 1 (piloting aircraft) to the management system level (Systems 2–5). Within the VSM framework, understanding environmental change is essential to analyze aggregated variability. Systems must continuously monitor and respond to changes in their environment to remain viable and achieve their goals. 'Viable' can be synonymous with 'safe' in this sense. Therefore, VSM emphasizes the need for systems to be sensitive to environmental signals, capable of adapting to changing conditions, and resilient in the face of uncertainty or disturbances from the environment.

Step 4: Manage the Variability

Based on the obtained model, this step explored both actual and hypothetical scenarios within JFK's operations to promote learning and adaptation rather than enforce rigid tasks. From the perspective of VSM, variability management consists of dealing with potential organizational pathologies in terms of structural, functional, and communication.

For example, the runway incursion at JFK demonstrated specific organizational pathologies directly related to communication breakdowns and decision-making bottlenecks. Critical factors included inadequate handling of high-workload situations, particularly in critical phases of operation, distracting the American crew from their primary task of safely navigating the aircraft.

This is consistent with the NTSB final report of the incident, which stated:

> ... the probable cause of this incident was... due to distractions caused by the performance of concurrent operational tasks during the taxi, which resulted in a loss of situational awareness. Contributing to the incident was... American Airlines' lack of adequate risk controls to prevent concurrent flight crew tasks from leading to distraction, loss of situational awareness and deviation from an authorized taxi clearance.

(NTSB, 2024)

To conclude this section, the application of both thematic and deductive analysis, in conjunction with advanced NLP techniques, allowed the researchers to investigate the complexities of the aviation near-miss incident at JFK. By partitioning the data through Beer's VSM and developing a FRAM model, we were able to identify functions and the variabilities that contributed to the event. This iterative process revealed the importance of understanding how systems, particularly in high reliability industries like aviation, interact dynamically with their environment. The study now moves to results, conclusions, and suggestions for future research.

RESULTS

Following the convergent mixed-method approach, this section presents the results of the qualitative analysis, followed by findings from the quantitative analysis employing NLP. An FRAM model is then introduced based on the converged data.

RESULTS OF PHASE 1: CODES AND THEMES

From the analysis of the official investigation documents, this phase allowed us to obtain codes and themes at three different levels.

At the first level of analysis, the researchers conducted an initial review of the data to establish the scope of the JFK near-miss case study. This began with the identification of an initial set of codes from the data sources outlined in Table 6.1. Each author independently worked in this stage and then collaborated to improve the thoroughness of the findings. Through this analysis, a total of 368 comments and phrases were identified as important for contextualizing the near miss. This effort resulted in the creation of 48 first-order codes.

Moving to the second level of analysis, first-order codes were employed to identify recurring themes relevant to the research purpose. This step led to the refinement of these codes into 28 second-order codes, distributed across the five systems. The process was iterative, involving a thorough review and adjustment of codes to ensure that they effectively captured the commonalities and aligned with the study objectives. This process ensured that the codes remained relevant and accurately reflected the data. For example, codes such as 'distraction', 'shift in focus', 'recognition of cognitive biases', and 'operational condition' were grouped under the second-order code 'situational awareness'.

Building on the 28 second-order codes, we further combined these with insights from NLP and conducted multiple rounds of analysis to enhance their precision. This refinement process allowed us to more accurately capture the nuances within specific quotes. The resulting third-order codes were then applied in FRAM modelling, yielding 11 third-order codes that served as essential functions in our framework.

RESULTS OF PHASE 2: VSM SYSTEMS

This section explores the system partitioning through the VSM as illustrated in Figure 6.2, which details the components of the five systems associated with the JFK near miss, including their associated environments. The subsequent analysis explores each system, where the themes from the classification of first-order codes highlight the important components of the incident.

System 1

Using the VSM perspective, System 1 (S1) are operational units at the front line, which represents the operation in the flight deck in this context. Piloting an aircraft encompasses operational activities and decision-making processes both in the air and on the ground. Our findings reveal several instances in which the flight deck crew made critical decisions and took action during taxi out.

Domain Expert

This included the American pilots' handling of changing runways and the acknowledgement of taxi clearances, which underscore System 1's involvement in executing the fundamental tasks that ensure the viability of the operation within the larger system, in collaboration with ATC:

> '…obviously there's a runway change; we're on 4 Left now, because that's the procedure for 4 Left departure' (D4, p.14) 'And then we got our taxi clearance, which we acknowledged'.

> (D4, p.17)

The instance where the pilot recognized the red glow on the runway serves as a direct reflection of System 1's capacity for self-regulation:

> …all of a sudden I saw that red glow and I just – right away I said something – that ain't right.

> (D4, p. 18)

Completing all checklists and obtaining a closeout before proceeding is mentioned as a critical procedure. The 'closeout' refers to the final weight and balance numbers sent by the company, and its data have been uploaded to the flight management system:

> I always tell the FO we're not going anywhere until we get a closeout, and we can take care of the checklist. And that way we're both…paying attention where we're going….

(D4, p. 30)

This practice kept both the pilot and the first officer (FO) well informed about their duties and the aircraft's status, boosting safety and efficiency. The swift execution of these processes underscored a well-coordinated approach to flight preparation, crucial for operational integrity and safety.

Additionally, passenger communication, particularly during emergencies, is a critical function of System 1. After the aborted takeoff, the Delta pilot mentioned:

> We stopped on the runway…and I directed FO Rodriguez to make a public address (PA) to the passengers to remain seated […].

(D5. p. 2)

This action ensures that all passengers are informed of the situation and are instructed on how to act to maintain safety and order. The prompt decision to reject the takeoff demonstrates the autonomous and responsive nature of System 1 as described by the VSM. This action was central in averting what could have been a disastrous incident, thus maintaining adequate separation of the two aircraft.

Referring to the red in-pavement lights that illuminated, the American captain stated:

> …I saw that red glow and I said…have someone…call the tower and ask them why they put those red lights on.

(D4, p. 17)

This sentence highlights the pilot's instant response to an unusual visual cue (red lights), which is a highly salient cue to elicit a reaction to unexpected signals. The decision to communicate with the tower also indicates the beginning of an interaction aimed at clarifying potential problems.

Pilot interview data indicate a shift in the fleet harmonization procedure, where the responsibility for the pre-takeoff passenger announcement had been transferred from the captain to the first officer (FO):

> She's supposed to make a prepare for takeoff PA roughly 2 minutes before departure. I mean, it's something the captain's done and they now they switched it over to the FO.

(D4. p. 37)

However, while this distribution of critical tasks from the captain to the first officer (FO) might enhance the efficiency of the operation, it inadvertently might add a layer of oversight for the captain, who must now ensure that the FO completes the task:

> Now you're telling the first officer to do it, … but then technically you got to make sure the first officer is doing it. If they're not doing it, you got to do it. So it's…it's just an added burden to…taxiing, where before you're just doing it [yourself].

> (D4, p. 39)

These moments from the JFK incident highlight the essence of System 1's role within the VSM: to ensure the *continued piloting of aircraft safely* in the face of routine challenges and unexpected hazards.

From our data, the following themes emerge to show the critical role of System 1 in navigating the complexities of aircraft operations.

- S1.1: Handling runway changes: adapting navigation and taxi paths in response to changes in runway assignments
- S1.2: Acknowledging taxi clearances: confirming and complying with taxi instructions to maintain safety and coordination on the airport surface
- S1.3: Complying with ground movement instructions: ensuring adherence to all ground movement directives to facilitate safe and efficient airport operations
- S1.4: Completion of all checklists
- S1.5: Communicating with passengers
- S1.6: Observing visual cues: Detecting visual indicators on the runway
- S1.7: Responding to visual cues: reacting to visual indicators on the runway that may affect the safety of takeoff or landing operations
- S1.8: Conducting safety checks
- S1.9: Making a pre-takeoff announcement

System 2
Through the lens of VSM, S2 is designated as ATC, which provides coordination services and manages conflicts and interfaces within the organization. During the incident, initial communication shortcomings were evident.

Domain Expert

A pilot (Traci) described crossing the runway and then receiving a notification of a possible deviation, indicating a lapse in proactive communication.

> … we cross over, and when we crossed to the other side then…Traci looked down and said we're still on ground. So, then she switched tower, called tower. And then they said we had a possible deviation, pilot deviation, and she kind of shut up […].

> (D4, p.17)

This scenario underscores the breakdown of timely information sharing and coordination, which is critical in managing the safe movement of aircraft on active runways. The incident was further compounded by delays in decision making, as reflected when the tower required the crew to call back, thereby indicating a pause in the immediate response:

> ... the tower...gave me a phone number to call. So, I called the tower, and the tower said, hang on, they're pulling their tapes, to call back in 5 minutes.
>
> (D4, p. 18)

This delay not only demonstrates the inefficiency of conflict resolution mechanisms but also shows the potential risk of information asymmetry. While ATC dealt with a ground proximity alarm and called for the abort, the American crew may not have realized the loss of separation created on the runway, as evidenced by their continued flight to London. Additionally, there was confusion about the clearance for movement on the runway, which led to unnecessary hold-ups that affected the operation.

> After a short delay holding on the runway due to sequencing behind a 'heavy' aircraft, Tower cleared us for takeoff.
>
> (D5, p. 2)

ATC also asked if the Delta crew needed time to run checklists after the rejected takeoff, which showed expert-level awareness from the controllers. After a rejected takeoff, the brakes can become hot, and a cooling period is necessary, often exceeding 30 minutes (Piľa et al., 2017). This incident also exhibited moments of critical decision-making under pressure, where a supervisor initiated a runway change, trying to stay ahead of operational demands:

> ...a supervisor said that they wanted a runway change... start coordinating runway change... call N90, and I try to stay ahead of the game.
>
> (D2, p. 14)

This exemplifies the proactive nature of coordination that sometimes precedes without full situational awareness. Additionally, the findings indicate instances of miscommunication and confusion in System 2:

> [American] might have thought that he is going to runway 3-1 Left [...]. I don't know why they would... go to the full length. I don't know why they would ignore the hold bar lights. I can't make sense of it, [...].
>
> (D2, p. 184)

This miscommunication led to reduced safety margins and also stressed the importance of a shared mental model, coordinated communication, and questioning of other actors' actions.

The above analysis of System 2 activities reveals the following themes:

- S2.1: Coordinating communication
- S2.2: Resolving conflicts[3]
- S2.3: Facilitating decision-making
- S2.4: Monitoring operational integrity
- S2.5: Adapting to dynamic situations

System 3 & 3*

In the VSM, System 3 serves as the operative management of a collective of subsystems. It coordinates and regulates internal operations to ensure that the system functions efficiently and effectively. In our case, System 3 is considered the technological control and support system. Using advanced technology to manage workflow and allocate resources, it plays a key role in overseeing the safety of day-to-day air traffic operations. Moreover, System 3* is the auditing mechanism of behaviours to ensure compliance with System 3's objectives and System 2's regulations.

By analyzing data from multiple sources, including input from System 2, System 3 ensures that all operational units run effectively together, ensuring that operations run seamlessly and autonomously. In addition, it filters essential information to higher levels, supporting decision-making in complex situations and ensuring the stability and effectiveness of the ATC.

Domain Expert

Our findings indicate instances where System 3's role as a technological control and support system is put to the test. For instance, in the interview with the NTSB, the American crew mentioned:

> We weren't getting a closeout. So, then I started thinking is there something that…
> is causing us not to have a closeout? Because usually the company is pretty good
> on getting it closed out …

(D4, p. 28)

This situation shows a potential weakness in System 3's capability to promptly diagnose and rectify technical failures, which is important for maintaining real-time situational awareness. The delay in resolving the closeout issue reveals challenges in the system's real-time responsiveness and problem-solving efficiency.

Our findings demonstrate aircraft inspection as a technological control of System 3. After the rejected takeoff:

The aircraft was inspected by DAL maintenance at approximately 2130L.

(D5, p. 2)

However, the reliance on manual inspection points to potential areas for improvement, leveraging automated diagnostics and real-time data analysis capabilities within System 3.

We find successful examples in which System 3 ensured that necessary resources were accessible for immediate use during critical moments, supporting the crew's ability to manage high-pressure situations effectively:

Holding short… the Captain instructed me to get the Quick Reference Handbook (QRH) while he was making a PA to the passengers.

(D5, p. 2)

These data underscore the role of System 3 in maintaining operational readiness and providing critical support when needed.

After returning to the gate, the passengers left the aircraft. Per Maintenance personnel, no damage was noted. The flight was delayed until the next morning.

(D5, p. 3)

This decision to delay the flight despite the absence of observed damage emphasizes a cautious approach that prioritizes safety and thorough checks. The tower cab coordinator at JFK discussed monitoring tasks, which are aligned with System 3*'s auditing objectives:

If clearance delivery needs help, it's really general assist, that's why we have it in our SOP that in the absence of higher priority duties, Local 1 is monitored, because that's usually where the most things can go wrong.

(D5. p. 22)

Additionally, the role of the cab coordinator exemplifies the auditing functions of System 3*, being described in the transcript as:

[…] an extra set of eyes to look out the window.

(D2, p. 22)

Similarly, the controller's active monitoring and response to potential errors further demonstrate System 3*'s auditing capabilities:

She's reading back where she's crossing…because those are safety things… that's the one I got to listen to, because I got to make sure she's not crossing the wrong runway.

(D2, p. 62)

Vigilance addressed in this statement is an integral part of maintaining safety and operational integrity.

The data collected for System 3 outlines the following themes. Note that for model parsimony, we have combined Systems 3 and 3* together:

- S3.1: Guidance for accurate decision making
- S3.2: Allocation of resources to maintain safety
- S3.3: Analyzing data from multiple sources
- S3.4: Supporting air traffic by providing technological resources
- S3.5: Providing capacity for managing high-workload situations
- S3.6: Diagnosis of technical failures and issues
- S3.7: Monitoring pilot responses

System 4

In the VSM view, System 4 ensures adaptability to external challenges, complementing Systems 1 to 3. It scans the environment, identifies opportunities, and proposes strategies. In our case, System 4 is labelled as the 'Socio-technical Intelligence' system, critical for strategic decision-making and proactive planning in aviation incidents. This subsystem links internal operations with external factors, such as weather conditions, changes in runway configuration, or even regulatory changes. Although its role in the immediate management at JFK may be less obvious, it typically involves senior controllers or designated quality assurance personnel who analyze operational data and trends to optimize safety and efficiency.

The findings reveal several instances in which the attitude of System 4 affected the operations of the aircraft.

Domain Expert

For example, a controller interviewee used the Swiss cheese model to illustrate the complexities of operating aircraft as a sociotechnical system:

> ... I think that the factors [were]... the Swiss cheese holes all lined up at the wrong time, unfortunately.

(D4, p. 58)

This reflection again reveals the controller's expert understanding of safety theory and his effort to highlight the complex factors where vulnerabilities can align to create mishaps. It emphasizes the role of identifying and mitigating these risks to enhance aviation safety. The controller outlined the decision-making in ATC, requiring quick, confident actions and the integration of technical and social input to enable a swift, informed decision:

> When it comes down to it, and when you got to actually hit that button, when shit's going on, it's me.

(D2, p. 88)

The controller emphasized his personal responsibility in operations. System 4 supported this by synthesizing data, anticipating outcomes and suggesting actions, and helping to prevent decision paralysis from second-guessing at critical moments.

> …if you start second-guessing what you got going on, then you're not going to be able to react next time. And then I'll be sitting down here at a desk doing nonsense and not working airplanes. So no, my response is not to hit that button because she's in the middle of the runway […].

> (D2, p. 88)

The mention of the decision to paradoxically not act ('not to hit that button') because of the situation on the runway is remarkable. Based on experience, the controller felt that if he told the American aircraft of the incursion, the captain would stop the aircraft immediately (on the runway). The 'button' here serves as a sociotechnical metaphor, similar to an emergency brake on a train. Despite the incredible severity of the situation, not hitting the button actually resulted in a safer outcome.

> We still had not received the closeout and were concerned it had something to do with the NOTOC [notice to captain] issue.

> (D1, p. 3)

The NOTOC system is in place to inform pilots of dangerous goods or hazardous materials on board. The pilots had previously experienced an issue where not receiving the NOTOC inhibited the transmission of performance data. This system very well illustrates how both technical systems (NOTOC transmission) and social systems (operational routine and checklist cadence) intersected during this event.

The data collected for System 4 outlined the following themes:

- S4.1: Identifying and addressing threats
- S4.2: Personal accountability
- S4.3: Pilots balancing sociotechnical systems

System 5

System 5 serves as the overarching governance and policy-setting layer. It is essential to maintain coherence across various operational systems and ensure that policies and regulations align with national and international aviation standards. In our case, it involves the Port Authority and the FAA, which establish the framework within which JFK operates. They provide strategic direction and ensure that their actions contribute to the overall safety and efficiency of ATC.

The findings demonstrate multiple examples of how System 5's approach influenced aircraft operations.

For example, when interviewees were asked whether marking a hotspot in that location might have changed the outcome of the taxi, the informant answered the following.

> It definitely would have been a factor for sure. I don't know that we would have continued that takeoff briefing at that point if we knew we were approaching a hotspot.

(D4, p. 24)

The discussion about potentially marking a hotspot underlines the importance of System 5 in managing the rules for critical markings. Recognizing a hotspot could change how pilots manage their operations, emphasizing the need for System 5 to provide clear and easily understood information. It is worth noting that after this incident, a hotspot was created at this intersection[4] ().

Furthermore, the pilot's uncertainty regarding the red glow and the decision to depart the runway illustrate the responsibilities of S5 in ensuring that navigational aids, such as lights and markings, are clear and functional.

> I'm seeing that red glow. And then I also decided I'm getting off this runway because I don't know why those lights came on.

(D3, p. 11)

This reflection reveals a potential fault in airport safety systems (false positive) or a deficiency in training and protocols that guide pilot response to runway status lights (true positive). It also underscores the need for System 5 to supervise the accuracy of airport lighting systems and the utility of training, thus fostering better safety margins and operational efficiency (subsequent analysis by the NTSB indicated that the red lights appeared earlier in the event than the American pilots recollected). In other words, there was a latency between the time the lights came on and the time they were perceived by the crew.

> The CA [captain] asked me per the new AOM [airplane operating manual] procedures to make 'the prepare to take off PA [passenger address].'

(D6, p.3)

This statement reflects the role of System 5 in enforcing new operating procedures, which require pilot actions to ensure compliance with standards. Although most hard skills are trained in the simulator, soft skills are routinely disseminated by the electronic flight bag. Pilots sometimes lament a 'train by memo' approach, as it can be opaque and relies solely on prospective cognitive memory.

The data collected for System 5 outline the following themes:

- S5.1: Lack of a safety policy in place for hotspot identification
- S5.2: Complex taxiway and runway layout
- S5.3: New policy at American Airlines may not be fully absorbed by pilots
- S5.4: Detection of runway status lights by pilots

Applying the VSM served as a framework to systematically identify functions, taking into account the holistic aspects of the system. The 28 themes identified in the five systems detail the roles of each system in the context of a near-miss incident at JFK, from routine air traffic operational management to strategic adaptation when the runway incursion occurred. These themes illustrate how elements work synergistically to maintain safety and efficiency.

RESULTS OF PHASE 3: NLP OUTPUTS

The following table shows the converged topics as a result of the NLP analysis. We utilized NLP techniques to reveal relationships and topics that may not have been apparent with a traditional qualitative approach. This integrated methodology helped to build a comprehensive mapping of operational roles across different systems.

Via TF-IDF, it was possible to extract specific operational tasks (e.g., departure start weather, runway and situational awareness factors like 'taxi captain think kilo delta awareness', while LDA identified broader contextual themes such as runway pile, radio, say, panel and paperwork).

Some of the terms may be strange to a non-expert. For example, 'pile' and 'panel' refer to printouts that were sent to the American pilots during taxi, which ended up making a pile of papers on the centre panel. This caused a distraction.

Examining Table 6.3, we see each VSM system in the left column and a set of converged topics in the right column, based on the TF-IDF score and LDA topic distributions.

To elucidate, System 1 converged on seven topics (to conduct pre-takeoff checklists, anticipate flight path, maintain situational awareness, communicate with ATC, communicate to cabin, load performance data, and cross a runway). Determining the phrase for the converged topic was based on the researcher's domain expertise and previous examples in the literature (see, e.g., Kim & Gil, 2019). System 2 converged on four topics (managing changes, resolving disparities, oversight of aircraft, and coordinating workload); System 3 on two topics (activation of runway status lights and use of airport diagram); System 4 on three topics (visual scan of runway, event severity diagnosis, and shared mental model); and System 5 on three topics (training for new procedures, hotspot designation, and design of taxiway/runway).

The 19 converged NLP topics served to complement the findings from the 28 themes in the VSM partitioning, with overlap existing for most constructs. Of the 19 topics, seven corresponding qualitative themes emerged: flight preparation, ground operation, communication, communication ambiguity, quality assurance, training, and regulatory oversight.

When reviewing the converged topics presented in Table 6.3, we will now translate those topics into FRAM functions, nodes, and interdependencies.

TABLE 6.3

The Converged Topics as a Result of the NLP Analysis, and Comparison with the Data Partitioning with VSM

VSM System	NLP Method		Converged Topic	Corresponding Qualitative Theme
	TF-IDF	LDA		
System 1 Pilots	Brief briefing	Runway pile radio say panel paperwork pedestal	To conduct pre-takeoff checklists	Flight preparation
	Departure start weather runway	Start taxi weather head think company radio 31 distract bias expect	Anticipating flight path	Flight preparation
	Taxi captain think kilo delta awareness	Taxi recall brief briefing awareness short bravo clearance unaware position kilo 31	Maintaining situational awareness	Ground operation
	FO radio	Pilot 31 cross clear original know thought mush	Communication with ATC	Communication
	Departure awareness	Runway closeout tells departure idea time, pilot leave	Communication to the cabin	Communication
	FO receive closeout	Receive message, departure change, think close	To load performance data	Flight preparation
	Runway	Head thing think distract happen lose delay aircraft runway	To cross a runway	Ground operation
System 2 ATC	Runway change delta ground	Runway change, cancel takeoff clearance, hear the delta crew point	Managing changes	Ground operation
	Possible cancel catch hear tell	Time control, controller change, clearance ground catch aal106 taxiway	Resolving clearance disparities	Communication ambiguity
	Tower deviation clearance turn	Deviation pilot tower possible turn thing tell stop half dozen	Oversight of aircraft	Quality assurance
	Pilot stop cancel time takeoff	Pilot like_everybody ATC hotshot act amateur upstairs supportive	Coordinating workload	Ground operation

(Continued)

TABLE 6.3 (Continued)

The Converged Topics as a Result of the NLP Analysis, and Comparison with the Data Partitioning with VSM

	NLP Method		Converged Topic	Corresponding Qualitative Theme
VSM System	TF-IDF	LDA		
System 3 Technological controls	Light red runway ASDE alert illuminate	ASDE light red runway cross hotspot illuminates centerline eye glow alert	Activation of safety devices	Quality assurance
	Hotspot cross diagram	Airport diagram takeoff runway turn hotspot	Use of an airport diagram	Ground operation
System 4 Socio-technical intelligence	Runway scan think	Runway scan approach Swiss_cheese	Visual scan of the runway	Ground operation
	Hotline line Swiss_cheese	Hotline time tower	Information asymmetry	Ground operation
	Departure taxi change brief plan diagram	Change taxi confirm left	Shared mental model	Ground operation
System 5 Policy	Change procedure new QRH FOM	New procedure pilot harmonization QRH FOM follow	Training for new procedures	Training
	Policy clear reference	Incursion policy runway hotspot	Hotspot designation	Regulatory oversight
	Policy help follow	Runway check clear takeoff policy kilo runway_4_left	Design of taxiway/ runway	Regulatory oversight

RESULTS OF PHASE 4: FRAM MODEL

An FRAM model has been developed following a hybrid (qualitative thematic analysis first and quantitative data-driven NLP later) process. First, using ATA in the VSM framework, we have identified first-order codes (topics) and second-order codes (themes). Then, using NLP, machine learning-generated topics and themes were integrated into the VSM data to show third-order codes as FRAM functions. These range from System 1, which focuses on piloting aircraft, to System 5, which involves higher-level strategic management and oversight by policy bodies such as the FAA. This segmentation has allowed us to clearly delineate the roles and interactions across all organizational levels, enhancing our understanding of each level's contribution to the near-miss event.

These insights are depicted in Figure 6.3, which presents a graphical representation of the FRAM model aligned with the five systems. Figure 6.3 depicts 27

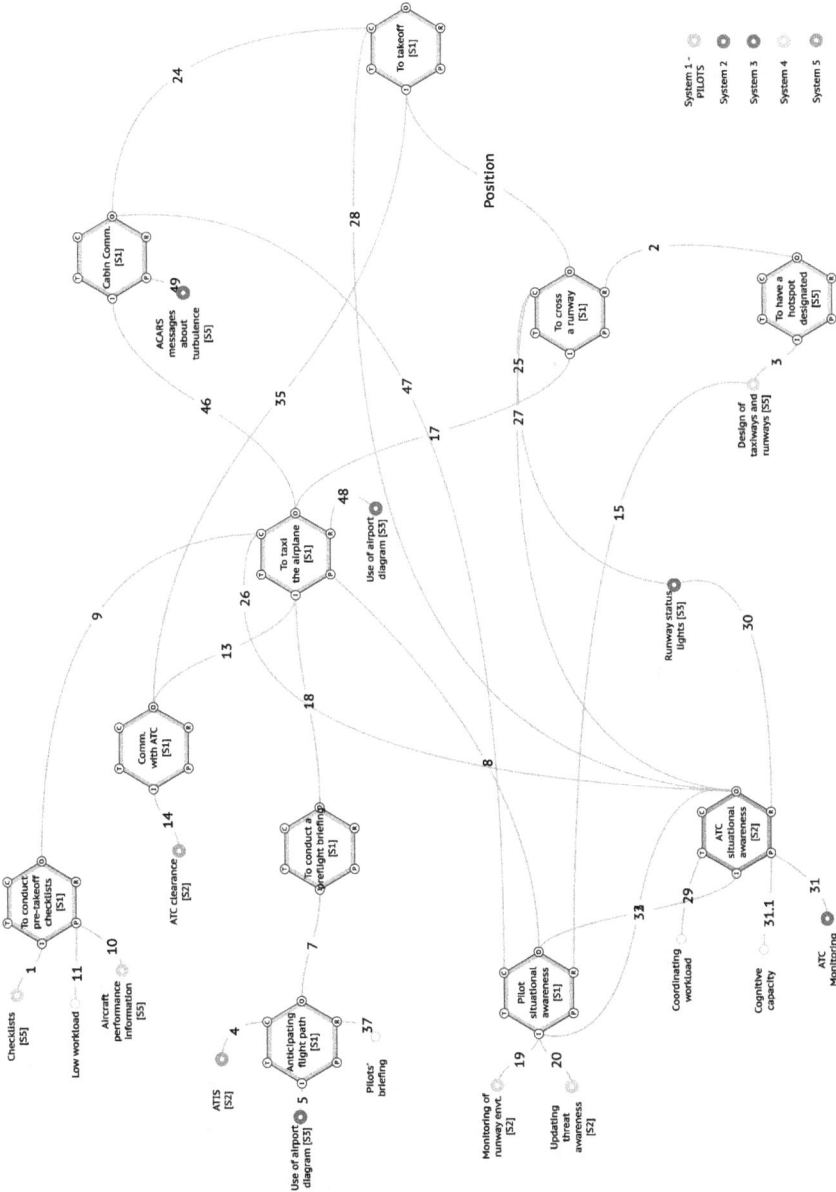

FIGURE 6.3 FRAM model for the JFK near miss incident, obtained by integrating VSM and NLP.

functions overall, 11 of which are foreground functions (input and output) and 16 of which are background functions. The difference between Table 6.3's content (19 topics and seven themes) and the final 27-function FRAM model takes into account features determined by the subject matter expert (SME) researcher, such as ATIS (weather and runway configuration for pilots) and ACARS messages that added workload. This nuanced difference shows the importance of SME examination of model refinement.

The above FRAM functions across different systems (1–5) interact within multiple recursive levels and tasks, each designed to *deliver specific value*, indicated by FRAM Outputs, in connected functions. The model reveals the dynamic interplay between these functions, where each level is interconnected through specific functions that may influence others across this recursive structure.

Once the model is built, it is possible to perform a variability analysis of its functions to understand functional resonance and design management actions, which will focus on real-time responsiveness and strategic foresight and conclude with enhancements in communication and response strategies.

Real-Time Responsiveness and Strategic Foresight

In the complex and dynamic field of aviation, direct operational tasks carried out by pilots in System 1 are key, as they constitute the genesis of the activity, directly impacting flight safety and efficiency downstream. The immediacy and critical nature of pilot decisions place them at the forefront of operational management and risk mitigation. The central role of pilots in managing these tasks is essential for maintaining the flow of operations and ensuring the safety of every flight.

An illustrative example related to real-time responsiveness comes from this study, where a pilot had to rapidly adjust his thinking due to unexpected operational shifts, noting, *'I just got distracted with all this happening and I stopped thinking about Kilo and started thinking about my original briefing to 3-1 Left'* (D3, p. 37). This highlights the need for pilots to maintain situational awareness and adapt quickly to changes.

Similarly, System 2's coordination capabilities are vital in managing and adapting to dynamic conditions. This is demonstrated when an ATC supervisor initiates coordination in anticipation of changes. *'Then a supervisor said that they wanted a runway change to start or, you know, start coordinating runway change, you know, call N90, and I try to stay ahead of the game'* (D2, p.15). Such anticipatory actions underscore the need for strategic foresight in frontline roles, allowing a preemptive rather than purely reactive approach to potential challenges.

Moreover, the VSM enhances these capabilities by providing a structured, yet adaptable framework that aligns well with the needs of real-time operational decision-making. Within VSM, System 1, which encompasses immediate operational activities such as those on the flight deck, is designed to facilitate swift and autonomous decisions. This was evident when a pilot responded instantaneously to an unusual situation: *'Well, all of a sudden I saw that red glow and I just - right away I said something - that ain't right'* (D6, p. 17). The pilot's immediate reaction to the red glow exemplifies the principle that System 1 should handle urgent operational decisions autonomously, allowing faster responses imperative for safety.

Through these examples, it becomes clear how a hybrid systemic approach, combining FRAM and VSM, significantly contributes to improving both real-time responsiveness and strategic foresight in complex operational environments.

Enhancing Communication and Response Strategies

Effective communication and response strategies are critical to managing complex operational details in any high-risk industry. While operators in System 1, pilots in this study serving as the operational core, not only respond to immediate environmental cues but also rely heavily on communication loops with ATC, for situational awareness and decision-making.

System 2 (ATC) operates as the nerve centre of communication, managing clearances and coordinating interactions across various units. Through the lens of FRAM, the resolution of disparities in ground-based anomalies underscores the importance of real-time communication channels facilitated by ATC. This aligns closely with VSM's emphasis on effective communication structures within System 2, essential for maintaining control and oversight over airport operations.

In Figure 6.2, the various communication channels are highlighted by lines connecting the five systems. System 3 (Technological Controls) displays integration of communication-enhancing technologies, such as runway illumination and hotspot alerts. These systems, depicted in FRAM as available resources such as safety devices, not only support pilot decision-making but also communicate vital information about runway conditions and hazards. This integration resonates with VSM's focus on communication channels within System 1, emphasizing the role of technology in facilitating clear and timely information exchange between stakeholders.

System 5's role, underpinned by policy, FAA oversight, and design considerations, sets the framework within which all operational decisions are made. Implementing new procedures often involves changes in crew responsibilities, demonstrating the need for clear and precise dissemination to ensure compliance and safety. For example, pilots noted a series of procedural changes and the need for constant updates, which occasionally added to their workload and were potentially distracting: *'The company was sending weather updates every few minutes, which was unusual; changes are happening now'* (D4, p. 96). This example illustrates the ongoing adjustments and the need for crews to adapt to new operational norms seamlessly. In addition to procedural adaptability, System 5's regulatory and design oversight ensures that operational strategies are not only responsive but also preemptive.

The proactive management during the JFK incident, such as coordinating runway changes and anticipating traffic flows, showcases this layer's crucial role. Nuanced knowledge of a blind spot highlighted this: *'...the intersection of Kilo and Juliet is actually blocked by the beams of the tower'* (D2, p. 19). Furthermore, ATC stated that they prefer to issue taxi clearances well in advance, allowing workload shedding; paradoxically, pilots perceived this 'dead air' as abnormal. Therefore, proactivity can have different effects.

Meanwhile, System 4's sociotechnical intelligence involves more than just handling immediate threats; it encompasses planning for future operations and integrating lessons learned into ongoing practices. Strategic discussions around taxi routes and departure runways, as well as the adjustments made following unforeseen events, underscore the importance of sociotechnical planning: *'We briefed the taxi route for 31L and discussed potential threats'* (D6, p.2).

These examples confirm that effective communication channels not only support decision-making and coordination but also improve situational awareness and mitigate risks, ultimately contributing to the overall resilience and effectiveness of airport management systems.

DISCUSSION

The discussion covers two dimensions: first, the theoretical synergies that emerge from a methodological perspective, and second, the larger implications for safety management later.

THEORETICAL SYNERGIES

The necessity for VSM's application becomes particularly evident when addressing the complexities involved in identifying and categorizing functions, a challenge often compounded by intricate interactions within system processes (Patriarca et al., 2018a, b).

VSM not only supports systems in continuously monitoring and responding to environmental changes but also emphasizes the need for sensitivity to environmental signals. Integrating this with NLP can enhance this approach by efficiently processing large volumes of data to extract and synthesize the core elements necessary for accurate system modelling. The comprehension of the operational context is essential for identifying functions in FRAM construction. Incomplete knowledge can obscure key interactions and constraints, critical to the analysis of system behaviour.

In the literature surrounding application of the FRAM, it is suggested that the identification of functions could begin with an initial step or function and then progress in a sequential manner, allowing for movement both forward and backwards, as well as a broader exploration across the system. This flexible approach extends beyond mere starting points, impacting the degree of resolution of the descriptions and the delineation of the system boundaries (Hollnagel, 2012, p.42). Consequently, the selection of the initial function, the resolution of descriptions and the establishment of system boundaries require a thoughtful approach, guided by an intricate understanding of the core operational dynamics and its environment. This ensures that the analysis is aligned with the inherent complexities of the system being studied.

In the previous sections, we outlined integrating VSM in FRAM model development, creating a replicable hybrid approach for identifying and analyzing system functions. This methodology synergistically combines multiple theories for safety management, resilience engineering, and cybernetics under a common knowledge representation framework developed by Hollnagel in 2012.

In addressing this complexity, traditional safety approaches such as the System-Theoretic Accident Model and Processes (STAMP) (Leveson, 2012) and AcciMap (Rasmussen, 1997) utilize a hierarchical approach. A hierarchical method structures the analysis through layers of command and control, aiming to systematically manage and mitigate complex system behaviours. However, it may also simplify control into linear chains of command, potentially overlooking the cascading effects and the multifaceted nature of control interactions within the system. This reduction can make the true aspect of control appear as an illusion (Hoverstadt, 2020), where the model structure conceals the subtleties of actual system dynamics and emergent properties.

In contrast, the VSM offers a different perspective on autonomy and control, distinguishing itself by recognizing that different organizational levels handle varying types of complexity. This clear differentiation ensures that as an organization is structured from its foundational operations upwards, there is a distinct focus for management decision-making at each level. Crucially, concerns at one level generally differ from those at another level. This framework allows for a more nuanced understanding and management of control, accommodating the complexities of real-world operations without oversimplifying the interactions between system components.

The combined model acknowledges the critical roles of every system element, humans, machines, and organizations. It emphasizes the importance of each function, including seemingly minor ones like illumination signals, from both safety and saliency perspectives. Dynamic analysis facilitated by this hybrid approach allows for real-time visualization of time constraints, resource availability, and precondition fulfilment. This enhanced visibility enables more effective operational decisions, leading to optimized duty assignment and authority management based on the time and control needs of each function.

What VSM adds to the 'foreground and background' functions in FRAM is structured labelling, which is defining organizational identity. Identity arises from the relationships between the components and stakeholders (Espejo & Reyes, 2011), such as pilots and air traffic controllers in our case. However, for observers, such as investigators and analysts, it is the actions of the organization that make this identity visible. The adoption of naming systems from System 1 to System 5 in the modelling of aviation systems is a strategic choice. This methodology provides a structured framework that allows for clear categorization and systematic exploration of various organizational perspectives. For example, these perspectives might include aircraft movement procedures, operational efficiency tradeoffs, or communication standards between pilots and air traffic controllers. This clarity is instrumental when applying FRAM, as it relies on the identification of functions and their potential variability, which affects the overall performance of a system. Essentially, VSM lays the groundwork for a more targeted and effective implementation of FRAM by elucidating the system's structural and functional architecture.

IMPLICATIONS FOR SAFETY MANAGEMENT

This study squarely confronts the issue of the industry's situation of being 'data rich and information poor'. As incidents are receiving nearly the same attention as accidents (at least in the aviation domain), they are producing voluminous amounts of data.

The methodology introduced in this study provides a methodical way to delve into a large textual data set, identify FRAM functions, and produce a FRAM model utilizing an existing theoretical approach.

Second, it offers a holistic and comprehensive approach to incident analysis. Although airlines have formalized accident response plans and procedures, regulations such as EU 376/2014 stipulate a robust analysis component after incidents, or 'occurrences' in EU parlance.

In our view, incident analysis receives less attention than accident analysis at the local level, and our framework offers a solution in that regard. Like safety management systems (SMSs), incident investigation must be scalable to meet the needs of entities. Our method offers a bespoke set of choices for company investigators to identify system levels and develop a FRAM model for near-miss incidents.

Third, this study continues a steady march down the road that departs from the axiom that 'human error is the cause of 80% of accidents'. Some may call this an illusion or even a myth (Besnard & Hollnagel, 2014; Wróbel, 2021). We believe that the five-tier framework, through the lens of VSM, as well as the interrelationships from the FRAM model, reveal complex interrelationships and interactions that go well beyond the frontline worker. To recall, American Airlines had updated a procedure weeks before the near-miss, which occupied space in their working and prospective memory while taxiing, possibly contributing to the event.

Fourth, we provide a toolkit for safety-critical organizations to implement a robust method in their SMSs. An SMS is mandatory in many aviation, maritime, oil and gas, and nuclear industries, while it is voluntary in rail, healthcare, construction, and manufacturing. However, organizations are gradually adopting SMS, or, in the case of aviation, it is expanding to areas beyond the airlines (e.g., maintenance organizations, charter, and corporate aviation). The VSM framework fits well within the four pillars of an SMS. For example, the first component of SMS is policy, which is identical to Beer's fifth system. The other SMS components of management, assurance, and promotion are also more easily discernable when partitioning the VSM and producing a FRAM model.

Fifth, we strived to demonstrate the usefulness of NLP when it comes to discerning topics and themes from a large textual data set, such as an NTSB docket. The pipeline utilized for this study is well established and can be used with an intermediate level of coding knowledge. With the rise of AI-based assistants such as ChatGPT, Copilot, and Gemini (as explored in other chapters of this book), it is possible that coding will not be necessary for safety researchers in the future. Rich data sets exist for accident investigation via official report dockets, which are now evolving into incident investigation. For example, a similar near miss produced more than 3500 pages of data. This new 'incident genre' presents a great opportunity for future researchers to expand the scope of analysis by creating small or large language models from the report data sets.

Finally, public and professional mindsets may revert to accident frameworks when it comes to incidents, which may be suboptimal. For example, the cockpit voice recorder (CVR) was overwritten after the American 777 continued its flight to London and was therefore unusable for the investigation. This caused consternation among the public and the regulator. Although the CVR was intended as a forensic analysis tool after a nonsurvivable accident, it may now be seen as just another data source for incident analysis, which, like the flight data recorder, diverges from its original intention and greatly concerns pilot unions.

It is clear that existing paradigms used after an accident may not necessarily be easily applied to incidents. Therefore, this research aims to provide a theoretical framework for the investigation of standalone incidents and see them as a separate genre in themselves.

CONCLUSIONS

At the outset of employing the FRAM, two seemingly simple questions emerge: How do we determine the starting point and depth of our analysis? And how to identify functions? These initial steps are not merely procedural; they are a strategic choice that deeply influences the trajectory of the entire analysis.

This study utilized NLP in conjunction with VSM to augment the capabilities of the FRAM, establishing a hybrid systemic approach for the identification and analysis of FRAM functions. NLP provided a means to deconstruct large data sets to identify themes and patterns. By systematically breaking down large datasets, NLP identified key themes and patterns, while expert interpretation ensures that these insights are meaningful within specific operational contexts. The VSM offers a guideline for identifying functions. The integration of VSM with FRAM not only enriches the representation of each function within a multidimensional space but also deepens our understanding of how function variability can impact system safety, particularly in identifying potential near misses. This methodological synergy provides a comprehensive framework for examining systemic interdependencies and improving the adaptability of systems to environmental changes.

The methodology of our work is designed to be transparent and replicable. Detailed descriptions of the NLP algorithms used, the selection criteria for incident reports, and the thematic analysis process are provided on a GitHub page maintained by the author.[5] This ensures that other researchers can replicate the study with the same or similar data sets to verify the findings or explore additional cases. Furthermore, future researchers can use the novel method with their own data sets.

Looking ahead, more empirical studies are needed to validate and refine hybrid FRAM and VSM models. Future research should explore how this approach can be used in different industries, determining the adaptations needed to align with various organizational cultures and technological environments. It is suggested that official accident or incident reports are referenced to increase the internal validity of researchers' efforts. Furthermore, the development of training materials and guidelines remains central to helping practitioners navigate effectively the complexities of this innovative approach, ensuring its successful implementation and maximum safety benefit.

NOTES

1   We refer to the tragic Germanwings Flight 9525, accident of 2015.
2   https://github.com/mesaba-james/Hybrid-Approach-FRAM-Chapter.

3   An interesting aspect of this incident was the air traffic controller's disclosure that he coun-
    terintuitively chose not to say anything to the American aircraft as it crossed the runway.
    He stated that based on his experience, the aircraft would have likely stopped if notifed of
    an incursion, exacerbating the severity of the event.
4   Personal correspondence, (2024).
5   https://github.com/mesaba-james/Hybrid-Approach-FRAM-Chapter.

REFERENCES

Arnold, R. D., & Wade, J. P. (2015). A definition of systems thinking: A systems approach.
    Procedia Computer Science, 44, 669–678.
Beer, S. (1984). The viable system model: Its provenance, development, methodology and
    pathology. *The Journal of the Operational Research Society, 35*(1), 7–25. https://doi.
    org/10.1057/jors.1984.2
Besnard, Denis & Hollnagel, Erik. (2014). I want to believe: Some myths about the manage-
    ment of industrial safety. *Cognition, Technology & Work* 16. https://doi.org/10.1007/
    s10111-012-0237-4
Blei, D. M., Ng, A. Y., & Jordan, M. I. (2003). Latent Dirichlet allocation. *Journal of Machine
    Learning Research, 3*, 993–1022.
Clarke, V., & Braun, V. (2017). Thematic analysis. *The Journal of Positive Psychology, 12*(3),
    297–298. https://doi.org/10.1080/17439760.2016.1262613
Creswell, J. W., & Creswell, J. D. (2017). *Research design: Qualitative, quantitative, and
    mixed methods approaches.* Sage publications. http://fe.unj.ac.id/wp-content/uploads/
    2019/08/Research-Design_Qualitative-Quantitative-and-Mixed-Methods-Approaches.
    pdf
Crocetti, G. (2015). Textual spatial cosine similarity. arXiv preprint arXiv:1505.03934.
Dekker, S., Cilliers, P., & Hofmeyr, J.-H. (2011). The complexity of failure: Implications of
    complexity theory for safety investigations. *Safety Science, 49*(6), 939–945. https://doi.
    org/10.1016/j.ssci.2011.01.008
Dekker, S., Hollnagel, E., Woods, D., & Cook, R. (2008). Resilience engineering: New direc-
    tions for measuring and maintaining safety in complex systems. Final report.
Diop, I., Abdul-Nour, G., & Komljenovic, D. (2022). The functional resonance analysis
    method: A performance appraisal tool for risk assessment and accident investigation
    in complex and dynamic socio-technical systems. *American Journal of Industrial and
    Business Management, 12*(02), 195–230. https://doi.org/10.4236/ajibm.2022.122013
Espejo, R., & Reyes, A. (2011). *Naming systems: Tool to study organizational identity.*
    Springer, https://doi.org/10.1007/978-3-642-19109-1
Espinosa, A., & Walker, J. (2011). *Complexity approach to sustainability, A: Theory and appli-
    cation. Series On Complexity Science.* https://doi.org/10.1142/p699#t=toc
Fereday, J., & Muir-Cochrane, E. (2006). Demonstrating rigor using thematic analysis: A
    hybrid approach of inductive and deductive coding and theme development. *International
    Journal of Qualitative Methods, 5*(1), 80–92.
Guest, G., MacQueen, K. M., & Namey, E. E. (2012). Introduction to applied thematic analy-
    sis. *Applied Thematic Analysis, 3*(20), 1–21.
Hollnagel, E. (2012). *FRAM: The functional resonance analysis method: Modelling complex
    socio-technical systems.* Ashgate Publishing
Hollnagel, E. (2016). Resilience engineering: A new understanding of safety. *Journal of the
    Ergonomics Society of Korea, 35*, 185–191.
Hollnagel, E. (2017). *Safety-II in practice: Developing the resilience potentials.* Routledge.
Hollnagel E., Hill R. (2020). Instructions for use of the FRAM Model Visualiser (FMV).
    Available at: https://zerprize.co.nz/Content/FMV_instructions_2.1.pdf

Hoverstadt, P. (2020). The viable system model. In M. Reynolds & S. Holwell (Eds.), *Systems approaches to making change: A practical guide* (pp. 89–138). Springer. https://doi. org/10.1007/978-1-4471-7472-1_3

Hox, J. J., & Boeije, H. R. (2005). Data collection, primary vs. secondary. *Encyclopedia of Social Measurement, 1*(1), 593–599.

Huang, A. (2008). Similarity measures for text document clustering. In *Proceedings of the sixth new Zealand computer science research student conference (NZCSRSC2008)*, Christchurch, New Zealand (Vol. *4*, pp. 9–56).

IATA. (2024). *Air passenger market analysis – February 2024*. International Air Transport Association. https://www.iata.org/en/iata-repository/publications/economic-reports/air-passenger-market-analysis-february-2024/

International Civil Aviation Organization. (2020). *Annex 13 to the convention on international civil aviation: Aircraft accident and incident investigation* (11th ed.). ICAO.

International Civil Aviation Organization. (2023). *Procedures for air navigation services: Air traffic management (PANS-ATM)* (ICAO Doc 4444, 16th ed.). ICAO.

Kim, S. W., & Gil, J. M. (2019). Research paper classification systems based on TF-IDF and LDA schemes. *Human-centric Computing and Information Sciences, 9*, 1–21.

Leveson, N. (2012). *Engineering a safer world: Systems thinking applied to safety*. MIT Press.

Leveson, N. G. (2011). Applying systems thinking to analyze and learn from events. *Safety Science, 49*(1), 55–64. https://doi.org/10.1016/j.ssci.2009.12.021

Li, W., He, M., Sun, Y., & Cao, Q. (2019). A proactive operational risk identification and analysis framework based on the integration of ACAT and FRAM. *Reliability Engineering and System Safety, 186*, 101–109. https://doi.org/10.1016/j.ress.2019.02.012

Madsen, P., Dillon, R. L., & Tinsley, C. H. (2016). Airline safety improvement through experience with near-misses: A cautionary tale. *Risk Analysis, 36*(5), 1054–1066.

Milch, V., & Laumann, K. (2016). Interorganizational complexity and organizational accident risk: A literature review. *Safety Science, 82*, 9–17. https://doi.org/10.1016/j.ssci.2015.08.010

Muecklich, N., Sikora, I., Paraskevas, A., & Padhra, A. (2023). Safety and reliability in aviation – A systematic scoping review of normal accident theory, high-reliability theory, and resilience engineering in aviation. *Safety Science, 162*, 106097. https://doi.org/10.1016/j.ssci.2023.106097

NTSB. (2023a). Air traffic control factual report. https://data.ntsb.gov/Docket?ProjectID=106577

NTSB. (2023b). ATC interview. https://data.ntsb.gov/Docket?ProjectID=106577

NTSB. (2023c). Operational & human performance factual report (p. 195). National Transportation Safety Board Offices.

NTSB. (2023d). AAL pilot interview transcripts. https://data.ntsb.gov/Docket?ProjectID=106577

NTSB. (2023e). Delta Air Lines pilots' statements. https://data.ntsb.gov/Docket?ProjectID=106577

NTSB. (2023f). AAL pilots' statements. https://data.ntsb.gov/Docket?ProjectID=106577

NTSB. (2024). Runway Incursion and Rejected Takeoff American Airlines Flight 106 Boeing 777-200, N754AN, and Delta Air Lines Flight 1943 Boeing 737-900, N914DU. Available at: https://www.ntsb.gov/investigations/AccidentReports/Reports/AIR2401.pdf

Palmer, B. (2013). *Understanding Air France 447*. William Palmer.

Patriarca, R., Bergström, J., & Di Gravio, G. (2017). Defining the functional resonance analysis space: Combining abstraction hierarchy and FRAM. *Reliability Engineering and System Safety, 165*, 34–46. https://doi.org/10.1016/j.ress.2017.03.032

Patriarca, R., Di Gravio, G., Costantino, F., Falegnami, A., & Bilotta, F. (2018a). An analytic framework to assess organizational resilience. *Safety and Health at Work, 9*(3), 265–276. https://doi.org/10.1016/j.shaw.2017.10.005

Patriarca, R., Falegnami, A., Costantino, F., & Bilotta, F. (2018b). Resilience engineering for socio-technical risk analysis: Application in neuro-surgery. *Reliability Engineering and System Safety*, *180*, 321–335. https://doi.org/10.1016/j.ress.2018.08.001

Pfiffner, M. (2022a). *The neurology of business: Implementing the viable system model.* Springer Nature.

Pfiffner, M. (2022b). Quick diagnoses. In *The neurology of business: Implementing the viable system model* (pp. 247–269). Springer.

Piľa, J., Korba, P., & Hovanec, M. (2017). *Aircraft brake temperature from a safety point of view. Zeszyty Naukowe.* Transport/Politechnika Śląska.

Preece, G., Shaw, D., & Hayashi, H. (2013). Using the viable system model (VSM) to structure information processing complexity in disaster response. *European Journal of Operational Research*, *224*(1), 209–218. https://doi.org/10.1016/j.ejor.2012.06.032

Preece, G., Shaw, D., & Hayashi, H. (2015). Application of the viable system model to analyse communications structures: A case study of disaster response in Japan. *European Journal of Operational Research*, *243*(1), 312–322. https://doi.org/10.1016/j.ejor.2014.11.026

Pruchnicki, S. (2018). *Improving facilitated debriefings – How are barriers to learning recognized by instructors and mitigated during post-simulator debriefings?* The Ohio State University.

Rasmussen, J. (1997). Risk management in a dynamic society: A modelling problem. *Safety Science*, *27*(2), 183–213. https://doi.org/10.1016/S0925-7535(97)00052-0

Ruiz-Martin, C., Pérez Rios, J. M., Wainer, G., Pajares, J., Hernández, C., & López-Paredes, A. (2017). The application of the viable system model to enhance organizational resilience. *Advances in Management Engineering*, 95–107, https://doi.org/10.1007/978-3-319-55889-9_5

Schwaninger, M., & Scheef, C. (2016). A test of the viable system model: Theoretical claim vs. empirical evidence. *Cybernetics and Systems*, *47*(7), 544–569. https://doi.org/10.10 80/01969722.2016.1209375

Shorrock, S. (2019). QF 32 - How it went right. *Hindsight*, *29*, 52–55.

Steen, R., Norman, J. E., Bergström, J., & Damm, G. F. (2024b). Dark knights: Exploring resilience and hidden workarounds in commercial aviation through mixed methods. *Safety Science*, *175*, 106498.

Steen, R., Roud, E., Torp, T. M., & Hansen, T.-A. (2024a). The impact of interorganizational collaboration on the viability of disaster response operations: The Gjerdrum landslide in Norway. *Safety Science*, *173*, 106459. https://doi.org/10.1016/j.ssci.2024.106459

Syed-Yahya, S. N., Idris, M. A., & Noblet, A. J. (2022). The relationship between safety climate and safety performance: A review. *Journal of Safety Research*, *83*, 105–118. https://doi.org/10.1016/j.jsr.2022.08.008

Taarup-Esbensen, J. (2023). Distributed sensemaking in network risk analysis. *Risk Analysis*, *43*(2), 244–259. https://doi.org/10.1111/risa.13895

Thoroman, B., Goode, N., & Salmon, P. (2018). System thinking applied to near misses: A review of industry-wide near miss reporting systems. *Theoretical Issues in Ergonomics Science*, *19*(6), 712–737. https://doi.org/10.1080/1463922X.2018.1484527

Von Bertalanffy, L. (1950). An outline of general system theory. *The British Journal for the Philosophy of Science*, *1*(2), 134–165.

Wróbel K. (2021). Searching for the origins of the myth: 80% human error impact on maritime safety, *Reliability Engineering & System Safety*, *216*, 107942, ISSN 0951-8320, https://doi.org/10.1016/j.ress.2021.107942

7 Building a FRAM Model Using Large Language Models

A Dream or a Nightmare?

David Slater and Josué E. Maia França

CONTEXT: LARGE LANGUAGE MODELS FOR FRAM

When teaching potential users the FRAM approach, they readily appreciate the ideas of thinking about safety as ensuring a system works safely and that most systems are not simple but need more thought as to how they work in practice. Also, that this means that we need to understand all the interactions and interdependencies among the activities involved, and that this, in turn, requires a more open, adaptive mindset. It further requires a new way of visualizing these nonlinear behaviours. And now we can do better than portray them as simple boxes and arrows. In FRAM, we have a better way to model them, to trace their workings and predict unexpected emergence, or resonance, among the functions identified as necessary to operate the process.

So, we easily "get" the theory, but to apply the approach, we need to build these models in practice. But having just been shown how to model something like making a cup of noodles and a cup of tea, the next step is quite daunting. The more expert readers may recall these examples from one of the FRAM tutorials, traditionally delivered on the first day of the annual FRAM meeting and workshop.

Now build your own model of a real system! Naturally, as with most things (e.g., riding a bike?), learning is painful, and until comprehension dawns, with the first set of hexagons that gives insights that you had not appreciated, it is difficult to accept that it is worth the struggle of trying to build these funny-looking pictures. As with painting, the first brush stroke is the most difficult. In building FRAM models, the first model is a mental struggle, but, as always, it only needs to be a first approximation. An accurate, "correct" first draft is not necessarily crucial, as it is inevitably incomplete and will get modified and improved as our understanding develops.

This chapter concentrates on describing some novel suggested approaches to getting started (but not necessarily finishing) that have proven successful in producing these first models. It looks at the essential steps needed. Then, at the complementary methods available for helping achieve them, and finally discusses the pros, cons, and pitfalls that the analyst needs to be aware of when employing them. For example, anyone can prompt one of the new chatbots to produce text on a given application.

DOI: 10.1201/9781003518167-9

But is it sound, satisfactory, and safe to do so? We will try to answer this question in the coming pages, also presenting lessons learned from testing this approach and the experience of using this method of "Kwikstarting" building FRAM models. Please note that by Kwikstarting, we refer to a rapid and pragmatic way for initiating the process of building a FRAM model. It means that we do not have to start the analysis with the full and complete model. We start with an iterative approach where the initial model is seen as a rough draft but rich in contributions, creativity and tacit knowledge. The goal is to get something on paper (mostly on screen, e.g., with the FRAM Model Visualizer software) that serves as a foundation for refinement and deeper understanding through its development.

The reflections in this chapter arise from a number of applications attempted. To better understand the application of LLMs in a FRAM analysis, one main case study will be presented related to an offshore oil well drilling. A standard prompt to be applied will be finally suggested to be used as a basis for various domains and, indeed, iterated in the appendix of this chapter in a healthcare case.

It is highly likely that users looking to try FRAM as a systems analysis approach have been used to other approaches that have proved less satisfactory but yield more familiar results. In our areas of interest in complex socio-technical systems, such as healthcare, O&G, civil aviation, maritime shipping, and nuclear, Jens Rasmussen (1997) was probably the first to put forward ideas on looking at what is influencing the behaviour of the system, such as in his Accimaps visualizations. Leveson (2011) extended this approach to model systems as a set of controllers/controlled processes. It took Hollnagel's realization that complex systems were not fixed networks or just control/influence diagrams, and as such, they needed a way of modelling capable of tracing the often-random variations in real-life operations that could give rise to emergent and unexpected behaviours (out of control?).

The more technology advances and the more complex the systems that make up the environments we circulate in – workplaces, homes, entertainment, etc. – become, the clearer the need to look at the system as a whole and then to discover and document all the key activities and their sequences and details required to operate the systems themselves.

This chapter aims to provide aids and suggestions for how to get started with a FRAM analysis supported by artificial intelligence. LLMs are advanced machine learning models that utilize deep learning techniques, particularly those based on the transformer architecture, to generate, understand, and manipulate human language. LLMs are trained on diverse and extensive datasets comprising text from books, articles, websites, and other digital text sources. Through training, these models learn intricate patterns and structures in natural language, enabling them to perform a wide range of language-related tasks.

THE RATIONALE FOR LLMs IN FRAM MODELLING

Whether the would-be FRAM user is a domain expert or not, we now live in an age where system information and descriptions are but a click away. Not only that, but by means of conversational agents, a semi-intelligent dialogue can now be held with various LLMs, trained on immense amounts of data, making them capable of understanding and generating natural language and other types of content to perform a wide range of tasks, including describing and explaining system details. So, whether this is a discovery or a safety net, this option is worth considering as it is expected to get better over the coming years. The development of LLM represents a significant milestone within the recent resurgence of AI, the latter having been driven primarily by advancements in deep learning techniques and applications, for example, in high-profile healthcare imaging and diagnostics research studies (Ebrahimian et al., 2023; Roberts et al., 2023). User-facing LLMs, such as OpenAI ChatGPT and Google Bard (since February 2024, called Gemini), represent the current peak of public interest in AI, showcasing unprecedented capabilities that have the potential to permeate all facets of society.

The rapid adoption of LLMs is evident from the exponential growth in their user base, as seen, for example, with ChatGPT, which is said to have accumulated 100 million users within just two months of its public release at the end of 2022 (Wu et al., 2023). LLMs have sparked widespread interest and generated considerable hype across many work domains with their potential to transform how people interact with technology and how work is done. In fields like medical decision-making and education, LLMs have demonstrated their potential to augment human capabilities, leading to discussions about their integration into everyday practices (Qi et al., 2023; Stechly, 2023; Oviedo-Trespalacios et al., 2023; Brundage et al., 2018).

Beyond healthcare, industries spanning finance, manufacturing, and entertainment are also leading the exploration of LLM in enhancing productivity and innovation (Kocoń et al., 2023; Deng et al., 2023; Badini et al., 2023). Numerous standards and guidance documents have been developed, including integrative standards, such as BS30440 in the healthcare domain (Sujan et al., 2023), which synthesizes best practices and guidelines into an auditable validation framework for healthcare AI. It is timely to additionally consider how safety practitioners might benefit from employing LLMs, especially, for the scope of this book, those interested in FRAM modelling.

Method Geek

The key to using LLM lies in the careful design of the prompts and follow-up questions.

For example, one needs to specify:

- Role – What's the context, and at what level of expertise do you want it to phrase its answers?
- Task – What exactly do you want it to do, research, imagine, analyze, etc.?
- Tone – Write using a formal, expert, critical, etc., tone.
- Format - Give details, be concise, and output as bullet points, tables, etc.

It is possible to use one single mega prompt, but in general, a series of follow-up prompts probing for specific details, asking to explain in more detail or more fully, or correcting unclear or incorrect assertions or hallucinations.

METHODOLOGY

Since the standard steps to build a FRAM have been explained in the previous chapter about "The FRAM essentials", the focus here will be on the phases concerned with the LLM usage within the identified steps. Of course, for any and all analyses, it is important, as Step 0, to discuss what exactly the system is to be studied, delineating what its boundaries are and known issues are, namely, the scope of the analysis. Besides this preliminary step, here we present our suggested phases to include LLM in a FRAM model building.

So, as a precursor for a non-domain specialist in FRAM, some outline description, before diving into the details that are, of course, almost too well known to those involved on an everyday basis, is a prerequisite. Can an LLM fulfil that role? Let's see.

PHASE 1 – IDENTIFICATION OF THE LLM

Select, from all LLM applications available, the most suitable for the system under study. Notice that, depending on the amount of data you may have, you will need an LLM with high processing capacity. When writing this chapter, several LLMs existed (GPT, Claude, Gemini, Grok, Mistral, PaLM, etc.). One should also consider the possibility of using a Small Language Model (SLM), especially in those cases where the domain is covered by one of those highly specialized ontological frameworks.

Method Geek

Large Language Models (LLMs) and Small Language Models (SLMs) serve distinct roles, depending on scale, complexity, and the application they are meant for. LLMs feature hundreds of millions to billions of parameters and are trained on vast, diverse datasets. These models excel in a wide range of natural language processing tasks due to their deep understanding of language nuances, enabled by advanced transformer architectures. Conversely, SLMs are compact, containing thousands to millions of parameters, and are usually trained on smaller, domain-specific datasets. They are designed for operational settings where computational resources, power, and response times are limited, making them ideal for embedded systems or case-specific applications. On the one hand, LLMs offer broader capabilities and superior performance, especially for complex tasks; SLMs provide practical solutions where efficiency is critical, and the knowledge needed to build the analysis can be limited ex-ante.

PHASE 2 – SETTING A PROMPT

Once in the traditional Step 0 of a FRAM application, you define which part of the system you will study and its boundaries. Here, you set a prompt that reflects the aim of your analysis to be used in the conversational agent linked to the LLM. For instance, if your analysis is how an engine works, looking for the development of a troubleshooting guide, your prompt must have terms, acronyms, and verbs related to the engine functioning.

Following experience in various case studies, it has been considered helpful to use within the prompt the Hierarchical Task Analysis (HTA) as a means to deconstruct a process via its constituents' functions (or tasks), leveraging the potentialities of the LLM.

Method Geek

The HTA (Stanton, 2006) is a systematic methodology used in the field of human factors and ergonomics to decompose a task into its constituent sub-tasks in a structured, hierarchical manner. This approach helps in understanding the complex processes involved in task execution by breaking them down into more manageable and analyzable components. HTA begins with identifying the overall goal of the task and then divides it into a hierarchy of sub-tasks and operations, each described by its specific goals, required actions, and any procedural or conditional dependencies. The analysis includes specifying the sequences and conditions under which sub-tasks are to be carried out. It is widely applied in various domains, including aviation, healthcare, military operations, and interface design, making it a versatile tool for addressing human–system interaction issues. By providing a clear, visual, and logical structure of tasks, HTA supports the identification of potential points of failure, training needs, and opportunities for technological interventions.

Asking the LLM to use the FRAM structure and terminology, presenting it in a table. It will help you build the model.

Phase 3 – Analyzing the LLM Answer

No matter how "intelligent" an AI agent is, it still depends on a database and prior information to function. Its greatest asset is its processing capacity. And this ability, if it processes one or more incorrect data, could generate a scenario that does not make sense. Therefore, analyzing the LLM response and checking whether it makes sense is essential.

Method Geek

This analysis should be performed by the analysts, and it should account for (at least):

- Contextual Relevance: Assess whether the output of the LLM is relevant to the prompt or question asked, without veering off-topic.
- Logical Consistency: Ensure that the output maintains consistency in its arguments, claims, and facts throughout, without contradicting itself.
- Factual Accuracy: Assess the accuracy of facts, especially in the case of detailed or technical content, by cross-referencing with reliable sources.
- Specificity: Be cautious with overly generic responses, as well as extremely detailed ones. The former could be a sign of misunderstanding the prompt.

PHASE 4 – DESIGNING THE FUNCTIONS AND THE COUPLINGS

Taking the LLM's answer, after your confirmation that it makes sense, start modelling the system. One option could be to use the FMV (FRAM Model Visualizer), but other options are viable, too. The functions and their couplings presented in tables, with their respective aspects (tip above), provide a differentiated performance at this stage, but can serve as a starting basis for a complete analysis. This phase corresponds to traditional Step 1 of a FRAM analysis.

PHASE 5 – ANALYZING THE MODEL

In this last phase, your FRAM modelling, supported by an LLM, is done. You and your team will discuss it, providing the analysis of the system under study. Note that from the function identification phase (cf. Phase 4), it is also necessary to draw information about phenotypes of variability, which allows you to gain a complete picture of the system at hand.

Notice that the discussion over the model is the analysis aimed at by the methodology and cannot be omitted. When a FRAM is built without an LLM, the discussion starts in Step 0 and goes through the entire process, so this phase should bear in mind the need to cover an extensive discussion over the model's peculiarities.

DATA COLLECTION

For the development of the FRAM model in the case study proposed by this chapter, a group of two experts in drilling and one FRAM expert (with particular knowledge of drilling) was set (Table 7.1).

This group had two rounds of online meetings for the development of the initial model and one round of online meetings for the validation of the final model. The location of their working operations is the same: an oil rig placed in the Brazil pre-salt area.

The stakeholders involved are the drilling crew, composed of the Rig Manager, Driller, Assistant Drillers, among others. They work in a 12-hour/day shift, having shift changes at 06:00 and 18:00, where they usually have a few minutes to discuss

TABLE 7.1
Experts Involved in the Study

	Driller #1	Driller #2	FRAM Expert #1
Age	46	33	45
Gender	Male	Male	Male
Nationality	Brazilian	Brazilian	Brazilian
Background	Mechanical technician; Mechanical engineer	Petroleum engineer	Electronics technician, Safety engineer, PhD
Years of experience in O&G industry	22	12	16
Years of experience in drilling	20	12	3

and pass the work for the next crew. Besides these professionals, one should consider the company operator and third-party contractors, among others.

For one oil rig – an offshore oil platform dedicated to drilling – there are from 150 to 300 POB (People on Board), depending on a series of factors: contract type, place of drilling, Country Regulation, age (generation of the oil rig), availability of space on board, etc.

These numbers serve as a confirmation of the complexity of the socio-technical system at hand.

RESULTS

This section describes an example of how we have tested this approach in a real application, to look at the intense activities involved on the drill floor of an offshore oil exploration installation. We initially used the LLM to try to describe the operation and system features involved, producing a list of the critical functions and the inter-actions/interdependencies involved. Then, we built the FRAM model using the FMV as a "novice" helped by AI, generating the Work-As-Imagined model that, in this case, will be more accurately called the "Work-As-AI-Imagined", WAAII. After the initial WAAII was built, we then asked domain experts in drilling offshore oil wells, guided by an experienced FRAM practitioner, to analyze the same set of activities and then produce a FRAM model of the Work-As-Done, WAD. For simplicity, we assume this model to be indeed the WAD, while it would actually be a Work-As-Disclosed variety about the WAD itself. The comparison between the two varieties then forms the basis of the discussion of the usefulness and validity of the approach.

Before starting, we recognized that to drill an offshore oil well is not an easy or risk-free task. It involves extreme process parameters, such as high temperature, flow and pressure, in an environment that is highly hostile – an industrial plant on the high seas.

In this environment, some (and only some!) of the activities needed to have a well ready for production, from the initial drilling till the installation of the WCT (Wet Christmas Tree), are:

- To control the drilling
- To install new drilling pipes
- To monitor drilling parameters
- To maintain constant communications

Domain Expert

Also, it is important to mention that a failure to control these processes, which are intrinsic to the workplace, can lead to significant losses, worker injuries, and fatalities (França et al., 2019). Offshore drilling, particularly in deep-water reservoirs such as the Brazilian pre-salt, represents a high-risk and high-cost operation, and they are very aware that a most severe accident can occur, that is, a blowout, an uncontrolled flow up the drill pipe from the reservoir, which poses a grave threat to the drilling rig and its crew.

Oil spills resulting from offshore blowouts (and resulting fires and explosions) can cause extensive damage to the environment and marine ecosystems, sometimes taking several years for them to recover (Xue et al., 2013). Understanding this activity in detail, knowing how the complex dynamics involving risks are established, is not only important to preserve the lives of workers but also the environment and business sustainability.

The prompt starts with defining a list of functions that describe the process. But we need more than just a list of functions to build a FRAM model. It needs a discussion on how the work is actually done in the dynamic conditions of a complex reality. This case presents a comparison between the model generated by the LLM, the WAAI, and a model based on real testimonies of experienced professionals, the WAD.

WAAII (Work As AI Imagined)

A FRAM analysis offers a unique chance to comprehend in more detail the complexity of the combination of all these elements that compose the activities to drill an offshore oil well. It sounds exciting, but before starting to discover the WAD (Work as Done), can ChatGPT provide a reasonable, at least simplified, version of it, as for the WAAII?

Let's ask.

You are a professional safety engineer in the oil and gas industry. You are also a system modeller, familiar with systems thinking. You will be using the Functional Resonance Analysis Method (FRAM) to model a system and identify the tasks that need to be carried out to deliver the output of the process the system is designed to facilitate.

In this case, writing in the style of a succinct technical report, describe the processes needed for the operation of an offshore drilling platform drilling a subsea well.

Use the language of a Hierarchical Task Analysis representation of how and in what sequence these tasks are carried out. Identify each of these tasks as achieved by a function, which you will label as <To do something>. Set out a list of these functions and any interactions or interdependencies with other functions. Apply the syntax of the functions and the criteria of the functions, coupling with each other as demanded by the FRAM methodology.

List all the functions and subfunctions identified in the Hierarchical Task Analysis as FRAM functions with labelled inputs, outputs, preconditions, resources, control, and time interdependencies with other function outputs as necessary, or discovered. Present this modelling as a table, showing which aspects of the FRAM – Input, Output, Time, Precondition, Control and Resource – are coupled with each other function, determining the connections between functions that model the system.

Continue to do the full process for all of the functions; do not use shortcuts or the tag – continue as before, shortcut. Once the functions have been listed, think and show thinking about their completeness, relevance, and adequacy, and point out any reservations or if this misses anything.

Here comes the outcome, as obtained from ChatGPT 4.0 (August 2024)

Domain Expert

Technical Report: Functional Resonance Analysis Method (FRAM) for Offshore Drilling Platform Operations

INTRODUCTION

This technical report applies the Functional Resonance Analysis Method (FRAM) to model the processes needed for the operation of an offshore drilling platform drilling a subsea well. The Hierarchical Task Analysis (HTA) representation is used to identify and sequence the necessary tasks. Each task is represented as a function, and its interactions and interdependencies are analyzed. The table provided details the FRAM functions, including inputs, outputs, preconditions, resources, controls, and time interdependencies (Tables 7.2 and 7.3).

HIERARCHICAL TASK ANALYSIS (HTA)

1. Mobilize Drilling Platform
2. Position Platform at Drilling Location
3. Prepare Drilling Equipment
4. Conduct Pre-drilling Safety Checks
5. Start Drilling Operations
6. Monitor Drilling Parameters
7. Manage Drilling Fluids
8. Handle Drilling Cuttings
9. Perform Well Logging
10. Cement Well Casing
11. Complete Drilling Operations

COMPLETENESS

The identified functions encompass the key operations required for the drilling of a subsea well on an offshore platform. The sequence from mobilization to completion covers the essential stages.

RELEVANCE

Each function is crucial for the drilling process, ensuring the safety, efficiency, and success of the operations.

ADEQUACY

The functions are adequate for the primary drilling process. However, the model may benefit from additional details in certain areas, such as emergency response procedures, maintenance routines, and coordination with support vessels.

TABLE 7.2
List of FRAM Functions as Obtained by the LLM

Function	Input	Output	Pre-Condition	Resource	Control	Time
Mobilize Drilling Platform	Mobilization Order	Mobilized Platform	Mobilization Plan	Logistics Team	Mobilization Protocol	Scheduled Start Date
Position Platform at Drilling Location	GPS Coordinates	Positioned Platform	Mobilized Platform	Navigation Team	Positioning Protocol	Immediate
Prepare Drilling Equipment	Equipment List	Prepared Equipment	Positioned Platform	Drilling Crew	Preparation Checklist	Before Drilling
Conduct Pre-drilling Safety Checks	Safety Checklist	Safety Approval	Prepared Equipment	Safety Officer	Safety Protocol	Before Drilling
Start Drilling Operations	Safety Approval	Drilling Initiated	Safety Approval	Drilling Crew	Drilling Plan	Scheduled Start
Monitor Drilling Parameters	Drilling Data	Drilling Status Report	Drilling Initiated	Monitoring Equipment	Monitoring Guidelines	Continuous
Manage Drilling Fluids	Fluid Inventory	Managed Fluids	Drilling Initiated	Fluid Management Team	Fluid Management Plan	Continuous
Handle Drilling Cuttings	Cuttings Data	Handled Cuttings	Drilling Initiated	Waste Management Team	Waste Management Plan	Continuous
Perform Well Logging	Drilling Data	Well Logs	Drilling Progress	Logging Team	Logging Protocol	During Drilling
Cement Well Casing	Casing Plan	Cemented Casing	Drilling Completed	Cementing Crew	Cementing Guidelines	After Drilling
Complete Drilling Operations	Drilling Data	Drilling Report	Drilling Completed	Drilling Crew	Completion Protocol	After Drilling

TABLE 7.3
List of FRAM Couplings, as Obtained by the LLM

From Function	To Function	Input/Output/Control/ Resource Interdependencies
Mobilize Drilling Platform	Position Platform at Drilling Location	Output (Mobilized Platform)
Position Platform at Drilling Location	Prepare Drilling Equipment	Output (Positioned Platform)
Prepare Drilling Equipment	Conduct Pre-drilling Safety Checks	Output (Prepared Equipment)
Conduct Pre-drilling Safety Checks	Start Drilling Operations	Output (Safety Approval)
Start Drilling Operations	Monitor Drilling Parameters	Output (Drilling Initiated)
Start Drilling Operations	Manage Drilling Fluids	Output (Drilling Initiated)
Start Drilling Operations	Handle Drilling Cuttings	Output (Drilling Initiated)
Start Drilling Operations	Perform Well Logging	Output (Drilling Initiated)
Monitor Drilling Parameters	Perform Well Logging	Input (Drilling Data)
Manage Drilling Fluids	Monitor Drilling Parameters	Input (Fluid Inventory)
Handle Drilling Cuttings	Monitor Drilling Parameters	Input (Cuttings Data)
Perform Well Logging	Monitor Drilling Parameters	Input (Well Logs)
Monitor Drilling Parameters	Cement Well Casing	Output (Drilling Status Report)
Cement Well Casing	Complete Drilling Operations	Output (Cemented Casing)

RESERVATIONS AND POTENTIAL OMISSIONS

1. Emergency Response Procedures: Specific functions for emergency response and contingency planning should be explicitly included.
2. Maintenance and Inspection: Regular maintenance and inspection routines for equipment and infrastructure might need separate functions.
3. Coordination with Support Vessels: Functions related to the logistics and coordination with supply and support vessels should be detailed.
4. Environmental Monitoring: Continuous monitoring of environmental impact and compliance with regulations should be incorporated.

CONCLUSION (BY THE PROMPT)

The FRAM analysis of the offshore drilling platform operations provides a detailed overview of the necessary functions and their interdependencies. While the current model is comprehensive and relevant, continuous evaluation and updates are essential to ensure adequacy in various scenarios and incorporate all necessary aspects of the drilling operations.

This report serves as a foundation for understanding the complex interactions in offshore drilling and can guide improvements in safety, efficiency, and environmental compliance.

The FRAM analysis above provided us with a sufficiently comprehensive understanding of the offshore drilling process and the critical functions involved. With this LLM response, a draft FRAM model can be obtained, showing its functions, aspects and, mainly, its couplings that characterize the functioning and resonance of the system under study. Figure 7.1 shows this FRAM model, which is the WAAII of offshore well drilling operations, that is, a simplified version of operations, obtained with the assistance of the AI.

WAD (WORK AS DONE)

To produce a genuine FRAM model of real operation, a domain expert team from the offshore oil and gas industry was invited to contribute to this study. The composition, characteristics and other information of this group are in the section "Data Collection" of this chapter. The FRAM developed by this group, representing the WAD of offshore oil well drilling operations, is shown in Figure 7.2.

DISCUSSION

The results obtained in this study showed that the prompt illustration was helpful and provided a reasonable basis on which to build a FRAM model. We can agree that in the case at hand, the LLM served as an aid to providing not just a sensible, if generic, overview of the system being studied but a respectable set of starter functions with which to build, explore and develop the actual FRAM model. For instance, the function <To perform completion tests> was not covered by the WAAII, but it is the most critical for the completion phase. Also, the function <To cement the casing of each section of the well>, and its aspects Time <To cure the cementation process> were not considered, even though it is what in fact builds an offshore well. Since the Macondo accident (França et al., 2022), it is clear the need for having a proper cementation in an offshore oil well. On another hand, the beginning of the drilling process, represented in the WAAII by the functions <Mobilize Drilling Platform> and <Position Platform at Drilling Location> was adequately considered, showing coherence and relative accuracy of the LLM.

POINTS OF DIVERGENCE

The domain experts added a front end to get the drilling rig in place, while the LLM was only asked to build a FRAM model of the actual drilling and completion. Obviously, the drillers know a lot more about the details of the process, and this is reflected in the plethora of background functions shown, supplying necessary support to the drillers.

But their FRAM model simplified the drilling to a column of simultaneous activities that do not interact. Because they see and feel it as obvious and straightforward, they do it naturally – it is normal to them.

The LLM, being a non-expert generalist, does highlight the sequence and interactive aspects involved in the operation, which to the analyst is very interesting and useful, particularly when you look at the variabilities that can occur in personnel,

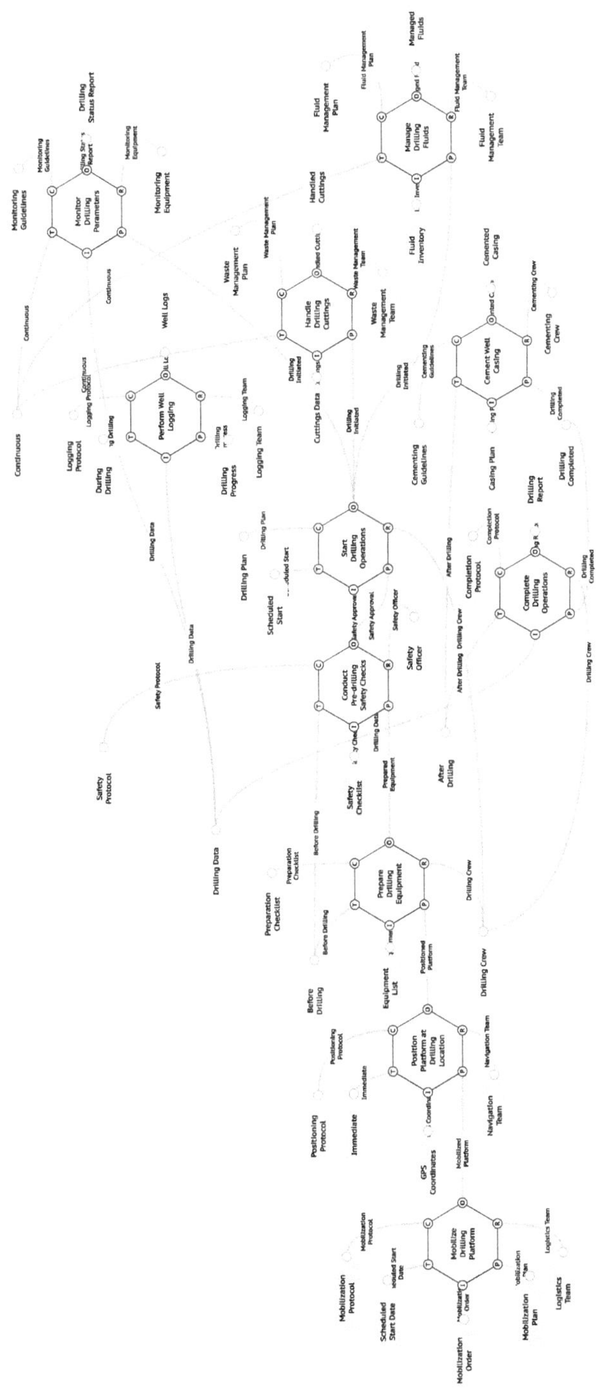

FIGURE 7.1 FRAM model of the WAAII of the offshore drilling operations.

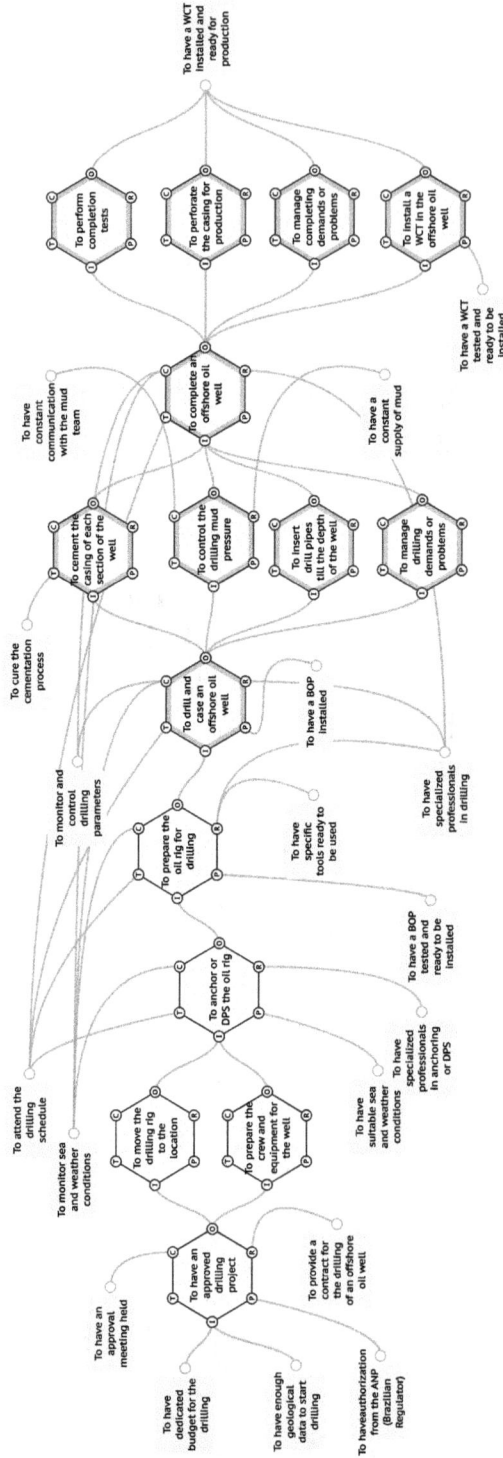

FIGURE 7.2 FRAM model of the WAD of the offshore drilling operations, where new groups of functions are highlighted with a darker border.

maintenance, supplies, mud control, etc. The Macondo incident (França et al., 2022) highlights this case very well.

So, for a naïve data analyst, the LLM FRAM model looks helpful. From the drillers' point of view, we think their FRAM model gives them a chance to identify all the functions that are supposed to be involved. An expert data analyst knows that this is the core for avoiding getting biased information into their investigation.

Based on these early results, we believe that the LLM can prove useful for analysis, but will always remain an element of support, helping the analyst build a starter FRAM model and, for the drillers, perhaps helping to explain and to deal with the non-trivial happenings that may appear. For instance, we think that the LLM will most likely never assign a function <To deal with pelicans on the drill floor> or <To keep replacing the silver tape on the auxiliary mud hose>, both functions that appeared in the conversation about the function <To manage drilling demands or problems>.

So, while the LLM is a great tool to plan, design, and develop preliminary steps of a drilling operation, represented by the WAAII, the operators remain an (the?) essential source of how operations actually happen, how they make tough choices (under both normal and emergency circumstances), and essentially how they keep things going. However, it is possible to perceive that, if we feedback to the LLM these real scenarios and non-planned activities in the field, it could possibly identify FRAM functions more aligned with WAD.

But there will always be a "gap" of uncertainties not mapped by LLM, resulting from the complex combinations of our increasingly technological offshore work environments. Quite contrary to what many managers would think, the insertion of more technological artefacts/controls (automation) in offshore drilling does not reduce complexity or increase control but rather brings more data and possibilities for action in the drilling system, which masquerades as pseudo-control of complexity. And why? While automation is designed to minimize human intervention and improve efficiency, it can inadvertently create systems so intricate that unforeseen problems beyond the scope of automated responses can occur. In such complex sociotechnical workplaces, human adaptive variability becomes an appropriate response to the natural unpredictability that will emerge. Humans possess the ability to respond to non-trivial and unpredictable situations by drawing on expertise, experience, and non-technical skills – competencies that automation, by its rigid design, cannot replicate. This underscores the necessity of balancing automation with human intervention to manage the dynamic and emergent outcomes from complex sociotechnical systems effectively. This can be demonstrated with a FRAM model to help the management of change more effectively, a clear necessity for safety professionals.

The comparison of these two results, the WAAII and the WAD of the offshore drilling, despite showing clear differences, also shows the possibility of an integrated development, where an LLM can support the processing, design, and development activities of drilling offshore wells, while the human brings the necessary perspective to make what was planned occur productively and safely in real life. It is often observed that all plans seem perfect and doable until the first exposure to reality (paraphrasing von Clausewitz) (Schön, 1983). So, if we can give operational feedback to LLM, we may have the possibility of updating our WAI with WAD, or better yet, WAAII with WAD. But an indispensable reference point is still the drillers who handle the real demands on adaptability that occur in the daily routine of off-shore drilling.

The Need to Validate the Prompt

Initially, we found that with the earlier ChatGPT versions (e.g., ChatGPT 3.5), it was necessary to educate the model via prompts as to exactly what a FRAM analysis was. It needed to know about the functions, the aspects, the couplings, etc. When this was provided, it produced the results shown. The LLM had a better knowledge of the more conventional tools, HTA, BPM, etc., but nevertheless this was also reinforced.

Method Geek

However, by the end of the study (August 2024), the available version (ChatGPT-4) probably had access to more up-to-date literature, which obviously had more relevant information on the FRAM processes. ChatGPT-4, developed by OpenAI, is a chatbot based on the LLM GPT-4. It excels in understanding and generating human-like text across various contexts and languages. Key features include enhanced contextual understanding, improved coherence in longer texts, and superior problem-solving capabilities compared to its predecessors. ChatGPT-4 has access to an extensive range of data, but its knowledge is limited to information available up to December 2023.

A suggested prompt based on the use case of this chapter, that is, offshore drilling operations, as well as on many other experiences by the authors (see another iteration of it in a healthcare domain in Appendix 7.A). As such, it can be applied by a FRAM expert to have comparative data or a starting point and by newcomers who wish to start using FRAM as an analysis tool.

*You are a **[put here the profession/activity under study]**. You are also a system modeller, familiar with systems thinking. You will be using the Functional Resonance Analysis Method (FRAM) to model a system and identify the tasks that need to be carried out to deliver the output of the process the system is designed to facilitate.*

*In this case, writing in the style of a succinct technical report, describe the process of **[put here the process or subprocess of the system you are studying with as much detail as possible – several iterations may be necessary to ensure contextual reference, logical consistency, factual accuracy and specificity]**.*

Use the language of a Hierarchical Task Analysis representation of how and in what sequence these tasks are carried out. Identify each of these tasks as achieved by a function, which you will label as <To do something>. Set out a list of these functions and any interactions or interdependencies with other functions. Apply the syntax of the functions and the criteria of the functions coupled with each other, demanded by the FRAM methodology.

List all the functions and subfunctions identified in the Hierarchical Task Analysis as FRAM functions with labelled inputs, outputs, preconditions, resources, control, and time interdependencies with other function outputs as necessary or discovered. Present this modelling as a table, showing which aspects of the FRAM – Input, Output, Time, Precondition, Control, and Resource – are coupled with each other's function, determining the connections between functions that model the system.

Continue to do the full process for all of the functions; do not use shortcuts or the tag – continue as before, shortcut. Once the functions have been listed, think and show thinking about their completeness, relevance, and adequacy, and point out any reservations or if this misses anything.

This is only a suggested prompt! Only you and your team will know what information, data and peculiarities your system under study really needs and is relevant. This prompt is a starting point, a suggestion, from which you can improve.

IS THIS REASONABLE TO USE?

One must always look at the output of a dialogue very sceptically. In trying to string intelligent sentences together, the LLM can sometimes become too imaginative, confused or sometimes contradictory. The function list produced is a valuable base from which to probe further for veracity and accuracy. But above all, remember that the dialogue can be limited or even substantially different each time.

For example, the list of functions produced in a similar study on how a microprocessor system works, repeated a month later from the same prompts, produced only six main functions compared to the more detailed 16 originally produced. Crucially, the missing functions were mostly focused on error handling, which an analyst might think was essential. Consistency as well as confusion can be an issue.

The limitations of LLMs and their potential to contribute to risk management in safety-critical domains have rightly received considerable scrutiny, even if further work is required to provide a more comprehensive understanding. For example, "hallucinations", that is, false or imagined outputs of the LLM, which are not based on source data, are a frequently described phenomenon. Hallucinations have significant safety implications because they may sound plausible and can be hard to detect (Bruno et al., 2023). A review examining the risks of using ChatGPT-4 as a chatbot in healthcare also found that the underlying LLM, GPT-4, was prone to hallucinations and could mislead users (Lee et al., 2023). In addition, LLMs are trained on publicly available data and do not have access to restricted data such as patients' medical records, which can affect their relevance and accuracy. These hallucinations typically occur because the LLM, relying solely on patterns in data, lacks true understanding and grounding in real-world facts. This is something that a FRAM analyst, expert or newbie should always bear in mind when trying to replicate the experiments discussed in this book.

CONCLUSIONS

The application of LLMs to assist in building FRAM models presents a significant advancement in system modelling, particularly for novice practitioners but also for experts, especially for comparisons. The comparative analysis between the Work As AI Imagined (WAAII) and Work as Done (WAD) models highlights both the potential and limitations of LLMs. While an LLM offers a valuable starting point by generating comprehensive system descriptions and identifying critical functions, it lacks the nuanced understanding of domain experts. It is important to remember that no matter how "intelligent" an AI is, it still depends on the way a human analyst prompts it and on the database and prior information it uses to function. Its greatest asset is its processing capacity. And this ability, if it processes one or more incorrect data, could generate a scenario that does not make sense.

Integrating human expertise with LLM-generated models enhances the accuracy and relevance of FRAM analyses. Continuous feedback and iterative refinement are crucial to bridge the gap between AI-generated models and real-world operations, underscoring the importance of human oversight in leveraging AI for safety-critical tasks.

The key principle here is balance: do not rely entirely on LLM to build a FRAM model, but also recognize human limitations in doing so. The integrated use seems to be the most appropriate, especially for methodological developments linked to safety.

Indeed, the use of LLM affords safety analysts the opportunity to enhance the FRAM analysis by facilitating initial model generation and offering different perspectives. Responsible and effective utilization of LLM requires careful consideration of their limitations as well as their abilities. Human expertise is crucial both with regard to validating the output of the LLM as well as in developing meaningful interactive prompting strategies to take advantage of LLM capabilities such as self-critiquing from different perspectives.

Finally, it is vital to remember that a computer-aided first impression is not only, inevitably, Work as Imagined (WAI) but, before employing it as in the methods suggested here, to be even more aware that it could be even less accurate/realistic – Work as Artificial Intelligence Imagined! (WAAII) or WA^2I^2 – Work as exponentially imagined?

APPENDIX 7.A

For demonstration purposes, another use case is presented here, leaving an experienced reader to answer the perennial questions on the usability of the LLM for FRAM purposes. The case is on purpose completely different from the Accident and Emergency (A&E) arrivals system in a healthcare facility. One of the challenges in many hospitals is the more efficient and effective operation and management of the very unpredictable arrival stream of patients at the A&E Department. This is even more crucial, of course, in the event of a mass casualty incident. A FRAM model can offer the chance to examine in more detail the process, sub-processes, and actors involved.

PROMPT

You are a healthcare professional. You are also a system modeller, familiar with systems thinking. You will be using the Functional Resonance Analysis Method (FRAM) to model a system and identify the tasks that need to be carried out to deliver the output of the process the system is designed to facilitate.

In this case, writing in the style of a succinct technical report, describe the process of admitting a patient to the accident and emergency department of a hospital.

Use the language of a Hierarchical Task Analysis representation of how and in what sequence these tasks are carried out. Identify each of these tasks as achieved by a function, which you will label as <To do something>. Set out a list of these functions and any interactions or interdependencies with other functions. Apply the syntax of the functions and the criteria of the functions coupled with each other, demanded by the FRAM methodology.

List all the functions and subfunctions identified in the Hierarchical Task Analysis as FRAM functions with labelled inputs, outputs, preconditions, resources, control, and time interdependencies with other function outputs as necessary, or discovered. Present this modelling as a table, showing which aspects of the FRAM – Input, Output, Time, Precondition, Control and Resource – are coupled with each other function, determining the connections between functions that model the system.

Continue to do the full process for all of the functions, do not use shortcuts or the tag - continue as before, shortcut.

Once the functions have been listed, think and show thinking about their completeness, relevance, and adequacy and point out any reservations or if this misses anything.

Domain Expert

Technical Report: Functional Resonance Analysis Method (FRAM) for Patient Admission in Accident and Emergency (A&E) Department

INTRODUCTION

This technical report applies the Functional Resonance Analysis Method (FRAM) to model the process of admitting a patient to the Accident and Emergency (A&E) department of a hospital. The Hierarchical Task Analysis (HTA) representation is used to identify and sequence the necessary tasks. Each task is represented as a function, and its interactions and interdependencies are analyzed. The table provided details of the FRAM functions, including inputs, outputs, preconditions, resources, controls, and time interdependencies (Tables 7.A.1 and 7.A.2).

HIERARCHICAL TASK ANALYSIS (HTA)

1. Receive Patient Arrival Notification
2. Initial Triage Assessment
3. Patient Registration
4. Detailed Triage Evaluation
5. Allocate Treatment Area
6. Assign Medical Team
7. Conduct Medical Assessment
8. Develop Treatment Plan
9. Administer Initial Treatment
10. Monitor Patient Condition
11. Decide on Admission or Discharge

TABLE 7.A.1
List of FRAM Functions

Function	Input	Output	Pre-Condition	Resource	Control	Time
Receive Patient Arrival Notification	Notification from EMS	Patient Arrival Record	EMS Notification	EMS Communication	Hospital Protocol	Immediate
Initial Triage Assessment	Patient Arrival Record	Initial Triage Category	Patient Arrival	Triage Nurse	Triage Protocol	Within 10 minutes
Patient Registration	Initial Triage Category	Patient Record	Initial Triage	Registration Clerk	Registration Protocol	Concurrent with Triage
Detailed Triage Evaluation	Patient Record	Detailed Triage Category	Initial Triage	Triage Nurse	Triage Protocol	Within 15 minutes
Allocate Treatment Area	Detailed Triage Category	Treatment Area Assigned	Detailed Triage	Hospital IT System	Allocation Protocol	Immediate
Assign Medical Team	Treatment Area Assigned	Medical Team Assigned	Treatment Area Allocation	Department Coordinator	Assignment Protocol	Immediate
Conduct Medical Assessment	Medical Team Assigned	Medical Assessment Report	Medical Team Assignment	Medical Team	Assessment Guidelines	Within 30 minutes
Develop Treatment Plan	Medical Assessment Report	Treatment Plan	Medical Assessment	Medical Team	Treatment Protocol	Immediate
Administer Initial Treatment	Treatment Plan	Initial Treatment Outcome	Treatment Plan	Nursing Staff	Treatment Guidelines	Within 60 minutes
Monitor Patient Condition	Initial Treatment Outcome	Patient Condition Report	Initial Treatment	Monitoring Equipment	Monitoring Protocol	Continuous
Decide on Admission/ Discharge	Patient Condition Report	Admission/Discharge Decision	Patient Monitoring	Medical Team	Decision-Making Criteria	After Monitoring

TABLE 7.A.2
FRAM Interactions and Couplings

From Function	To Function	Input/Output/Control/ Resource Interdependencies
Receive Patient Arrival Notification	Initial Triage Assessment	Output (Patient Arrival Record)
Initial Triage Assessment	Patient Registration	Output (Initial Triage Category)
Initial Triage Assessment	Detailed Triage Evaluation	Output (Initial Triage Category)
Patient Registration	Detailed Triage Evaluation	Output (Patient Record)
Detailed Triage Evaluation	Allocate Treatment Area	Output (Detailed Triage Category)
Allocate Treatment Area	Assign Medical Team	Output (Treatment Area Assigned)
Assign Medical Team	Conduct Medical Assessment	Output (Medical Team Assigned)
Conduct Medical Assessment	Develop Treatment Plan	Output (Medical Assessment Report)
Develop Treatment Plan	Administer Initial Treatment	Output (Treatment Plan)
Administer Initial Treatment	Monitor Patient Condition	Output (Initial Treatment Outcome)
Monitor Patient Condition	Decide on Admission/ Discharge	Output (Patient Condition Report)

COMPLETENESS

The identified functions cover the critical steps in the patient admission process in the A&E department. The sequence from patient arrival to the decision on admission or discharge appears comprehensive.

RELEVANCE

Each function is relevant to the process of admitting and initially treating patients in the A&E department. These functions are necessary for ensuring patient safety and providing timely medical care.

ADEQUACY

The functions identified are adequate for the core process. However, additional details may be necessary for specific cases, such as handling critically ill patients, coordinating with other departments, or managing high patient volumes.

RESERVATIONS AND POTENTIAL OMISSIONS

1. Coordination with Other Departments: Explicit functions for coordination with radiology, laboratory, or other specialties may be needed, especially in complex cases.
2. Handling of Critical Cases: Adaptations for critically ill patients or mass casualty events should be considered to ensure the model's robustness.

3. Patient Handover: The transition from A&E to inpatient wards or other departments is not detailed and should be addressed.
4. Patient Information Management: Continuous updating and accessing of patient records should be explicitly included to ensure accurate and timely information flow.

CONCLUSION

The FRAM analysis of the patient admission process in the A&E department provides a comprehensive overview of the necessary functions and their interdependencies. While the current model is complete and relevant, continuous evaluation and updates are recommended to ensure adequacy in different scenarios and incorporate all necessary aspects of patient care.

This report serves as a foundation for understanding the complex interactions in the A&E department and can guide improvements in efficiency and patient care quality.

REFERENCES

Badini, S., Regondi, S., Frontoni, E., Pugliese, R., 2023. Assessing the capabilities of ChatGPT to improve additive manufacturing troubleshooting. *Advanced Industrial and Engineering Polymer Research, 6*, 278–287.

Brundage, M., Avin, S., Clark, J., Toner, H., Eckersley, P., Garfinkel, B., et al. 2018. The malicious use of artificial intelligence: Forecasting, prevention, and mitigation. arXiv preprint arXiv:180207228.

Bruno, A., Mazzeo, P.L., Chetouani, A., Tliba, M., Kerkouri, M.A. 2023. Insights into classifying and mitigating LLMs' hallucinations, arXiv preprint arXiv:231108117.

Deng, X., Bashlovkina, V., Han, F., Baumgartner, S., Bendersky, M.. What do LLMs know about financial markets? A case study on Reddit market sentiment analysis. *Companion proceedings of the ACM web conference 2023.* Austin, TX: Association for Computing Machinery; 2023. pp. 107–110.

Ebrahimian, M., Behnam, B., Ghayebi, N., Sobhrakhshankhah, E. 2023. ChatGPT in Iranian medical licensing examination: evaluating the diagnostic accuracy and decision-making capabilities of an AI based model. *BMJ Health & Care Informatics, 30*, e100815.

França, J., Valle, I. L. C., Hollnagel, E. 2022. Reanalysing deepwater horizon accident with FRAM: Enhancing learning and understanding complexities to improve safety. *Proceedings of the Rio oil & gas expo and conference*, Rio de Janeiro, RJ, Brazil, https://doi.org/10.48072/2525-7579.rog.2022.479

França, J. E. M., Hollnagel, E., Luquetti dos Santos, I. J. A., Haddad, A. N. 2019. FRAM AHP approach to analyse offshore oil well drilling and construction focused on human factors. *Cognition, Technology & Work, 22*, 653–665. https://doi.org/10.1007/s10111-019-00594-z

Kocoń, J., Cichecki, I., Kaszyca, O., Kochanek, M., Szydło, D., Baran, J., et al. 2023. ChatGPT: Jack of all trades, master of none. *Information Fusion, 99*, 101861.

Lee, P., Bubeck, S., Petro, J. 2023. Benefits, limits, and risks of GPT-4 as an AI chatbot for medicine. *New England Journal of Medicine. 388*, 1233–1239.

Leveson, N. G. 2011. *Engineering a Safer World – Systems Thinking Applied to Safety.* Massachusetts: MIT Press.

Oviedo-Trespalacios, O., Peden, A. E., Cole-Hunter, T., Costantini, A., Haghani, M., Rod, J. E., et al. 2023. The risks of using ChatGPT to obtain common safety-related information and advice. *Safety Science*, *167*, 106244.

Qi, Y., Zhao, X., Huang, X. 2023. Safety analysis in the era of large language models: A case study of STPA using ChatGPT. arXiv preprint arXiv:230401246.

Rasmussen, J. 1997. Risk management in a dynamic society: A modelling problem. *Safety Science*. *27*, 183–213.

Roberts, R. H., Ali, S. R., Hutchings, H. A., Dobbs, T. D., Whitaker I. S. 2023. Comparative study of ChatGPT and human evaluators on the assessment of medical literature according to recognised reporting standards, *BMJ Health & Care Informatics*, *30*, e100830.

Schön, D.A. 1983. *The reflective practitioner: how professionals think in action*. New York: Basic Books.

Stanton, N. 2006. Hierarchical task analysis: Developments, applications, and extensions. *Applied Ergonomics*, *37*.

Stechly, K., Marquez, M., Kambhampati, S. 2023. GPT-4 doesn't know it's wrong: an analysis of iterative prompting for reasoning problems. arXiv preprint arXiv:231012397.

Sujan, M., Smith-Frazer, C., Malamateniou, C., Connor, J., Gardner, A., Unsworth, H., et al. 2023. Validation framework for the use of AI in healthcare: overview of the new British standard BS30440. *BMJ Health & Care Informatics*, *30*, e100749.

Wu, T., He, S., Liu, J., Sun, S., Liu, K., Han, Q. L., et al. 2023. A brief overview of ChatGPT: The history, Status Quo and potential future development, *IEEE/CAA Journal of Automatica Sinica.*, *10*, 1122–1136.

Xue, L., Fan, J., Rausand, M., Zhang, L. 2013. A safety barrier-based accident model for offshore drilling blowouts. *Journal of Loss Prevention in the Process Industries*, *26*, 164–171. https://doi.org/10.1016/j.jlp.2012.10.008

8 A Semantic Lifting of Language Models to Support FRAM Models Building

Antonio De Nicola and Maria Luisa Villani

CONTEXT: SEMANTIC REASONING VIA AI

The advent of Industry 4.0 and 5.0 demands features such as digitization, automation and adaptation, and optimization (Lu, 2017). A prerequisite for them is the availability of model representations of operations and processes. This can be accomplished using various modelling notations and methods. This chapter, in line with the scope of the book, orients its interest towards the Functional Resonance Analysis Method (FRAM), which in recent literature appears as a method capable of merging both qualitative (Cantelmi et al., 2021) and more quantitative dimensions (Patriarca et al., 2017).

Nonetheless, a further advancement in leveraging knowledge encompassed in a FRAM process model is to provide an explicit formal specification that can be processed by a software application. This would facilitate reasoning mechanisms capable of inferring new knowledge from existing ones. To achieve this objective, ontologies and knowledge graphs can provide valuable support in representing the semantics behind FRAM knowledge and support semi-automatic reasoning. Indeed, ontology is a conceptual model aimed at providing a shared understanding of a specific domain. Linking process models with an ontology would clarify their meaning, enhancing both human comprehension and machine computation.

To facilitate the semantic enrichment of FRAM process models, two preconditions must be met. First, there is a requirement for a shared semantic model in the form of an ontology, encompassing both foundational knowledge inherent in FRAM process models and domain-specific and application-specific knowledge. Second, a method for establishing a connection between the aforementioned process models and the ontology needs to be developed and applied.

Therefore, two main research objectives are addressed in this chapter: i) to demonstrate how to construct an ontology for FRAM process models, and ii) to establish a semantic annotation, which entails linking a FRAM process model with a semantic model. Indirectly, the chapter pertains to demonstrating the utility gained from transitioning from operational knowledge contained within a FRAM model to semantic

DOI: 10.1201/9781003518167-10

knowledge encapsulated within a FRAM domain ontology. In simpler terms, the chapter leverages formalized semantics to fully empower FRAM analysts in extracting value from their FRAM models.

We define semantic lifting of a FRAM model as the transformation of its information content into an ontological representation (De Nicola et al., 2008). This semantic lifting facilitates the integration of a FRAM model with an ontology, establishing the connection mentioned earlier. It is worth mentioning that foundational knowledge inherent in FRAM models is not limited to providing model components with semantics. Indeed, each FRAM model can represent a specific view on the process itself, such as the Work-As-Imagined, Work-As-Done, Work-As-Disclosed, and Work-As-Observed, among others (Patriarca et al., 2021).

Although this chapter is not related to a particular domain, as its usefulness is general, a case study is presented concerning a possible use of a chatbot following a minor incident in a hospital.

Some additional background about ontological modelling and semantics is needed before delving into the proposed methodology.

ONTOLOGIES

An ontology defines the concepts, relationships, and instances within a specific domain or pertaining to a specific application. It serves as a structured conceptual framework for organizing information in a way that can be understood by both humans and machines. Ontologies are used to facilitate knowledge sharing, data interoperability, and reasoning; to standardize terminology and semantics; to enable interoperability, and to facilitate communication between different systems and domains.

Method Geek

In formal terms, an ontology is an explicit formal specification of a shared conceptualization (Borst, 1997; Gruber, 1993). They are often specified using formal languages such as RDF (Resource Description Framework) or OWL (Web Ontology Language). Ontologies can provide a foundation for intelligent systems, supporting tasks such as reasoning, inference, and decision-making based on structured knowledge representation. An ontology is at the core of the semantic lifting proposed in this chapter.

Notably, other FRAM ontologies have already been reported in the literature (De Nicola et al., 2023; Lališ et al., 2019).

SEMANTIC ANNOTATION

The semantic annotation of a business process involves enriching the process description with additional semantics to clarify its meaning and allow process interoperability (Liao & Zhao, 2019). It typically includes annotating elements such as activities, resources, inputs, outputs, and relationships within the process model by building links to a domain ontology.

Formally, a semantic annotation <c, e> pairs an element from an ontology (c) with an element from a business process (e). For example, considering the process activity *Give the alert* and the ontology concept *Alert*, the pair <Alert, Give the alert> represents a semantic annotation of the process activity.

This annotation adds semantics to the process elements, facilitating automated reasoning and decision-making. Semantic annotations use ontologies to ensure consistency across different processes. By providing a structured representation of the business process semantics, semantic annotation enables better understanding, integration, interoperability, and reuse of process knowledge across organizational boundaries. It supports various tasks such as process discovery, analysis, optimization, and automation, ultimately contributing to improved business process management and efficiency.

CHATBOTS BASED ON LARGE LANGUAGE MODELS

A large language model (LLM) is a type of artificial intelligence (AI) system trained on vast amounts of text data to understand and generate human-like language. These models, such as OpenAI's GPT (Generative Pre-trained Transformer) series, consist of deep neural networks with millions or even billions of parameters. They learn the statistical patterns and structures of language by processing massive datasets, enabling them to generate coherent and contextually relevant text. A chatbot based on an LLM uses this advanced AI technology to converse with users. By leveraging the LLM capabilities, the chatbot can understand and respond to user queries, provide information, or engage in dialogue on a wide range of topics. A chatbot mimics human-like conversation, offering users a natural and intuitive interface for communication. Chatbots continuously improve their language understanding and generation abilities through ongoing training on new data, allowing them to adapt and evolve over time. In a CSTS, a chatbot can serve multiple roles. For example, it can act both as an agent executing a process and as an analyst offering insights on that process. By using an ontology, the chatbot's roles and activities within the system can be clearly defined and represented in an unambiguous manner.

WAX FRAMEWORK AND WAX ONTOLOGY

The WAx framework (Patriarca et al., 2021) is a conceptual framework tailored to offer a systematic organization of the diverse work varieties and work properties in operations. Here, we focus solely on the essential concepts crucial for comprehending the proposed method for semantic annotation of business processes, conscious that an interested reader can find full details in Patriarca et al. (2021).

Complex Cyber-Socio-Technical Systems (CSTS) require extensive knowledge dispersed among operators, integrated into technology, and delineated within organizational structures. The WAx framework facilitates the tracking of processes involving the creation, loss, amplification, transfer, and analysis of distributed knowledge within CSTSs. For instance, the framework introduces knowledge entities as specific manifestations of work (Work-As-x), hence deriving its name. These entities include Work-As-Imagined (WAI), Work-As-Prescribed (WAP), Work-As-Normative (WAN), Work-As-Done (WAD), Work-As-Disclosed (WADI), and Work-As-Observed (WAO).

WAI embodies the mental constructs surrounding human work activities, representing the idealized potential and belief-based perceptions of present, past, and future work execution. WAP encompasses the formalized organizational perspectives on work, articulated through procedures, checklists, standards, task descriptions, and training materials. WAN encapsulates external norms influencing the organization, ranging from formalized laws, regulations, and international standards to safety protocols and technical guidelines. WAD denotes the actual activities performed within the dynamic, unpredictable environment of the CSTS. It is only partially accessible and often diverges from imagined or prescribed work due to the real-world complexities. WADI reflects the conscious or unconscious presentation of work by system agents, shaped by interaction and intention. It conveys messages to specific audiences, influenced by both deliberate and inadvertent factors, thus carrying rich informational content.

WAO pertains to the mental models formed through work observation. Even in naturalistic observations, distortions can arise from the mental frameworks of both the observer and the observed, impacting the interpretation of the work being observed.

Adding to the body of knowledge of existing WAx varieties and considering the growing significance of LLMs (e.g., ChatGPT and Gemini) in productive contexts as peculiar AI artefacts, we hereby define the Work-As-AI Imagined (WAAII) to signify the activities as they should be performed in accordance with an LLM.

The WAx framework adopts a holistic and fractal viewpoint, conceptualizing systems as hierarchies comprising organizations, teams, individuals, and artefacts, each exhibiting agency characteristics intermittently. Individuals may assume diverse agency roles depending on the context. Given its focus on effectively describing knowledge within real systems, the WAx framework emphasizes the significance of influences. These influences, stemming from prior knowledge, can impact new knowledge elicitation processes. The framework acknowledges information loss and the deliberate pursuit of trade-offs between efficiency and effectiveness inherent in all real systems. These influences are considered not only in process realization and communication activities but also in meta-analytic endeavours such as modelling.

DATA COLLECTION

From the analyst's viewpoint, data collection for FRAM process modelling involves acquiring information about processes through documents, interviews, and/or direct observations. The data collection method indeed hinges on the specific WAx entity to be modelled, which essentially refers to the description of the process from a particular perspective.

In detail, the case study concerns a chemical spillage in the chemical laboratory of a hospital.

Domain Expert

The situation is as follows. A newly hired chemist at the hospital accidentally knocks over a bottle containing a hazardous chemical, causing it to spill onto the floor. Although the chemist recently underwent training on handling emergency situations, they struggle to recall the correct procedure to follow. In an effort to seek guidance, they consult an LLM chatbot for assistance. However, the chatbot provides a slightly different procedure than what should have been followed. This is adopted by the chemist, leading to confusion among the other team members involved in managing the incident. Indeed, the chatbot can be considered as a new actor performing activities in a CSTS, whether it has been customized for a specific industry or used in its standard form.

The final goal is to measure the allostatic load, a newly defined resilience indicator for a CSTS that allows for measuring accumulated tensions due to the different perspectives on the same process (De Nicola et al., 2023).

The case study involves the modelling of a work variety, which is here assumed to correspond to the WAP, a procedure to be followed by chemists of the laboratory to safely manage the emergency following the spillage of a chemical substance, and of a WAAII, that is, a procedure suggested by a chatbot fed by a LLM for the same emergency situation.

In light of the illustrative nature of the case study, an existing procedure has been used and validated by an experienced chemical scientist working at a laboratory of ENEA (Italian National Agency for New Technologies, Energy and Sustainable Economic Development). The following text describes the WAP FRAM modelling.

Domain Expert

WAP PROCESS DESCRIPTION

1. Alert and Evacuate:
   - *Inform everyone nearby about the spill.*
   - *If the spill is large, flammable or poses a serious health risk, evacuate the area immediately.*

2. Secure the area:
   - *If safe to do so, try to contain the spill by closing the container or using spill pads to prevent it from spreading.*
   - *If flammable liquids are involved, turn off heat sources and open windows for ventilation.*
3. Personal protection:
   - *Do not attempt to clean the spill without proper Personal Protecting Equipment (PPE). This typically includes gloves, safety goggles, and a lab coat, but the specific requirements will depend on the spilled chemical. Refer to the Safety Data Sheets (SDS) for the chemical for proper PPE recommendations.*
4. Assess and clean:
   - *Identify the spilled material using the container label and consult the Safety Data Sheet (SDS) to determine the hazards and appropriate clean-up procedures.*
   - *If the spill is small and you are trained in handling the specific chemical, proceed with cleaning up the spill using the spill kit available in the lab. Never mix chemicals during clean-up unless instructed in the SDS.*
5. Report and dispose:
   - *Report the incident to your supervisor or lab safety officer.*
   - *Collected spill waste should be disposed of according to hazardous waste disposal regulations.*
6. Additional recommendations:
   - *If the spill is serious or you are unsure how to handle it safely, call emergency responders immediately.*
   - *Consult the following resources for further information: "Chemical Spill Management and Response" and "How to Handle Chemical Spills in Laboratories".*

The first step for the analyst is to identify the process functions from the text and determine the appropriate level of abstraction for the modelling objective. Since the goal is to represent the various actions that the chemist must undertake and their dependencies, details such as the specific resources required for certain functions have not been included. Furthermore, it should be noted that the enumeration of the prescribed functions has not been interpreted as a sequencing of them, but the flow has been deduced from the overall description of the procedure.

Domain Expert

The FRAM functions identified for the model, the type (i.e., to whom they are attributed), and their description are presented in Table 8.1. Indeed, to be aligned as much as possible with the WAP information source, these descriptions required for the FRAM model and the semantic annotation method have been produced by querying Gemini.

TABLE 8.1

Functions, Function Types, and Descriptions of the Spill Generation Process as Specified in the Protocol

Function	Function Type	Description
Spillage generation	Technological	Spillage generation due to a technological device is the unintended leaking or overflowing of liquids or materials caused by a malfunction or design flaw.
Situation assessment	Human	A situation assessment is gathering information to understand the current state of a situation and identify factors that influence it.
Emergency management	Human	Emergency management is the practice of preparing for, responding to, and recovering from hazards that threaten communities.
Laboratory activity	Human	A laboratory activity is a controlled experiment or procedure conducted in a lab setting to test hypotheses, collect data, or develop new knowledge.
Evacuation	Human	Evacuation is the organized removal of people from a dangerous location to a safer one.
Wear PPE (Personal Protection Equipment)	Human	To wear PPE means to put on personal protective equipment, which shields your body from hazards in specific situations.
Equipment provision	Organizational	Equipment provision is the process of supplying the necessary tools and machinery for a task or operation to be completed.
Spill containment	Human	Spill containment refers to methods used to prevent the spread of a leaked liquid or material.
Ventilation	Technological	Ventilation is the process of bringing in fresh air and removing stale air from a space.
Training	Organizational	Training is equipping yourself or others with the skills and knowledge needed for a specific task or activity.
Protocol provision	Organizational	A protocol provision outlines a specific action or guideline to be followed within a larger set of procedures.
Clean up	Human	Cleaning up involves removing dirt, clutter, or a spilled substance to make something neat and orderly.
Incident report	Human	An incident report documents the details of an unexpected event, like a spill or injury, to aid investigation and future prevention.
Ask for help	Human	Asking for help involves requesting assistance from someone to address a difficulty or complete a task.
Turn off heat sources	Human	To turn off heat sources means to deactivate or extinguish anything producing heat, like a stove, oven, fireplace, or space heater.
Time check	Human	A time check is the act of finding out the current time.
Throwing away waste contaminants	Human	Throwing away waste contaminants involves identifying them and following specific disposal procedures to minimize risks to health and the environment.
Consult additional resources	Human	To consult additional resources means seeking out more information from various sources like books, websites, or experts.

The set of modelled couplings includes:

- Output-Input, that is, the Output of a function is needed to activate the connected function, for example, the situation assessment needs the spillage;
- Output-Resource, that is, the Output of a function is a resource for the connected function, for example, the clean-up requires availability of the spill kit in the lab;
- Output-Precondition, that is, the Output of a function is a condition that must hold true to execute the target function, for example, training in handling the specific chemical is a Precondition for the clean-up;
- and Output-Time, that is, a function provides a time constraint to the target function, for example, the time check function provides the time of activation for the ask for help function.

The whole description of the FRAM model realized for the WAP is omitted here for brevity, but it is available in an already published manuscript (De Nicola & Villani, 2024). As described earlier, we imagine that, during the emergency, the unexpert chemist consults ChatGPT for instructions on the management procedure through the prompt: "What should I do in case of chemical spillage in the laboratory?"

The text provided by ChatGPT is reported below and provides the basis for the FRAM WAAII model.

Domain Expert

WAAII PROCESS DESCRIPTION

1. *Assess the Situation: Quickly evaluate the extent and nature of the spill. Determine the type of chemical involved and whether it poses immediate hazards such as toxicity, flammability, or corrosiveness.*
2. *Alert Others: Notify nearby colleagues and evacuate the immediate area if necessary. Warn others to stay away from the spillage site to prevent exposure.*
3. *Personal Protective Equipment (PPE): Put on appropriate PPE, including gloves, goggles, a lab coat, and possibly a face mask or respirator if the chemicals are volatile or toxic.*
4. *Contain the Spill: If it's safe to do so, contain the spill by using absorbent materials, such as spill kits, sand, or commercial absorbent. Create a barrier around the spill to prevent it from spreading further.*
5. *Ventilation: Ensure adequate ventilation in the area to disperse any fumes or vapours. Open windows and doors and use fume hoods if available to remove airborne contaminants.*
6. *Neutralization: If you are trained and it's safe to do so, neutralize the spilled chemical using appropriate neutralizing agents. Follow established protocols and use caution to avoid creating further hazards.*

7. *Cleanup: Carefully clean up the spill using appropriate procedures and equipment. Dispose of the contaminated materials according to hazardous waste disposal guidelines.*

8. *Report the Incident: Inform your supervisor or safety officer about the spillage immediately. And report it to laboratory management for further investigation and follow-up.*

9. *Decontamination: After cleanup, decontaminate the affected area and any equipment used in the cleanup process. Follow established decontamination procedures to ensure safety.*

10. *Review and Prevent: Conduct a post-incident review to identify the cause of the spill and any opportunities for improvement in handling procedures. Take corrective actions to prevent similar incidents in the future.*

The same FRAM analyst, by proceeding similarly to the WAP, creates a FRAM model for the WAAII by identifying functions and dependencies from the text above. The function descriptions are generated by querying ChatGPT to mimic the interaction of the chemist with the chatbot to acquire specialized information. Where semantically appropriate, the FRAM analyst tries to re-use the names of the FRAM functions and of the couplings of the WAP model. The FRAM functions identified for the WAAII model, the type, and their description according to ChatGPT are presented in Table 8.2, while the complete FRAM model is available in a dedicated manuscript (De Nicola & Villani, 2024).

TABLE 8.2

Functions, Function Types, and Their Descriptions Mentioned in the Spill Generation Process Provided by ChatGPT without Prior Validation by the Human Expert

Function	Function Type	Description
Spillage generation	Technological	Accidental release or escape of substances, such as liquids or powders, from their intended containment, potentially leading to environmental contamination or safety hazards.
Situation assessment	Human	Situation awareness refers to the perception, comprehension, and projection of the status of elements in an environment and their future status. It is a crucial cognitive skill for emergency management.
Emergency management	Human	Emergency management is the systematic approach to preparing for, responding to, recovering from, and mitigating the effects of emergencies or disasters, aimed at protecting lives, property, and the environment.
Laboratory activity	Human	Laboratory activity encompasses experimental procedures and research conducted within a controlled environment to explore scientific phenomena, analyze data, and advance knowledge in various fields.

(Continued)

TABLE 8.2 (Continued)

Functions, Function Types, and Their Descriptions Mentioned in the Spill Generation Process Provided by ChatGPT without Prior Validation by the Human Expert

Function	Function Type	Description
Evacuation	Human	Evacuation is the process of moving people from a place of danger to a safer location.
Wear PPE (Personal Protection Equipment)	Human	Wear the PPE correctly, following the manufacturer's instructions. Ensure that all items are properly fitted and secured. For example, adjust straps and fastenings so that the PPE provides adequate protection without causing discomfort.
Equipment provision	Organizational	Equipment provision involves the procurement, distribution, and maintenance of tools, machinery, or instruments required for specific tasks or activities, ensuring their availability and functionality as needed.
Spill containment	Human	Spill containment refers to the methods and procedures used to prevent and manage the spread of hazardous substances when a spill occurs, ensuring that the spilled material is confined to a controlled area to minimize environmental impact and health risks.
Ventilation	Technological	Ventilation is the process of supplying fresh air and removing stale air from an indoor space to maintain air quality, control temperature, and ensure a healthy and comfortable environment.
To neutralize	Human	To neutralize (Neutralization) is a chemical reaction in which an acid and a base react to form water and a salt, effectively cancelling out each other's properties and resulting in a solution that is closer to a neutral pH (around 7).
Training	Organizational	Training is the systematic process of imparting knowledge, skills, and competencies to individuals or groups through instruction, practice, and feedback to enhance their performance in a particular area.
Protocol provision	Organizational	Protocol provision involves establishing and distributing guidelines or procedures detailing specific steps and standards to be followed in conducting tasks or activities, ensuring consistency and compliance with established protocols.
Clean up	Human	Clean-up the spill refers to the process of removing a substance that has been accidentally spilled onto a surface to restore the affected area to its original, clean condition, while also minimizing any potential environmental or health hazards.
Throwing away waste contaminants	Human	Throwing away waste contaminants involves disposing of hazardous or unwanted substances in a manner that risks environmental contamination or poses health hazards to individuals and ecosystems.
Incident report	Human	Incident reporting is the formal process of documenting any unexpected or unplanned event that occurs within an organization.

(Continued)

TABLE 8.2 (Continued)

Functions, Function Types, and Their Descriptions Mentioned in the Spill Generation Process Provided by ChatGPT without Prior Validation by the Human Expert

Function	Function Type	Description
Ask for help	Human	Asking for help involves requesting assistance, support, or guidance from someone else when you are unable to complete a task, solve a problem, or understand something on your own.
Decontamination	Human	Decontamination is the process of removing or neutralizing contaminants from a surface, object, person, or environment to make it safe for handling, use, or habitation.
Review and prevent	Human	Review concerns examining a process to identify any issues, inefficiencies, or areas for improvement. Prevention involves implementing measures to address these issues and prevent them from recurring in the future. Conduct a post-incident review to identify the cause of the spill and any opportunities for improvement in handling procedures. Take corrective actions to prevent similar incidents in the future.
Time check	Human	"Time check" typically refers to a quick verification of the current time, often requested to ensure punctuality or to synchronize activities with a specific schedule.

Broadly speaking, the main difference between the FRAM WAP and the FRAM WAAII models lies in the couplings (e.g., sequencing) between FRAM functions, thus impacting their definition. For example, the function <Clean up> requires consultation of the SDS procedures in the WAP (provided by the Training function), whereas procedures are not detailed in the WAAII, and therefore, the <Clean up> function is directly activated as one of the steps of the <Emergency Management> function. Instead, the function <Ask for help> is suggested to happen immediately in the WAAII, whereas the WAP suggests first assessing the situation and trying to take actions, and then eventually deciding to ask for assistance.

METHODOLOGY

This section describes the application phases, as proposed in this paper for the semantic lifting, starting from the automatic building of a FRAM domain ontology, its semantic annotation, and subsequent reasoning. To improve readability and facilitate connection between the theoretical and operational aspects of the proposed approach, each phase is described jointly with its respective results.

PHASE 1: AUTOMATIC BUILDING OF A FRAM DOMAIN ONTOLOGY

Ontology engineering is an activity aimed at building domain ontologies. There are several methodologies that support ontology engineers and domain experts in constructing an ontological model. Among the most used ones, we cite Methontology (Fernández-López et al., 1997), UPON (De Nicola et al., 2009), NeOn (Suárez-Figueroa et al., 2012), and UPON-lite (De Nicola & Missikoff, 2016).

Here, we use a gradual approach, beginning with the identification of relevant terminology and progressively increasing in complexity until we develop a full-fledged ontology. Initially, the mentioned terminology relevant to the domain of interest is compiled into a lexicon. Next, these terms and their associated textual descriptions are collected in a glossary to form a preliminary set of concepts. Following this, the glossary elements are organized into a specialization hierarchy and linked to upper-level concepts from ontological models like the Suggested Upper Merged Ontology (SUMO) (Pease et al., 2002) or the Basic Formal Ontology (BFO) (Arp et al., 2015). Ontological relationships, such as domain-specific connections or decompositions, are then used to interlink these concepts. Finally, the ontology is implemented using a formal language such as OWL or RDF. As mentioned, ontology engineering requires the collaborative work of domain experts and knowledge engineers.

Method Geek

To facilitate this activity, we have developed software that imports an FRAM model in .xfmv format and automatically extends a core ontology, including two upper ontologies that extend the SUMO ontology. These are the FRAM Upper Model (FUM) ontology and the WAx framework ontology, which provide ontological models, respectively, of the FRAM constructs and their mutual relationships and of the WAx agents, the knowledge entities, and their dynamics.

SUMO (the Suggested Upper Merged Ontology) is a foundational ontology that provides strong ontological foundations and enhances the precision of concepts and their definitions. Interoperability is crucial for achieving operational excellence in productive processes and increasing the competitive capacity of industries. SUMO facilitates interoperability between domain ontologies when they are mapped to it.

The FUM ontology is an upper ontology derived from the FRAM method. According to the FUM ontology, a FRAM model comprises couplings and functions. Functions are FRAM elements, as are aspects. Each function is characterized by a function type, indicating whether it is organizational, technological, or human. A function includes aspects, which can be inputs, outputs, controls, preconditions, resources, and times. Couplings specify the process flow by linking functions through the object properties *hasDownstreamFunction* and *hasUpstreamFunction*, and linking function aspects through the object properties *hasDownstreamAspect* and *hasUpstreamAspect*.

The WAx Framework Ontology is an upper ontology derived from the WAx framework, which formally defines the relationships between various work representations (e.g., Work-As-Imagined, Work-As-Done). These representations are considered knowledge entities generated by different agents, such as sharp-end operators, blunt-end operators, and analysts.

The import software is a Python application that parses .xmfv files and updates the ontology with the knowledge extracted from these files. Specifically, it extracts functions and aspects and extends the ontology by specializing the function, aspect, and coupling concepts. Similarly, the specific process, which relates to the CSTS process, specializes the process concept within the WAx framework ontology. The type of knowledge entity, such as Work-As-Imagined or Work-As-Done, refines the knowledge entity concept, while the agent's role, such as analyst, blunt-end operator, or sharp-end operator, refines the agent role concept.

Domain Expert

In the case study, the resulting ontology consists of 268 classes, 52 object properties and 1887 axioms.

Figure 8.1 presents a screenshot of the built ontology in the Protégé Ontology Management System.

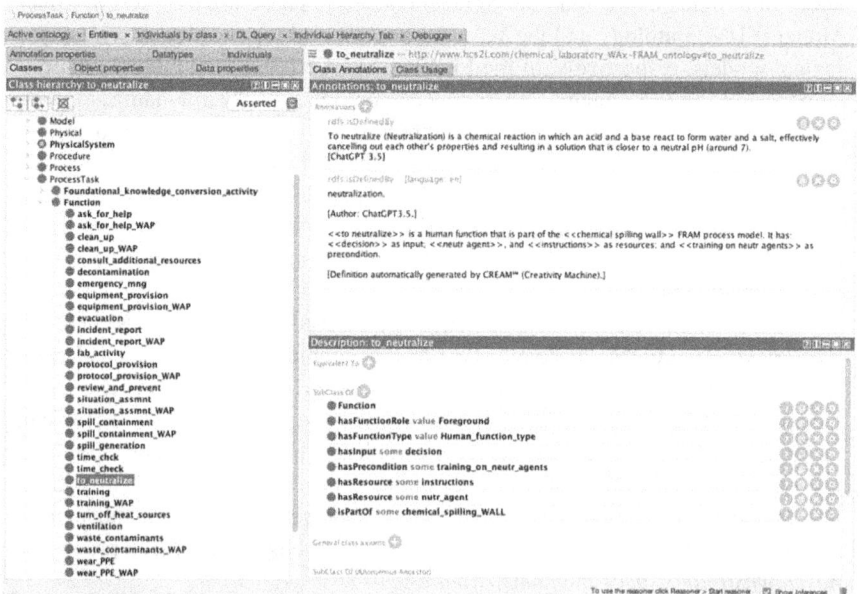

FIGURE 8.1 A screenshot of the built ontology in the Protégé ontology management system for the use case.

PHASE 2: SEMANTIC ANNOTATION OF THE FRAM MODEL

Semantic annotation Σ generally refers to a method to establish a link between any model and concepts of an ontology to provide its interpretation. The objective is to describe the significance of the model through an appropriate conceptual framework of the domain of interest so that it can be processed by a machine.

Method Geek

In detail, we envisage two types of semantic annotations for CSTSs. The former is devoted to semantically enriching components of the model. The latter aims at giving meaning to the model itself. Σ is defined as a set of predicates σ_i either of the form $\sigma_i\left(f_j, c_k^{FUM}\right)$ or $\sigma_i\left(f_j, c_k^{WAX}\right)$, where f_j is a concept and c_k is either a component of a FRAM model or a WAx entity.

For the sake of simplicity, Figure 8.2 is used to illustrate the semantic annotation method, distinguished in *Process-level semantic annotation* and *WAx-level semantic annotation*.

Process-Level Semantic Annotation

Given a process and a FRAM model representing it, the process level semantic annotation is finalized to establish a mapping from every component of the FRAM model (i.e., function, aspect, and coupling instance) to a concept of an ontology that describes its meaning.

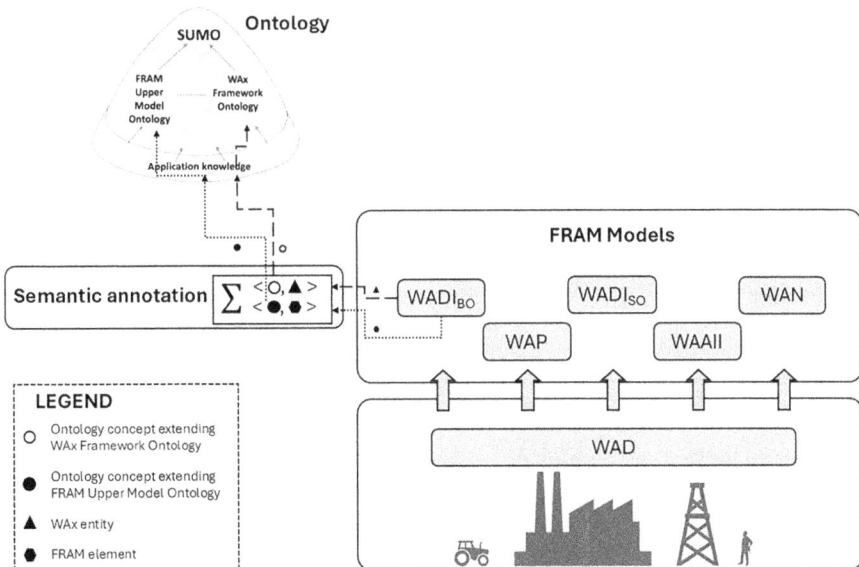

FIGURE 8.2   Semantic annotation of FRAM process models.

The semantic annotation requires a domain-specific ontology to be defined, that is, an ontology tailored to the real-world system where the modelling component is realized, and the FUM ontology, to represent the component's role in the FRAM model representation. To address these needs, an application ontology is created that extends the FUM ontology with system-specific concepts *as intended* in the FRAM model at hand. Such interpretation is provided in the description tags of the FRAM component instances and from the role of the components in the FRAM model.

In our methodology, for each FRAM model, say a FRAM WAAII, the creation/update of the application ontology is performed automatically as follows: (i) the parser of the FRAM model artefact (i.e., xmfv files) retrieves all the model components and their attributes from it; (ii) for each component, a new concept is created with the name of the component, and it is linked to the FUM ontology as a subclass of function and the label WAAII.

Figure 8.1 shows some of the concepts representing FRAM functions automatically created based on the FRAM WAAII model. As an example, the selected concept <To neutralize> from the WAAII from the function taxonomy on the class hierarchy of the panel is the correspondent of the function <To neutralize> in Table 8.2. The description of the concept reported in the Annotation tag of the ontology file, displayed in the upper right side of the panel, is also automatically generated based on the description of the function in the FRAM model (see Table 8.2) and its aspects. These are also created as new concepts and linked to the FUM ontology as subclasses of Aspect and with the label WAAII. The complete semantic representation of the FRAM function <To neutralize> in the ontology, shown in the bottom right side of the panel, is achieved through semantic associations to the aspect concepts, leveraging the object properties of the FUM ontology. The result of this automatic process is a full conceptualization of the FRAM WAAII model as part of the application ontology.

The method described above can be applied to any FRAM model. Accordingly, the automatically created application ontology provides formal semantic representations of all FRAM models of the same process, and it also replicates the FRAM structure to support many of the analyses that can be performed on the original FRAM files in addition to semantics-based analyses.

WAx-Level Semantic Annotation

A unique feature of the ontology is the capability to semantically describe FRAM model components, that is, their meaning and roles within a process under a specific agent's perspective, for example, ChatGPT in the case study, and the perspective itself or WAx entity, for example, the WAAII of ChatGPT. This is achieved by semantically annotating the various process descriptions (or knowledge entities) with the WAx ontology, as illustrated in Figure 8.2. The WAx ontology allows for enriching the process descriptions with (meta)concepts of the WAx framework, such as the links with other knowledge entities and the agent's roles providing them, following the holistic and fractal structure of the framework.

The conceptualization of instances of WAx entities, that is, the WAP and the WAAII chemical spill procedures, may support automatic identification of knowledge conversions and their formalization to handle the complexity of the analysis.

PHASE 3: SEMANTIC REASONING

The goal of the experimentation is to demonstrate how the semantic lifting of the Work-As-Artificial-Intelligence Imagined (WAAII) can enable safety-oriented reasoning. Specifically, we compute the allostatic load to measure the level of organizational tension caused by misalignment between the WAAII and the Work-As-Prescribed (WAP) model.

As mentioned earlier, and as expected, the WAP for managing a chemical spillage in the hospital laboratory is different from the WAAII. Indeed, some functions of the WAAII are not included in the WAP and vice versa (e.g., <To consult additional resources>, which is included only in the WAP). Even for shared functions, there can be specific differences. For example, concerning <To throw away waste contaminants>: the WAAII instructs on using designated waste equipment and following specific instructions, while the WAP directs users to follow general waste disposal procedures.

In total, 18 functions are included in the WAP and 19 in the WAAII. For a CSTS, a business process (i), and a pair of WAx entities (j,k) representing it, the allostatic load $\mathcal{A}_{i,(j,k)}$ can be measured as 1 minus the semantic similarity between the WAx entities (De Nicola et al., 2023). Values close to 1 indicate a high allostatic load and, therefore, a high level of perceived organizational tension in the CSTS being analyzed, while values close to 0 indicate a low level of perceived organizational tension.

Method Geek

From an analytical point of view, De Nicola et al. (2023) defined the allostatic load both in a global view dimension and in a local dimension.

$$\mathcal{A}_{gv} = \frac{1}{n} \cdot \sum_{i=0}^{n} \sum_{j=0}^{m} \frac{\left(1 - waxss_{i,j}\right)}{m}$$

Here, n represents the total number of CSTS processes, m represents the total number of WAx entity pairs available for the process i, and $waxss_{i,j}$ denotes the semantic distance between the pair of WAx entities labelled j for process i. A_{gv} is defined within the *[0,1]* range.

The allostatic load in the local view can be assessed by multiple indicators, including the aforementioned $A_{i,(j,k)}$, as well as others derived from variations of the algorithm that compute semantic similarity between WAx entities. These variations may focus on specific functions or particular aspects.

Accordingly, it is possible to measure semantic similarity between two WAx entities considering only specific function types (i.e., human, technological, or organizational) or only specific function aspects (i.e., input, output, precondition, resource, constraint, and time).

The corresponding indicators for allostatic load are *allostatic load by function type* denoted as $A_{n,(j,k)}^{FT}=1-waxss_{i,(j,k)}^{FT}$ and *allostatic load by aspect*, denoted as $A_{i,(j,k)}^{ASP}=1-waxss_{i,(j,k)}^{ASP}$. The algorithms to compute the WAx semantic similarity by function type and by aspect are reported in De Nicola et al. (2023). These algorithms can serve as a basis for other cluster assessments, that is, load by a specific agency (i.e., only about functions performed by agent i-th) or any other metadata available.

This paper exemplifies the local allostatic load assessment. Based on the use case details, the allostatic load $A_{1,(WAAII, WAP)}$ is 0.38. This value can be used to prioritize additional processes if needed. The high semantic similarity value between WAAII and WAP indicates only minor differences between them.

We then computed the value of allostatic load by analyzing one function type and one aspect at a time. First, we considered human functions, technological functions, and organizational functions separately.

Domain Expert

The results (see Figure 8.3) show that allostatic load due to Technological functions (see $A_{1,(WAAII, WAP)}^{Tech}$) in the two WAx entities is zero, while the highest load is due to Organizational functions ($A_{1,(WAAII, WAP)}^{Org}$). Human functions, also, contribute to allostatic load (see $A_{1,(WAAII, WAP)}^{Hum}$). Therefore, an analyst could decide to focus on the Organizational functions since they have the highest corresponding allostatic load values. As shown in Figure 8.3, most misalignments in the process are due to different perceptions of the Output (see $A_{1,(WAAII, WAP)}^{Out}$) and required Resources ($A_{1,(WAAII, WAP)}^{Res}$) and Time ($A_{1,(WAAII, WAP)}^{Tim}$), while there is more agreement on the inputs (see $A_{1,(WAAII, WAP)}^{Inp}$), and preconditions (see $A_{1,(WAAII, WAP)}^{Pre}$).

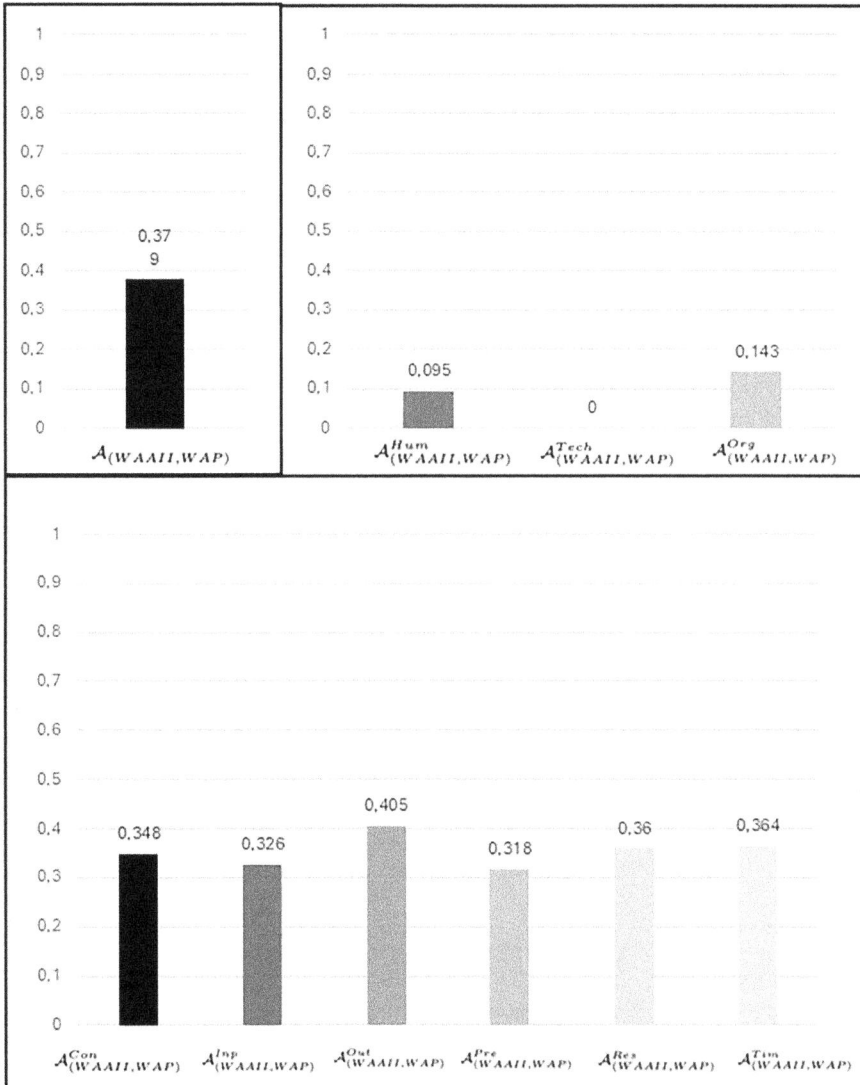

FIGURE 8.3 Allostatic load values in the local view between WAAII and WAP entities.

These values can be used to recognize the need for potential interventions to reduce organizational dissonance (Hinojosa et al., 2017), where relevant. Accordingly, recommended actions should be oriented towards clarifying the rationale for such deviations (e.g., not up-to-date procedures, training not aligned with practices, organizational drifts, etc.) and acting accordingly (Falegnami et al., 2020).

CONCLUSION

In this chapter, we demonstrated how semantics can lift an FRAM model effectively. We provided a concise overview of ontologies, including their construction, and discussed chatbots based on LLMs. Additionally, we outlined the process of semantically annotating FRAM models and explored the benefits of these annotations for enhancing safety analysis.

We presented a case study focused on managing chemical spills in a hospital laboratory to illustrate our approach. In this context, we showed that LLMs could function as actors within CSTSs, highlighting their potential to introduce organizational tension due to misalignment in process perceptions. This tension can be quantified by a dedicated indicator: the allostatic load.

Our findings suggest that the semantically enriched FRAM process models can significantly enhance safety analysis through advanced reasoning techniques. While our study presents a viable application, further research could explore deeper analyses, such as utilizing newly defined SPARQL queries to exploit the ontological relationships within the FRAM and WAx-based ontology. This approach promises to unlock additional insights and capabilities for safety analysis and organizational resilience, especially in a "data rich and information poor" environment.

REFERENCES

Arp, R., Smith, B., & Spear, A. D. *Building Ontologies with Basic Formal Ontology*. MIT Press, 2015.

Borst, W. *Construction of Engineering Ontologies. PhD thesis, Institute for Telematica and Information Technology*, University of Twente, Enschede, 1997.

Cantelmi, R., Di Gravio, G., Patriarca, R. Reviewing qualitative research approaches in the context of critical infrastructure resilience. *Environment Systems and Decisions*, 41(3), 341–376, 2021. https://doi.org/10.1007/s10669-020-09795-8

De Nicola, A., Di Mascio, T., Lezoche, M., Taglino, F. Semantic lifting of business process models. *Proceedings - IEEE International Enterprise Distributed Object Computing Workshop*, EDOC, art. no. 4815008, 120–126, 2008. https://doi.org/10.1109/EDOCW.2008.55

De Nicola, A., Missikoff, M., A lightweight methodology for rapid ontology engineering. *Communications of the ACM*, 59(3), 79–86, 2016. https://doi.org/10.1145/2818359

De Nicola, A., Missikoff, M., Navigli, R. A software engineering approach to ontology building. *Information Systems*, 34(2), 258–275, 2009. https://doi.org/10.1016/j.is.2008.07.002

De Nicola, A., Villani, M. L. Case study of work-as-large language FRAM models for safety-oriented analysis. *Mendeley Data*, 1, 2024. https://doi.org/10.17632/z2vxnndp97.1

De Nicola, A., Villani, M.L., Sujan, M., Watt, J., Costantino, F., Falegnami, A., Patriarca, R. Development and measurement of a resilience indicator for cyber-socio-technical systems: The allostatic load. *Journal of Industrial Information Integration*, 35, 100489, 2023. https://doi.org/10.1016/j.jii.2023.100489

Falegnami, A., Costantino, F., Di Gravio, G., Patriarca, R. Unveil key functions in socio-technical systems: Mapping FRAM into a multilayer network. *Cogn Tech Work*, 22, 877–899, 2020. https://doi.org/10.1007/s10111-019-00612-0

Fernández-López, M., Gómez-Pérez, A., Juristo, N. Methontology: From ontological art towards ontological engineering. In *Proceedings of the AAAI Spring Symposium Series* (Stanford, CA, Mar. 24–26). AAAI Press, Menlo Park, CA, 33–40, 1997.

Gruber, T. R. A translation approach to portable ontologies. *Knowledge Acquisition*, 5(2):199–220, 1993.

Hinojosa, A. S., Gardner, W. L., Walker, H. J., Cogliser, C., & Gullifor, D. A review of cognitive dissonance theory in management research: Opportunities for further development. *Journal of Management*, 43(1), 170–199, 2017. https://doi.org/10.1177/0149206316668236

Lališ, A., Patriarca, R., Ahmad, J., Di Gravio, G., Kostov, B. Functional modeling in safety by means of foundational ontologies. *Transportation Research Procedia*, 43, 290–299, 2019. https://doi.org/10.1016/j.trpro.2019.12.044

Liao, X., Zhao, Z. Unsupervised approaches for textual semantic annotation, a survey. *ACM Computing Surveys*, 52(4), 66, 2019. https://doi.org/10.1145/3324473

Lu, Y. Industry 4.0: A survey on technologies, applications and open research issues. *Journal of Industrial Information Integration*, 6, 1–10, 2017. https://doi.org/10.1016/j.jii.2017.04.005

Patriarca, R., Di Gravio, G., Costantino, F. A Monte Carlo evolution of the Functional Resonance Analysis Method (FRAM) to assess performance variability in complex systems. *Safety Science*, 91, 49–60, 2017. https://doi.org/10.1016/j.ssci.2016.07.016

Patriarca, R., Falegnami, A., Costantino, F., Di Gravio, G., De Nicola, A., Villani, M. L. WAx: An integrated conceptual framework for the analysis of cyber-socio-technical systems. *Safety Science*, 136, 105142, 2021. https://doi.org/10.1016/j.ssci.2020.105142

Pease, A., Niles, I., Li, J. The suggested upper merged ontology: A large ontology for the semantic web and its applications. *Working Notes of the AAAI-2002 Workshop on Ontologies and the Semantic Web*, 28, 7–10, 2002.

Suárez-Figueroa, M. C., Gómez-Pérez, A., Fernández-López, M. The NeOn methodology for ontology engineering. In *Ontology Engineering in a Networked World* (pp. 9–34). Springer, 2012.

Section II

Step 2
To Identify the Variability

9 Navigation Support from Shore
Analyzing Remote Pilotage Using the FRAM

Gesa Praetorius and Amit Sharma

CONTEXT: MARITIME PILOTAGE

Amongst the various modes of transportation, shipping forms a key component of the global movement of passengers, goods and services every day.

Domain Expert

As much as 80% of global trade by volume is carried through shipping (UNCTAD, 2021). It is estimated that this corresponds to approximately 11 billion tonnes of cargo transported every year, or 1.5 tonnes of cargo for each person annually (ICS, 2024). These goods are transported by a wide variety of ships. More than 100,000 ships of at least 100 gross tonnage operate worldwide, with their type depending on the primary cargo they carry (UNCTAD, 2023).

Today's shipping industry has evolved significantly in recent decades, with changes in size as well as means of propulsion. Such changes have also been observed in the development of ports, which act as the interface between maritime and land-based modes of transportation. It has resulted in an increasingly complex and interconnected maritime traffic system that comprises numerous ships often carrying sensitive cargo with a relatively constrained manoeuvring area in a time-bound manner. The efficiency of the system is ensured by the coordination of the ship's crew and shore personnel, communicating relevant navigation-related information and controlling the ship´s trajectory to navigate seamlessly in and out of busy waterways.

Despite the presence of advanced information and communication technology and regulations or protocols governing the actions of ships when traversing such areas, accidents often happen with dire consequences. The sinking of the passenger vessel *Costa Concordia* in January 2012 due to multiple underlying factors, including a navigational error, led to 32 personnel losing their lives and multiple casualties

DOI: 10.1201/9781003518167-12

in addition to the total loss of the vessel itself (Schröder-Hinrichs et al., 2012). The very next year, a cargo vessel, *Jolly Nero*, crashed into the port tower near Genoa, Italy, while departing its berth, leading to critical damage to port infrastructure and loss of nine lives (Vairo et al., 2017). More recently, in March 2024, the collision of the container vessel *Dali* with the Francis Scott Key Bridge, following a presumed propulsion failure during pilotage, at Baltimore, United States, led to the complete collapse of the bridge and six deaths (AP, 2024). While the causes of marine accidents could be varied, with multiple interdependent factors, the consequences of them are unfortunately lethal in terms of human lives involved and disruptive for the economy. Navigation of vessels in the coastal vicinity due to the risks posed by the very nature of such operations requires adequate considerations in terms of safety and reliability.

The navigation of vessels can be divided into four distinct phases, that is, (i) inland waterway phase, (ii) harbour/harbour approach phase, (iii) coastal phase, and (iv) ocean phase (Bowditch, 2012, p.2). In the first two phases (particularly in the first phase), services of an experienced navigator with localized knowledge, often referred to as the pilot, are required for safe navigation. When the pilot boards the ship and joins the bridge team (BT) for coordinating these phases of navigation, the vessel is said to be under pilotage.

Domain Expert

The pilots are experienced navigators who officially represent the coastal state and ensure that the bridge team led by the captain of the vessel can navigate to and from the harbour with the awareness of local conditions as well as the navigational constraints (Sharma et al., 2019). They are also monitored by the local Vessel Traffic Service (VTS), which, together with the vessel´s bridge team and the local pilotage service, forms the maritime traffic system (MTS) within the confines of a harbour area. The role of a pilot is unique in terms of exerting significant control over the movement of the vessel in such areas. Van Westrenen and Praetorius (2014, p. 66) have described a model of pilot control consisting of three levels. At the very first level, the pilot with a goal of maintaining traffic fluency and safety, formulates a strategic plan for the entire phase of the voyage. Afterwards, a track control plan with clear boundaries for manoeuvring and compliance with appropriate regulations is made. Finally, this tactical plan is translated into a set of course headings and engine thrust orders to follow the intended trajectory. The pilot receives information about the manoeuvring characteristics of the vessel and its other generic characteristics from the bridge team. Simultaneously, the tactical/strategic plans and control orders are shared by them as navigational advice upon which the vessel´s captain or master acts. Even though the control of the vessel always remains with the captain during the pilotage, in practice the pilot acts as the primary navigation officer who takes the lead in this phase of the vessel's navigation (Van Westrenen, 1999).

The act of pilotage involves boarding the vessel by the pilot in designated position coordinates while at sea or at the jetty while alongside in harbour. This is followed by a generic information exchange occurring between the captain and the pilot where the former shares the information about the vessel, its manoeuvring characteristics, and the state of machinery, whereas the latter communicates the sequence of the known traffic movement, recommended routes, and any navigational information specific to the local area. After signing a document confirming this information exchange, the pilotage officially begins with the pilot joining the vessel´s bridge team for a limited duration of the route, until the manoeuvring is completed with the vessel either at the limits of the harbour area while outbound or safely alongside when inbound.

The manoeuvring can be termed a critical phase of navigation for the vessel due to the presence of navigational hazards, limited room for manoeuvrability and considerable risk of human lives and material or environmental damage should an accident occur. An additional risk to consider is the act of pilot transfer to the vessel itself, where several casualties have happened over the years due to failure of deck machinery, miscommunication, or adverse weather conditions (Sakar & Sokukcu, 2023). The concept of remote pilotage has gained prominence over the past few years as an alternate option to provide navigational support from the shore by pilots without being physically present onboard. The practice of remote pilotage has gained some application in this period, especially in the European ports (Koester et al., 2007). The motivation for providing this type of service by the coastal state can be varied. The financial benefit from conservation of the pilotage resources, to safety considerations in the cases of adverse weather conditions, can be the reason for having the pilot guide the vessel from shore (Hadley, 1999). The vessels also need to be of a certain type and dimension to be able to be guided remotely. At the moment, the remote pilotage remains mostly a backup or limited option for the coastal state in the context of providing navigation support from the shore but presents an opportunity to further develop decision support systems and autonomous solutions for remote control of vessels in the near future (Lützhöft & Bruno, 2009).

NAVIGATIONAL SHORE ASSISTANCE (NSA)

Navigational shore assistance (NSA) is a type of service offered by the coastal administration in Swedish ports. The concept and its feasibility are currently explored within the frame of a research and innovation project titled "Navigation support from shore", led by the Swedish Maritime Administration. The project is a cooperation of regulators, academia, and industry to explore the feasibility of navigational shore assistance within limited testbed areas. The project, currently in its second phase, plans to explore the concept of shore-based pilotage by conducting field trials in dedicated testbed areas and simulations involving marine pilots from selected ports in Sweden.

The NSA service will be provided by the Swedish Maritime Authority (SMA). Licensed pilots will be located in a shore-based centre and utilize a variety of digital tools and decision support systems to provide the vessels, one or more at a time, with the support needed to navigate safely within a dedicated area (Swedish Maritime Administration, 2024). While the pilot may provide advice with regards to speed, course to steer and any other information important for the safety of navigation, the responsibility for a safe passage will remain with the bridge team on board.

Currently, the NSA concept is defined to consist of seven steps (Swedish Maritime Administration, 2024):

1. Concept conditions, methodology: Defines the requirements for participating in NSA (qualifications, ship specifics, navigational methodology and terminology, technology and communication requirements, etc.)
2. Ordering NSA: Downloading routes and passage plans, port planning (preliminary terminal, berthing, and manoeuvring plan), preparation of communication equipment and open lines, electronic information exchange, pilot card, and ship data
3. Information Exchange: Master–Pilot Exchange (weather, traffic information, task delegation, plans, agreements, etc.)
4. Start of NSA: Decision to start or abort the process, communication supported by message markers
5. NSA voyage management: Navigation and remote monitoring, tasks adapted to either green (ordinary), yellow (deviations from planning), or red (emergency) conditions
6. NSA manoeuvring and berthing: based on previously agreed-upon plans
7. Debrief (bridge team and remote pilot) and end of service

These steps are roughly defined, as the concept is still under development. At the time of writing this chapter, only technical testing runs have been conducted within the live testbed. Until now, all development efforts have been carried out through service simulations at the SMA's simulator centre. Domain experts, including navigators, researchers, and pilots, have actively contributed to exploring and shaping the concept.

As the concept of NSA is still under development, the current phase of the project aims to generate solutions in terms of redundancy, communication requirements, human-machine interface, common situational picture, and training needs analysis. However, many questions remain, such as what complexities arise when a marine pilot is moved to shore, what systemic issues may arise from changing the relationship between ship and shore, and what potential sources of risks arise.

AIM

The aim of this chapter is twofold. First, it will utilize the FRAM to enhance the understanding of the functional setup of the NSA service and to uncover systemic risks and opportunities associated with the overall system design. Second, and even more aligned with the rest of this book, this chapter attempts to serve as an example for how information concerning potential and actual variability, as well as potential aggregation of variability, can be identified in interview data to feed a FRAM analysis and model.

DATA COLLECTION

This study utilizes a qualitative approach combining different data sources and analysis techniques to identify the core functions of the model and to identify potential sources of variability, as well as potential aggregations resulting in functional resonance. Overall, the approach was exploratory, aiming to balance extensive expert input with a structured methodology. Thus, multiple data sources and relevant documentation (process description, teaching material, and earlier project reports) were used. These were complemented with semi-structured, structured and informal conversational interviews (Patton, 2002) with representatives of the Swedish maritime cluster (pilots, mariners, simulator personnel, and researchers from other organizations involved in the NSA-project), and unstructured observations during simulator exercises were used to collect data. Figure 9.1 below depicts the overall research process.

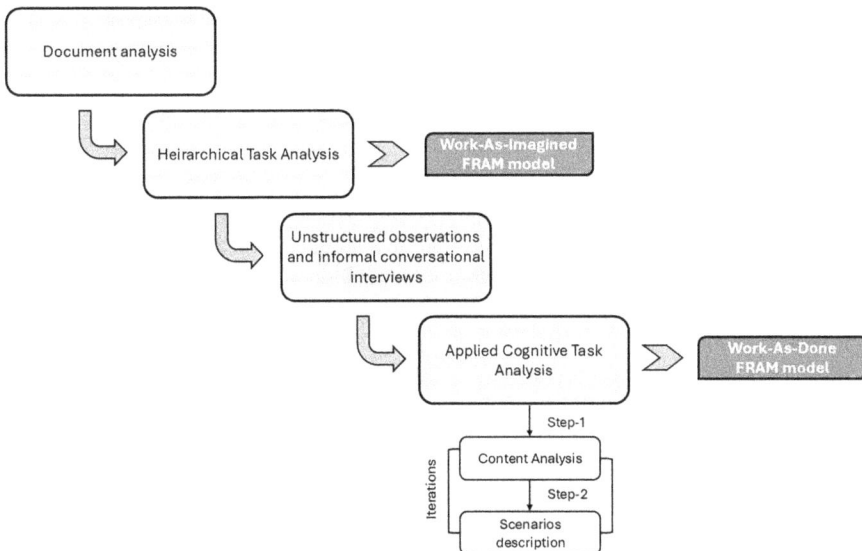

FIGURE 9.1   Overall research process for the study.

Document Analysis

Different types of documents and documentation available about the NSA service were collected from project partners and public sources, such as the homepage of the Swedish Maritime Authority (Swedish Maritime Administration, 2024). These documents represent the current state-of-the-art of the development of the NSA concept. Especially documentation about procedures and communication protocols, as well as the NSA service process as described to participating vessels, was studied. Further, teaching material for NSA pilots (NSAP) and service users produced by a maritime academy in Sweden was used to gain an overall understanding of the service, potential service levels, and requirements, as well as the complete process from ordering pilotage to being alongside the berth in port. The purpose of the document analysis was to help define the scope of the study, as well as to provide input to a Work-As-Imagined (Hollnagel, 2012) description of the overall process.

Hierarchical Task Analysis (HTA)

After the initial Work-As-Imagined model had been developed based on the available documentation, a hierarchical task analysis (HTA) was conducted in cooperation with a navigational expert to explore the different tasks of the NSA service in more detail. The expert is currently employed as a project researcher at one of the organizations participating in the NSA project. He is a master mariner by education and has previously worked as a navigator (navy, merchant marine fleet) and instructor for navigation and radio communication for the past 20 years. He has also been involved in the first phase of the navigational support from shore project, providing a background to how certain procedures for the NSA service had been developed. HTA is an analysis technique frequently used within human factors to establish which tasks, subtasks, and conditions need to be met to accomplish specified system goals (Kirwan & Ainsworth, 1992).

Method Geek

A top-level goal, such as safety and efficiency of the vessel's movement for the NSA concept, is broken down into sub-goals, which are then described in terms of specific tasks and conditions under which these are carried out. The method produces a hierarchical visualization of the analyzed system and thus may enable the analysts to reason about the potential organization of work (Stanton, 2006). Within the scope of this study, HTA was chosen for its ability to provide a formal description of the process, including hierarchies between the different types of tasks. As the NSA concept is still under development, the method also supports reasoning about relationships between tasks and the identification of potential gaps within the current process description. This also provided valuable input to decisions about the granularity of the functional description for the NSA service.

During the analysis, the developed HTA diagram was repeatedly shared with an active marine pilot and simulator instructor involved in the pilot education. He served as a subject matter expert (SME) to provide feedback on the current description and helped to fill gaps with regard to ordinary pilotage procedures and pilot work.

Unstructured Observations and Informal Conversational Interviews

To gain a deeper understanding of how the NSA process is carried out and to validate the HTA, informal exploratory observations were carried out during several NSA simulations conducted at the SMA simulator centre. Observations were carried out over the course of two full days of simulations with four NSAPs and BTs. During the observation, a researcher was placed in the simulator instructor room and could follow both the verbal communication between the NSAP and the BT in the simulators, as well as the vessel tracks and video feeds of the participants. Questions about the procedure, scenario and events were asked during and in between the different simulator runs in a non-structured manner, which can be characterized as informal conversational interviews (Patton, 2002). The aim of following the simulator runs in such a way was to gain a deeper understanding of the NSA process and observe how NSA is intended to be implemented. It also helped to refine the interview guide for the structured interviews.

Applied Cognitive Task Analysis

After the HTA had been conducted, Applied Cognitive Task Analysis (ACTA) (Militello & Hutton, 1998) was used as a guiding framework for structured interviews with four marine pilots. The aim was to gain a deeper understanding of how the NSAP perceives the service provision and to identify which potential functions are the most critical for the process. Furthermore, the intention was also to test whether the structure of the ACTA approach allows the identification of sources of variability (internal, external, and coupling).

ACTA can be described as a framework to extract information from subject matter experts on cognitively demanding aspects of work in high-risk operations domains, such as nursing or firefighting (Militello & Hutton, 1998). An ACTA consists of three stages: (1) task diagram, (2) knowledge audit, and (3) simulation interview.

Method Geek

ACTA, which is rooted in the naturalistic decision-making framework (Orasanu & Connolly, 1993), was originally developed in the late 1990s when the need for an increased understanding of decision-making under uncertainty had drawn attention towards fast-paced expert decision-making. Based on the critical decision method (Stanton et al., 2005), ACTA explores the use of information within dynamic and variable environments with the aim of providing guidance in the development of procedures, training, or the design of decision support. Rather than tasks, goals, and subtasks, this framework draws attention to the decision strategies and information needs of experts within highly uncertain contexts. Through the structure and probes, SMEs are encouraged to verbalize and explain what underlies their expertise and might normally not be uncovered (Militello & Hutton, 1998).

The three stages of ACTA can be summarized as follows:

- The task diagram aims to elicit an overall description of the task under study in three to six steps. The SME is asked to describe the task at hand and draw a diagram with no less than three but no more than six distinct sub-tasks. The purpose is to draw attention to the core sub-tasks that are conducted and provide an overview of the task structure. After the task diagram has been drawn, the SME is asked to identify the most demanding sub-task(s), which will then be the focus of the second stage, the knowledge audit.
- The knowledge audit focuses on the sub-task(s) which the participant considers to be particularly cognitively demanding, using probes to explore different areas associated with expertise (diagnosing, predicting, information perception, problem identification, improvisation, metacognition, compensation for equipment failures, and missing information) (Militello & Hutton, 1998). These areas are operationalized into eight distinct categories: past and future, big picture, noticing, job smarts, opportunities/improvising, self-monitoring, anomalies, and equipment difficulties. Within each category, probes are used to elicit information. As an example, the category "big picture" has the suggested probes *"Can you give me an example of what is important for the Big Picture for this task? What are the major elements that you have to keep track of?"* (Militello & Hutton, 1998, p. 1622)
- The final stage, the simulation interview, can be considered as a stepwise description of a scenario in which the participants are queried about what course of action they would take based on available information and are introduced to situations that require intentional decisions. As ACTA is rooted in the Critical Incident Method (Stanton et al., 2005), it has been common to use an incident as an example for the simulation interview.

In this study, an adapted version of ACTA developed by Berlin and Praetorius (2023) was used to elicit information on the work of the NSAP. This version was developed in a previous research project focused on the everyday work of marine pilots, thus designed to acknowledge domain and work-specific aspects. The interviews were conducted by one moderator in conjunction with simulator trials testing NSA pilotage. As the concept was still under development, only ACTA Step 1 and ACTA Step 2 of the method were conducted, that is, task diagram and knowledge audit.

Table 9.1 shows an overview of the participants. They all hold an unlimited pilot license for a port in Sweden and are currently actively working as pilots. Their engagement in the NSA project is as expert navigators and apprentice NSAPs. One of them had previously participated in a simulation; for the rest of the participants, the simulator trials were the first time they actively acted in the new role.

The interviews took approximately 45 minutes each and were both video and voice recorded. For the analysis of the data, all recordings were transcribed verbatim.

TABLE 9.1
Overview of the Participants in the ACTA Interviews

Participant ID	Total Years in Maritime Domain	Years as Pilot
1	29	14
2	28	5
3	33	18
4	18	3

METHODOLOGY

This section proposes the application steps followed during this study, in line with the traditional steps for applying FRAM.

STEP 0: RECOGNIZE THE PURPOSE OF THE ANALYSIS

The document analysis provided input to set the boundaries of the analysis and formulate the purpose of the modelling. The documents were reviewed with a focus on available details of the NSA process, agents involved and identified process steps. The boundaries can be selected, keeping in mind what would likely be the most variable parts of the process that require more detailed investigations.

STEP 1: IDENTIFY AND DESCRIBE THE FUNCTIONS

Based on the document analysis, the first description of the NSA process was produced to produce a Work-As-Imagined FRAM model. This was followed by the HTA to create an order and be able to reason about the granularity of the functions that would be needed for the modelling. The HTA was based on the process documentation but also used subject matter experts and marine pilotage procedures defined by the SMA to derive a detailed description.

Method Geek

The HTA was produced in MIRO, a web-based tool for visualization. While there are multiple tools for web-based sharing, MIRO was chosen as it offered an opportunity to repeatedly share the analysis in the research team and with the subject matter experts, that is, MIRO allows collaborative editing, which made it easy to structure and restructure the tasks, as well as to add new tasks and subtasks. Further, sharing the analysis with the SMEs allowed them to add and comment directly in the analysis.

To provide an overview of the HTA produced, Figure 9.2 shows one of the produced nodes, to make a turn under NSA service with the NSAP and BT closely collaborating and jointly monitoring the progress.

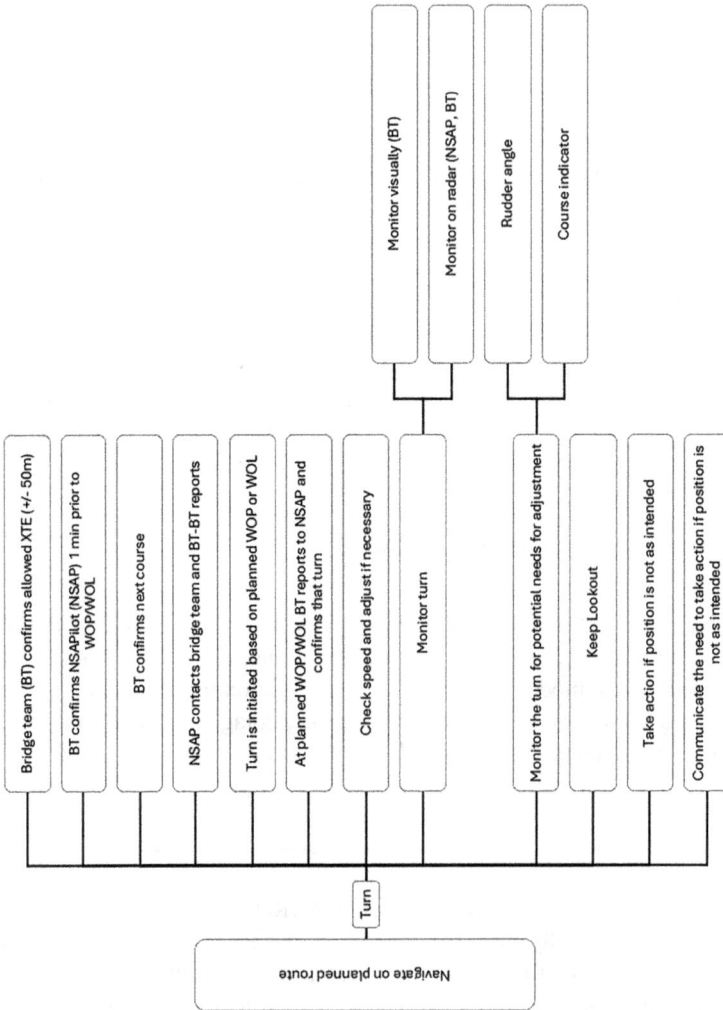

FIGURE 9.2 Excerpt of the HTA representing the NSA process based on documentation and expert feedback.

As the HTA was based on documents and SME input on the formal process, the knowledge gained from the simulations, informal conversational interviews and ACTA interviews was used to create a Work-As-Done model of the NSA service.

Method Geek

In a first step, the task diagrams from the ACTA were reviewed, and based on the interview transcripts, one task diagram for each participant was produced. The four diagrams were then merged into a single process description, which represented a surface overview of the NSA process and facilitated the identification of perceived core functions in the model, that is, those deemed most important by the participants.

In a second step, a content analysis (Patton, 2002) of the interview data was conducted to identify what tasks the NSA pilots perceived as part of the process, that is, what functions they and the bridge team perform to achieve a successful pilotage. In this iterative analysis process, the following codes emerged: Task distribution bridge-team-pilot, traffic planning, and tasks part of the NSA process (e.g., monitor external communications, establish contact, identification of potential risks). Based on this, a first functional model of the Work-As-Done (Hollnagel, 2012) process was developed. This model was iterated several times after being shared with and commented on by SMEs involved in the project.

Step 2: Identify Variability

The same content analysis (Patton, 2002) as described for Step 1 was applied to identify potential variability. The data collected during the knowledge audit of the ACTA interview were analyzed with a specific focus on content relating to potential sources of variability. The interview extracts were assigned a general code *variability*. Afterwards, all identified extracts were reviewed and coded as internal, external, or coupling variability. It was also possible for extracts to be classified with more than one code. Table 9.2 provides an example of how the following extract was coded.

> And the talk on the bridge, I couldn't really hear because of the technology. I couldn't really hear, because they were talking in … if you understand me correctly, in a messy way. I couldn't follow the thoughts simply. And they expected me as a NSA pilot to give courses or a lot of advises on speed et cetera, which put a lot of workload on me. But that is an adjustment I have to do according to the crew, how they are.
>
> (NSAP 2)

While the extract above provides an example for actual variability expressed by the respondent through an example from a simulator run, the extract below was coded as potential internal variability as it refers more generally to the work and a potential source for variations in how a function may be performed.

TABLE 9.2

Extract From NSAP 2 with Assigned Variability and Identification of Main Function That Is Affected

Extract	Variability (Code)	Justification
And the talk on the bridge, I couldn't really hear because of the technology. I couldn't really hear, because they were talking in … if you understand me correctly, in a messy way.	External variability	The NSAP explains the importance of being able to understand the BT's communication. Function(s) affected \<Provide NSA support>
I couldn't really hear, because they were talking in … if you understand me correctly, in a messy way. I couldn't follow the thoughts simply. And they expected me as a NSA pilot to give courses or a lot of advises on speed et cetera, which put a lot of workload on me	Coupling variability	The inability to clearly hear and understand the BT affects the NSAP's ability to \<Provide NSA support>
And they expected me as a NSA pilot to give courses or a lot of advises on speed et cetera, which put a lot of workload on me. But that is an adjustment I have to do according to the crew, how they are	Internal variability	Workload affects \<Provide NSA support>
	Coupling variability	Need for adjusting the NSA provision to the crew's needs

Like, it takes the most effort to figure out how to do it. Navigation with digital systems, it's also mixed up with experience that you have from doing it live, and try to apply that experience into doing it on two dimensional … Just a screen with a … Like, in a computer-based … Instead of seeing the whole picture live and try to put yourself in the situation where they are and from what information you have on the screen, so those are the two things that are most demanding from my side, that takes most workload.

(NSAP 4)

As the ACTA knowledge audit focuses on one or more sub-tasks that are cognitively demanding and fixed probes, the data collected provided rich descriptions of the various NSA tasks, how the respondents experienced their performance during the simulator runs and what cues and strategies they had developed to provide NSA service.

STEP 3: AGGREGATE VARIABILITY, ACTUAL, AND POTENTIAL

In this step, the content analysis focused on utterances that highlighted patterns, or combinations of functions, which were experienced as critical by the respondents. The focus of the analysis was on specific examples provided by the SME as part of the knowledge audit. Especially the four following sets of ACTA probes provided input into this analysis step:

- Noticing: Can you give an example when you had a look and realized something was not right about the situation?

- Past/Future: Is there a time when you looked at the task and were able to conclude how we reached this situation?
- Self-monitoring: Can you tell me about a time when you realized you needed to adapt your usual way of working in order to succeed at the task?
- Deviations: Can you describe a time when you noticed a deviation from the normal, or when you knew something was missing, something that should have been there?

As these probes triggered the respondents to provide examples and reflect upon these, the data derived was rich in descriptions and allowed the identification of extracts that described how variability had aggregated and strategies used by the NSAP to manage these situations. Extracts that described negative outcomes of aggregated variability were identified in the interview data and coded as functional resonance.

STEP 4: ASSESS THE CONSEQUENCES OF THE ANALYSIS

To identify measures to manage variability, the functional model derived in the earlier steps was reviewed with a specific focus on the scenarios and situations that were described by the participants. The analysis focused on what information gaps and needs the respondents identified, as well as potential novel relationships between the NSAP(s) and other actors in the maritime transport system. As strategies and cues had been highlighted through the probes in the knowledge audit, the analysis on measures to potentially manage, monitor or eliminate variability within and across the functional setup was facilitated.

Functions with a large number of couplings, as well as functions being perceived as critical by the respondents, were highlighted as a starting point. Further, the interview analysis was conducted with a focus on what was perceived as particularly well within the current setup of the NSA process by the participants.

RESULTS

The following section describes the results of the analysis based on the four FRAM building steps just described.

RESULTS OF STEP 0: THE PURPOSE

The purpose of the analysis was to explore emerging complexities and potential systemic issues, including new risks, that may affect and change the relationship between ship and shore. Therefore, the FRAM analysis aims to evaluate the current NSA system design with a specific focus on the four phases: ordering NSA, information exchange by VHF, start of NSA, and NSA voyage management.

Besides being highly variable phases, these were also the phases that were documented in the most detail and already tested by pilots in several simulator runs, providing a pragmatic scoping for the analysis at hand. Indeed, the case offered the possibility of recruiting interview participants for the ACTA who had one or more days of experience from the simulator on how to provide NSA as a pilot.

RESULTS OF STEP 1: THE FUNCTIONS

A total of 8 background functions and 20 foreground functions were identified from the data. Additionally, three functions were added for the sake of model completeness, that is, not having any orphans in the functions. They are background functions, but were not explicitly mentioned in the interviews or the current version of the process documentation (Figure 9.3).

Domain Expert

<Ordering NSA> can be described as a set of 4 background functions to the foreground activity of information exchange, the start of NSA, and NSA voyage management. The stage of the process consists of, among others, providing the ship's details and pilot card to the pilot service and receiving a suggested passage plan as well as a berthing and mooring plan. Thus, the functions included in this first phase are background functions providing control, resources, and preconditions for many of the foreground functions in the model.

The information exchange is initiated by either the vessel calling or the NSA pilot (NSAP) calling on the VHF radio as output of <Establish VHF contact>. Once contact is established, all necessary information for the NSA process is exchanged between ship and shore. This phase consists of nine foreground functions and focuses on creating a joint mental model between the bridge team and the NSA pilot. This includes what is normally conducted during a Master-Pilot Exchange (MPEx) and concerns agreements on plans (route, mooring, etc.) and monitoring procedure, task and distribution, and manning, as well as additional functions, such as a communication test and traffic information on potential meetings and traffic situations during the NSA.

After the information exchange is finished at a satisfactory level for both parties, the NSAP decides whether or not to initiate NSA, which represents the start of the NSA. The process consists of two foreground functions; the NSAP initiates NSA <Initiate NSA> through the use of the defined communication marker (Output) "NSA operation start", which is confirmed through closed-loop communication by the bridge team (BT) in the function <Confirm NSA initiation>.

NSA voyage management, the third phase under investigation in this study, represents the largest set of foreground functions within the model. In comparison to the functions within the previous phases that are carried out jointly, the functions resembling this phase are carried out either by the NSAP or the BT. The NSAP functions concern the management of communication with other actors in the maritime transport system, that is, VTS, other vessels, other services <Manage external comms> (VTS, other traffic, and other services), monitoring the progress of the vessel in the dedicated area and the NSA at large (<Monitor progress (shore)>), and <Provide NSA Support> when necessary, for example, when the vessel is deviating from the agreed route or misses to report to the NSAP. The functions carried out by the BT mainly concern the actual navigation (<Navigate on route>, <Prepare a turn>, <Make a turn>), and <Keep lookout> to support the monitoring function of NSA and to monitor the vessel's own progress (<Monitor progress (vessel)>).

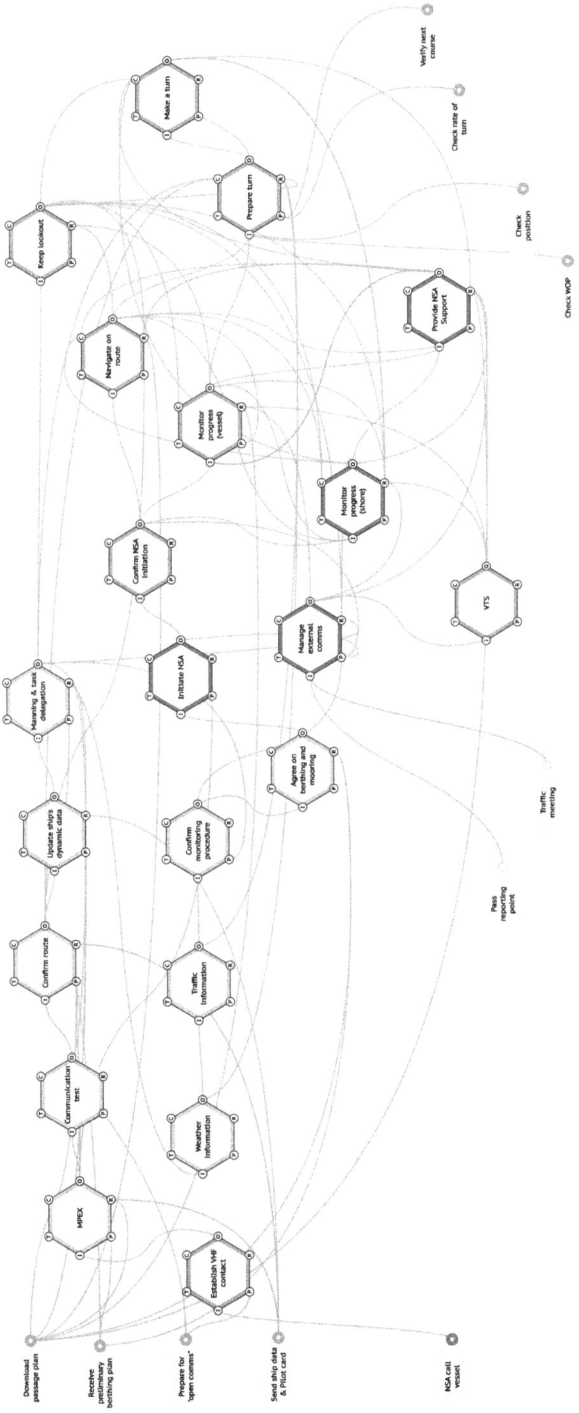

FIGURE 9.3 FRAM Model of the NSA Service System.

The following extract provides an example of how a subset of the foreground functions was identified, see Table 9.3.

> That I asked them to go more to starboard. I made the crew observant of this ship coming up from stern, giving the captain … First, I point out the danger, so to say. I make it clear how they will pass, and I advise them to go more to starboard. And when I see that it's getting closer and closer I advise the ship again to come even more to starboard to increase. And I call the vessel again to come to port, and they confirm this.
>
> (NSAP 2)

The respondent describes the process of how NSA service. In this sequence, the NSAP had been in contact with another vessel not under his pilotage to make meeting arrangements. Despite the meeting arrangement, the meeting will be a quite close encounter. He calls up the vessel under his pilotage and describes the situation, and provides advice. As he monitors the progress, the situation is not resolved to his satisfaction, so he contacts BT again to advise them to alter course. After that, he also contacts the meeting vessel to ensure that arrangements agreed upon earlier are kept.

TABLE 9.3
Examples of the Function Identification, Its Corresponding Extract and Justification as Outcomes of the Qualitative Study

Function	Extract	Justification
<Monitor progress (shore)>	*And when I see that it's getting closer and closer I advise the ship again to come even more to starboard to increase*	NSAP monitors the progress of the vessel's voyage in relation to other traffic and an upcoming traffic meeting
<Provide NSA Support>	*That I asked them to go more to starboard. I made the crew observant of this ship coming up from stern, giving the captain… First I point out the danger, so to say. I make it clear how they will pass, and I advise them to go more to starboard.*	NSAP provides support for the vessel to go more to starboard based on their judgement about the upcoming traffic meeting in the fairway. The decision is left to the BT (captain), but the NSAP motivates the decision by explaining his perspective on the upcoming vessel meeting
<Manage external comms>	*And I call the vessel again to come to port, and they confirm this*	NSAP calls the other vessel to make a traffic arrangement and confirm the intention

The following tables (Table 9.4–9.6) present an overview of the functions within the model, including which actors are responsible for carrying it out.

TABLE 9.4
Overview of Foreground Functions in the Model

Function Name	Description	Agent(s) Carrying Out the Function
<Establish VHF contact>	BT and NSAP establish VHF contact	BT & NSAP
<Conduct Master Pilot Exchange (MPEx)>	Exchange necessary information, and agree upon plan	BT & NSAP
<Conduct communication test>	Communication means, including open lines, are tested to ensure potential recovery.	BT & NSAP
<Confirm route>	The planned route is confirmed by both parties.	BT & NSAP
<Update ship's dynamic data>	Dynamic data is updated.	BT
<Agree on manning and task delegation>	Manning and task delegation are agreed upon.	BT & NSAP
<Inform about weather information>	Weather information about the current conditions is updated.	NSAP
<Provide traffic information>	NSAP informs BT on potential traffic meetings during pilotage and surrounding traffic	NSAP
<Confirm monitoring procedure>	The monitoring procedure for the NSA service is confirmed	BT & NSAP
<Agree on berthing and mooring>	Berthing and mooring plans are agreed upon	BT & NSAP
<Initiate NSA>	The pilot initiates the process of NSA through a distinct communication marker	NSAP
<Manage external communication>	NSAP manages all external communication (traffic, VTS, and other services) if otherwise agreed upon	NSAP
<Monitor progress>	NSAP monitors the vessel's progress in relation to the agreed plan with the help of technical equipment (AIS and RADAR)	NSAP
<Provide NSA support>	In case of deviations, upon request, or if a vessel is not responding accordingly, the pilot provides support (informs on deviation, advises on speed, course, and contact if non-responsive)	NSAP
<Confirm initiation of NSA>	BT confirms that NSA has been initiated using the dedicated communication markers	BT
<Navigate on route>	Vessel proceeds on agreed route	BT

(Continued)

TABLE 9.4 (Continued)
Overview of Foreground Functions in the Model

Function Name	Description	Agent(s) Carrying Out the Function
<Prepare turn>	Vessel prepares for turn using information available to the BT (technical equipment, visual, communication, and NSA support)	BT
<Make a turn>	Vessel turns, and BT communicates the next course to the NSAP	BT
<Keep Lookout>	BT keeps a lookout and provides information on the current situation to the NSAP	BT
<Monitor progress (vessel)>	BT monitors the progress of the vessel in relation to the agreed route	BT

TABLE 9.5
Overview of the Background Functions in the Model

Function Name	Description	Agent(s)
<Download passage plan>	Bridge team downloads the passage plan provided by the pilotage service	BT
<Receive preliminary berthing plan>	Bridge team receives a preliminary berthing plan from the pilotage service	BT
<Prepare for open communication>	Check communication equipment and provision of contact details, including mobile phone numbers	BT & NSAP
<Send ship data & Pilot card>	All necessary data about the vessel and the Pilot card are provided to the pilotage service	BT & NSAP
<Check WOP>	BT checks wheel over the point	BT
<Check position>	BT checks the position	BT
<Check rate of turn>	BT checks the rate of turn	BT
<Verify next course>	BT verifies the next course after a turn	BT

TABLE 9.6
Functions Created for the Completeness of the Model

Function Name	Description	Agent(s)
<Pass reporting point>	Vessel passes reporting line to the VTS area or a reporting point within the VTS area	BT
<Enter a traffic meeting>	Trigger for communication between NSAP and other traffic	NSAP
<Interact with VTS (Vessel Traffic Services)>	Provides traffic information and receives reports from the NSAP	NSAP

The model created is focused solely on the functions that involve human operators, that is, the pilot (NSAP) or the BT. Technology functions, such as those that provide the data to the various systems the BT and NSAP use, are not considered in greater detail in the identification of variability as these tend to be rather stable in their performance (Hollnagel, 2012).

Potential External Variability

Maritime operations are characterized by many external factors which are beyond the control of the joint system of NSAP and BT carrying out the joint action of navigation through the area under NSA. These factors include environmental aspects (e.g., weather, winds, and currents); other traffic (commercial, military, and leisure); other services (such as VTS, tugboat, or port services); or changes to the navigational space (e.g., military exercises, regattas, and maintenance in the fairway), which can impose constraints on the operation and may increase the variability of the functions' performance.

Further, potential external variability can also occur in the way that the NSAP and BT perceive each other and therefore adapt their communication and monitoring processes. As an example, one respondent emphasizes how the perceived competence of the BT and his experiences as a pilot help to anticipate the next interaction, which in turn impacts how he or she performs the pilot-related functions.

> ...they are experienced, so in many of the situations I can hear them talk, their talk makes sense, so I hear what they say internally, then I have a heads-up what they're thinking. And many times, I can anticipate what they will say. And they ... let's say they take a decision and double check it with me, and that makes it easier for me.
>
> (NSAP2)

Potential Internal Variability

In the FRAM model, the functions under investigation involve human operators, which creates the potential for internal variability since humans vary in their performance. However, what can be identified as an additional source of potential variability is the fact that the current NSA concept, procedures, and work arrangements are still under development.

> We are all new at this, the pilots here. So, I think that, when you have done this a few times, that you will find your own vocabulary that you will use, or that will make it easier for you to guide the ship.
>
> (NSAP 2)

The four respondents all defined the core tasks of the NSA service differently, and there was only a little overlap in what is perceived as tasks or components of the NSAP's work. Consequently, there might be differences in how the respondents execute and monitor the functions of the system. This may be due to the perceived role, including the interpretation of role and task distribution among the actors (BT and

NSAP), as well as to the novelty of procedures and communication markers, which are currently being developed in conjunction with the simulations.

> Do the traffic planning with only the digital equipment and this environment that you work in, when you're not there live.

(NSAP 4)

Further, the participants raised the fact that there is a need to interpret information in the technical systems, but not being on the bridge may cause an increased work-load, which in turn can lead to an increase in internal variability in the functions provided by the NSAP, such as <Provide NSA Support> (which in the current model incorporates the traffic planning), <Monitor progress (shore)>, or <Manage external comms>.

Potential Coupling Variability

The functions that show the most couplings to other functions are <Monitor progress (shore)>, <Monitor progress (vessel)>, <Provide NSA support>, <Keep lookout>, and <Prepare turn>. The functions show the most couplings to other units within the larger system and are therefore also those that can be considered the most critical functions for successful performance. However, through the multiple couplings to other functions in the system, they might also cause a spread of variability with risks for functional resonance, if timing or precision vary. The two monitoring functions, in conjunction with the <Keep lookout> function, are essential to ensure a safe sys-tem state. While the NSAP monitors the progress of the vessel in relation to the big-ger traffic picture and planned interactions with other services, such as VTS or port services, the BT monitors the vessel's progress with a specific focus on navigation.

> They take the workload for turning the ship. I take the load of the traffic. So this is what it will be in the future. A remote pilot or an NSA pilot, they can just focus on the traffic.

(NSAP 1)

The <Keep lookout> function is an insurance that important information that might be missed due to deficiencies in the technology, such as missing or wrongful Automatic Identification System (AIS) signals or a lack of radar information, and emphasizes that there is a need to combine visual information with information in the decision support systems. However, visual information can only be obtained on the vessel itself and needs to be communicated to the NSAP afterwards, creating both complex interactions with other functions and a potential for variability spreading if informa-tion is not communicated in time or if this communication takes too long.

Further, due to the nature of work being highly dependent on technologically mediated communication between NSAP and BT, working communication lines, such as those tested through <Conduct communication test> in the information exchange phase, are essential. Thus, if means of communication are affected, vari-ability can quickly arise within the couplings across the model. Without working communication, the NSAP cannot provide any support to the vessel.

Results of Step 3: The Potential Aggregation of the Variability

Even though the current analysis is based on interviews and concept tests in simulated environments, the participants' responses indicate the present actual variability. The following sections will thus show examples of how variability in NSA service provision is expressed and outline potential risks for aggregations of variability.

Prioritizing Among Information

The following quote shows an instance of internal variability. As the NSAP receives "Traffic information" (Output <Interact with VTS>), the BT is reporting 'Vessel turning (Output <Make a turn>). Both outputs are resources to <Monitor progress (shore)> and <Provide NSA support>, thus increasing the workload of the NSAP as they have to prioritize among the two different information sources (resources). This causes internal variability as the workload is exceeding what can be handled at one moment in time. The NSAP decides to prioritize the VTS information as that is judged to be more important as it will lead to an 'Updated traffic picture' (output <Monitor progress (shore)>), thus helping to maintain the overall safe state of the navigation for the vessel under his service.

> ...when the VTS is calling and you receive traffic information, the bridge team was discussing, and they were ignoring the channel one three [?? 0:10:35] for us. The traffic channel. Then I tried to deselect. I know what they said to me. They called me up during the time when the traffic was coming from the radio. Then I have to split myself in two. I go ... I take the bridge team later on. They will say to me that we are now turning. I know what they are doing, yes ... but the important thing is, what is happening in the traffic in the future? So then you have to split yourself in two for a short moment, and then you can just bring up the connection with the bridge later on and just say yes, I see your turn. It's good. You're going to 072 or something.
>
> (NSAP1)

The quote further shows external variability introduced by the fact that the NSAP realizes that the BT is not receiving the information that is transmitted publicly on the VHF channel. This may impact the output of <Monitor progress (vessel)>, <Monitor progress (vessel)>, as the BT is occupied with executing <Make a turn>. Thus, while stated procedures require the monitoring of progress from both shore and vessel, only one of the monitoring functions is active, putting more emphasis on receiving the traffic information than following up with the turn, as the vessel is currently behaving in an anticipated way.

While this situation is not critical, it needs to be acknowledged that, through the multiple couplings between <Monitor progress (shore)>, <Monitor progress (vessel)>, and <Provide NSA support>, there is a high risk for aggregated variability, which can affect when (timing) and to what extent (precision) monitoring and support are carried out. As a consequence, the NSAP may not be able to provide information in due time, or miss information essential for the vessel's safe conduct, nor have an "Updated traffic picture" when needed to make decisions on whether or not, as well as what support is needed by the BT.

Missing Transponder Data and Providing NSA Support

<Provide NSA Support> represents one of the core functions within the model. Depending on the situation, the Outputs ("Contact vessel if not reporting", "Inform on deviation", "Advise on course", and "Advise on speed") can serve either as control or resource for the vessel's navigation (<Navigate on route>), as well as being an essential control for when the vessel is turning (<Make a turn>). It is therefore important to note that the timing (how long a certain NSA advice is valid and meaningful) and precision (what information is relevant, needed and enough) are critical for the performance of the function. As the vessel under NSA service proceeds as part of the larger traffic system, other traffic may constrain the way in which the BT functions (<Navigate on route>, <Prepare turn>, and <Make a turn>) can be performed.

> I asked them to go more to starboard. I made the crew observant of this ship coming up from stern, giving the captain ... First, I point out the danger, so to say. I make it clear how they will pass, and I advise them to go more to starboard. And when I see that it's getting closer and closer I advise the ship again to come even more to starboard to increase. And I call the vessel again to come to port, and they confirm this. So I cannot say that I know what will happen, but it will improve the chances of a good outcome of this situation.

> (NSAP 2)

The quote of NSAP 2 shows that the time to react towards sudden changes is crucial. In the extract above, a vessel is approaching quickly, which imposes constraints on the ability to <Provide NSA support> in such a way that it can be of help for the BT when adjusting the performance of the functions associated with navigation (<Navigate on route>, <Prepare turn>, and <Make a turn>). Thus, delays in timing will increase the output variability for being able to provide necessary support and therefore, in turn, adjusting the navigation adequately based on the newly available information. As a result, such situations might require reformulating and adjusting plans for manoeuvring and navigation, as well as for traffic meeting arrangements and potential NSA support.

As the NSAP is located on shore with no visuals, having access to relevant data about other traffic is essential for the success of shore assistance. Data can either be obtained from the *VTS* (output of traffic information) or through making arrangements (<Manage external comms>). Data about the surrounding traffic state provides an important resource for <Monitor progress (shore)>, <Monitor progress (vessel)>, and <Provide NSA support>. The following quote exemplifies the risk of missing transponder data. As the respondent mentions, he is forced to make a traffic meeting arrangement with a vessel that is not visible in the decision support system, as she is not transmitting any AIS data; thus, the resource available to the monitoring functions, as well as to <Provide NSA Support> is imprecise.

> Well, for this time, there was no AIS signal. This first of all is very ... that's a risk you know. You can't really see her. I just heard [XXXXX] that she was leaving. I couldn't see her on the chart. So that's dangerous, I think. And she also had some problem replying to me, that I had to call her several times, and then when she finally answered she didn't give me an answer on my question.

> (NSAP 3)

The NSAP overhears a report from a vessel to the VTS recognizing (<Monitor progress (shore)>) that the vessel will affect the planned traffic situation. The function <Manage external comms> is activated as the need to make traffic arrangements is recognized by the NSAP. However, several times, the call is unsuccessful, which means that the output "Meeting agreement" is omitted. In this specific case, there was no major consequence, but it still shows the dependence on external communication and traffic information, especially that which is displayed in the decision support system. In this instance, the time to the meeting was allowing for variability within the output, but it can be anticipated that if less time is available, omitting "Meeting arrangement" can lead to close quarter situations or more severe incidents, as it is an essential resource for an adequate performance of <Provide NSA support>, as well as the functions that help ship and shore to monitor the vessel's safe progress.

While adjusting plans in a complex environment is expected, sudden changes may result in high-risk situations, as information about traffic not transmitting transponder data may be received late. As mentioned above, if transponder data is not available, the only way for the NSAP to receive relevant traffic information is either through the Vessel Traffic Service (VTS) or as output from <Keep lookout>.

> It's also hard to change over, when you get information that, okay, here is something big and challenging that I didn't know until now, to try to figure out a way to solve that problem. Which is in that situation is almost impossible to do when you are far away from … I mean, your distance becomes too far away when you're just on remote, as compared to if you were in real life. In real life I would have seen it a lot earlier visually

> (NSAP 4)

Whenever new information requires adjustments to plans, especially if this information is received with a certain delay, it may give rise to aggregations of variability across the functional system, that is, functions associated with the provision of the service and with navigation.

> I think, or communication. And one is the ordinary communication, where you are preparing for the next course, speed et cetera, together with the crew. And then it is somewhat unexpected advice, you could say.

> (NSAP2)

The quote of the respondent shows that ordinary communication, such as interactions about the next course, or heading, for example, as output from <Make a turn>, are normally executed according to procedure and are considered routine operations. However, in cases that unexpected traffic situations occur, such as a quickly approaching vessel, what the participant is calling unexpected advice is required, that is, the function <Provide NSA Support> will be executed with a less certain outcome, which requires a deviation from intended plans and a re-evaluation of the current traffic state.

Reconfiguration of Roles

Conducting NSA service successfully requires close cooperation between the NSAP and BT. Several of the respondents highlight that it is more difficult to determine and monitor the crew at a distance, as it can only be done through the communication between the ship and shore and judging the performance of the navigation functions of the BT. While ordinary pilotage allows the pilot to be in the situation and assess the task distribution and manning, as well as the state of the BT, NSA services are based only on interaction mediated by technology. This changes the role of the pilot as there is no opportunity to take control or directly delegate tasks in complex or risky situations.

> ...[you are] No longer in control, because this concept is heavily based on that you are proactive, not reactive, which means that when you go into reactivity, you are normally too slow. You are not able to get through quick enough. Normally, if you are onboard, then you can give orders or perform various tasks directly, sometimes that you don't have to tell anybody what to do, you have the controls by yourself, you can do it instantly, which means in a matter of seconds, now it may be a matter of minutes instead, and that in some cases is way too slow.

> (NSAP4)

Especially when new information becomes available and changes to plans may be required, technologically mediated communication is experienced as being slow. As an example for potential aggregations, it could happen that performing <*Monitor progress (shore)*> the NSAP has identified a problem, but there is no time to effectively execute <*Provide NSA Support*> omitting the output to create a situation, in which the BT can choose how to resolve the situation, as time to adequately evaluate options or generate advice is not available.

Further, delegation within high-workload situations in the novel NSA service system may create additional workload for the NSAP, thus affecting the output timing and precision in the NSAP functions.

> Depending on the density of the traffic, there can be too many ships for my brain to keep track of simultaneously, when they are moving at different speeds, and so on. But I can use the digital system to help me to keep track of that. Also, I can use the crew onboard to tell me things that they see which I cannot see. And I can tell them to keep track of things that I know that I would ... This is too much for me, I won't be able to keep track of that thing also. So, I can task delegate that to somebody else. Which is a problem, because they might not have the same view of it as me, so maybe they don't know how to do or they may not know what I mean when I say it, but ... It's like a last resource to do something

> (NSAP 4)

While delegating tasks to the BT would potentially be possible, the respondent stresses the risk of increased communication to align and create a joint perception of the current situation and what is needed, which in turn can lead to misunderstanding or misalignment of the perceived traffic situation. With regard to the functional model, the communication to align and create a joint perception would result in additional system functions, which in turn may delay the output of several other functions, such as <Monitor progress (shore)> or <Manage external comms>, as the alignment must be prioritized to ensure an adequate level of workload for the NSAP.

RESULTS OF STEP 4: THE MANAGEMENT OF PERFORMANCE VARIABILITY

Based on the modelling, the following considerations concerning the monitoring and management of performance variability on a system level could be obtained.

Experience as Key Resource

All participants raised the importance of having pilotage experience from real-world settings to be able to interpret and act upon the information available to them in various systems. It also helps to direct the focus of the NSAP and to determine what actions to take and when. Thus, experience from pilotage guides the attention and becomes a key resource and control for those functions carried out by the NSAP.

> We have some experience-based parts of the fairway where we know that … I know that if I'm here and this other ship is here, we're going to meet somewhere where it's not ideal to meet. So then I have to take … I have to look out for that ship. And I know … on the other hand, I know where … If a ship is somewhere in the fairway and I'm in a certain part of the fairway, I know that it's not going to be a problem, or most probably not. So I can see it already which ships I have to focus on.
>
> (NSAP4)

The experience from piloting vessels also provides a good knowledge of potential meeting points, different types of vessel behaviour and ways to determine when a change of plan is needed. It serves as an underlying resource for the NSAP functions and is essential to be able to interpret the traffic picture and patterns within the technology at hand, and enables quicker problem identification and solving, thus becoming an important enabler for safe and efficient service delivery.

Therefore, with regards to NSAP competence requirements, requiring several years operational experience, that is, having piloted different types of vessels through the fairway in which NSA is provided, is a way to ensure that the NSAP can quickly adapt to changing circumstances and find adequate ways to continue providing the service, even if this means prioritizing certain tasks or information, as seen in one of the examples above.

Further, two of the four participants discussed the advantage, having previously worked with a certain vessel or crew, as this may make the interaction and communication between NSAP and BT faster and easier. As communication between ship and shore is essential for a safe and efficient voyage, paying attention to pairings between those who provide a service and those who make use of it might be considered in the service planning stage.

Training and Procedures

The participants raised concerns about the sudden changes in workload during the simulation variations in workload based on traffic load, traffic scenario, or other circumstances, such as sudden shifts in weather conditions, may cause a swift increase in workload for the NSAP. The increase itself is not critical, but it can lead to changes in timing and precision for several core functions, such as <Monitor progress (shore)>, <Manage external comms>, and <Provide NSA support>. While workload might currently be associated with the novelty of the service, some of the issues are likely to remain. Therefore, means and measures to support the NSAP in these sudden workload shifts sure are of importance to consider.

All four respondents stressed the importance of swift, concise, and unambiguous communication for the success of the service. Thus, training the usage of message markers, such as specific Short Maritime Communication Phrases for this specific service, repeated training of swift changes in traffic density or complexity, and live tests in dedicated areas might help to lay the grounds for managing or dampening the effects of variability in functions carried out by the NSAP.

Procedures for NSA service are currently under development and are being tested. It might help to consider in what way these can help to support the new interactions and, consequently, the complexities that arise in the interactions between NSAP and BT. Especially procedures that support task delegation to either another shore-based service operator or to BT should be considered. As an example, currently, the NSAP manages all external communication.

While this is not particularly demanding in normal operations that proceed as planned, extracts from the interviews have shown the potential to give rise to coupling variability when too much information and communication needs to be handled at a single point in time. Thus, procedures for how to renegotiate the management of external communication, such as with the VTS or vessels in the proximity, should be explored.

Understanding the Role of NSA in the Maritime Transport System

Another line of development that should be pursued is a further exploration of the novel role of the NSA service within the maritime transport system. As described by the participants, providing NAS service is different from piloting a vessel.

> It's like to be in the backseat. It's like to pilot defensive. Definitely. Much more defensive than if you are on board.

> (NSAP1)

Therefore, it is important to understand how the service and those providing it relate and become part of the current transport system. One of the areas to explore further is, for example, the novel relationship between NSAP and the VTS services. As described in the quote below, VTS serves as an additional source of information, which may imply that collocating the two services within one centre can help to maintain a joint traffic picture, thus also complementing each other's perceptions on the current state of the traffic.

> But VTS, when he's in the room, you have the camera, you can see the camera. So then it … so it's actually a very good thing if the VTS and the NSA pilots is in the same room. You get very good information just talking to each other.
>
> (NSAP 3)

Traditionally, VTS has rather been a resource for pilots with regard to traffic information, but now that the system configuration is changing, NSA services and VTS operation are perceived as closer to each other.

> The cooperation with the VTS operation … VTS operator on a, like, person to person level is a quite strong … It makes a little bit of a team feeling and you can save each other a bit by just nudging each other to ask, "did you see this?" or "did you hear this?" or … Like, remind of things that are happening, which are hard to just keep in mind all by yourself.(…) I will not say similar, but just another task but in the same area, with the same point of view, and with the same interest of the result.
>
> (NSAP4)

In this example, rather than being a resource delivering traffic information, VTS is also a measure to ensure that essential information of the overall system state in the VTS traffic area is acknowledged by the NSAP. As recognized by NSAP4, both services are provided in the same area and have the same overall goal of guaranteeing safe and efficient traffic movements. Therefore, considering the NSA services' role and relation to other services can help to strengthen the service provision and help to manage performance variability that might be introduced by increased traffic volumes, unanticipated traffic meetings, or traffic without transponder information. Thus, in addition to information, VTS can now become an extra resource as backup in essential system functions, such as <Manage external comms> and <Monitor progress (shore)>.

CONCLUSIONS

This chapter has presented an approach to identifying system functions in a service and system that is still under development. Data sources ranging from formal documentation and training materials to qualitative data from interviews have been analyzed and synthesized to produce a comprehensive system model and to identify critical system functions and potential aggregations of variability. While the specific Navigational Shore Assistance is a service developed in specified areas within Sweden, the above can still serve as an example of how services, system functions and potential aggregations can be explored with FRAM, as well as how different data sources may contribute to the various steps in the FRAM process. While traditional risk analysis techniques have been applied as part of the project-related work, FRAM was chosen as a complement to be able to explore systemic issues that might arise when moving the pilot services provision ashore.

OPERATIONAL PERSPECTIVE

The results show that the NSA service creates a novel, complex system where BTs and NSAPs need to closely communicate with each other to be able to quickly adapt to sudden changes in the environment. Due to the many couplings related to the functions provided by the NSAP, it is likely that sudden shifts in traffic density or complexity might lead to problems in providing timely and adequate advice when quick responses are needed to avoid risky situations. While the current procedures are still under development, it was emphasized that the current way the service is provided requires a lot of preparatory work, such as planning traffic meetings, but that the NSAP can only offer little support when unanticipated events demand a coordinated response.

Further, the analysis has shown that the concept of NSA is not solely about moving the pilot ashore but also creates a new role for the NSAP within the maritime traffic system. The heavy reliance on working communication equipment and the fact that visual information is not available to the NSAP create a complex system in which new functions can quickly arise if there are discrepancies between what the BT and NSAP perceive as the current traffic situation. During the simulator trials observed and in the interviews, it became salient that participants utilized the VTS operator to manage an overload in information and to dampen variability arising from delayed outputs across the most critical system functions. Therefore, it is suggested to explore how this novel relationship between NSAP and VTS operator can be formalized in the working procedures, as it currently represents an essential resource for the safe NSA service provision.

Currently, the concept is not used in real-life operations, but there is a plan to move from simulator trials to a live testbed. Both simulator and live tests in the dedicated area offer an opportunity for learning about cues and strategies. As much of the service depends on the NSAP, which represents the coastal state and is the local expert, a knowledge bank surrounding useful cues and strategies, both to recover and to ensure safety but also to identify how to monitor progress, is needed. Monitoring variability and finding ways to manage or dampen it is rooted in an understanding of work-as-done. Therefore, reviewing and debriefing participants from the ship and shore side on how sudden shifts in workload have been handled or finding examples for how complex situations were resolved can serve as a basis for learning from the positive and can, in turn, later inform the revision of procedures, as well as this can become a baseline for training developments.

METHODOLOGICAL PERSPECTIVE

The model of the NSA service was developed in close collaboration with the subject matter experts and based on what is currently developed within the joint research and innovation project. While this has been an essential input for the work, it needs to be acknowledged that the documentation, training, and concept itself are still under development and therefore might not present a complete picture of all risks and opportunities associated with the introduction of NSA. It is therefore to be considered that what is presented here corresponds to the state-of-the-art information available in early 2024. Depending on future developments, such as the implementation of a testbed area and further simulations, many of the challenges that the participants in this study experienced might be changed or resolved.

In addition to the documentation, the analysis has been based on four structured interviews conducted with the help of the ACTA framework. This helps to identify many important aspects of how variability is experienced during operations and how it might aggregate, but using a combination of HTA and ACTA also puts a focus on the perspective of the NSAP, not the BT. It might be a fact that the analysis based on interviews from the shipside may emphasize other challenges and create a vastly different functional model. Future studies should therefore focus on the balancing ship and shore, also considering using a similar approach of combining a structured interview methodology with functional modelling.

This chapter aims to explore arising complexities and systemic issues in the concept of Navigational Shore Assistance as currently developed in Sweden. The analysis has shown that the system design as is has created complex interactions across ship and shore-side that mainly depend on and are conducted through technologically mediated communication. The short decision frame to be able to produce an adequate and timely output in several core functions creates a challenge for BT and NSAP to be able to align their perceptions of the situation and agree upon a response, which makes recovery in the face of unanticipated events very challenging. Therefore, measures to support adaptability, monitoring and response capabilities within the current system should be explored further, such as reiterating the current FRAM model after living tests of the service have been conducted.

ACKNOWLEDGEMENTS

The authors would like to acknowledge that the work presented here has been funded by the Swedish Transport Administration within the "Navigationsstöd från Land" project. We would like to thank Johanna Larsson for her substantial support during the data collection and the initial analysis steps. We would also like to thank Anders Johannesson, Andreas Edvall, Erik Sandberg and the marine pilots who participated in the ACTA interviews for their support of our work. The second author would like to acknowledge support from the Shaping European Research Leaders for Marine Sustainability (SEAS) programme under the European Union's Horizon 2020 Framework for Research and Innovation, with the Marie Skłodowska-Curie agreement No. 101034309.

ACRONYMS

ACTA   Applied Cognitive Task Analysis
AIS   Automatic Identification System; a transponder that transmits certain static and dynamic information about a vessel (e.g., name, call sign, destination, draught, etc.)
BT   Bridge team in charge of the vessel functions
HTA   Hierarchical Task Analysis
MPEx   Master–Pilot Exchange
NSA   Navigational Shore Assistance service: a novel form of providing remote pilotage to merchant vessels

NSAP   Navigational Shore Assistance Pilot; a pilot providing NSA service from a
shore centre
SMA   Swedish Maritime Authority providing pilotage services in Swedish waters
SME   Subject matter expert
VHF   Very High Frequency radio that is used for communicating between ships
and ship and shore. All communication on the VHF is public
VTS   Vessel Traffic Service; information services in the port approach area
WOP   Wheel Over Position/Point

REFERENCES

AP. (2024). Baltimore Key Bridge Collapse. https://apnews.com/article/baltimore-bridge-collapse-53169b379820032f832de4016c655d1b. Date accessed: April 20th, 2024.

Berlin, C., & Praetorius, G. (2023). Applied Cognitive Task Analysis (ACTA) of marine piloting in a Swedish Context. *Human Factors in Transportation*, 95, 709–718.

Bowditch, N. (2012). *The American Practical Navigator. An Epitome of Navigation. Originally by Nathaniel Bowditch (1802) Volume 1*. Compiled and Published by National Geospatial Intelligence Agency.

Hadley, M. (1999). Issues in Remote Pilotage. *The Journal of Navigation*, 52(1), 1–10.

Hollnagel, E. (2012). *FRAM: The Functional Resonance Analysis Method - Modelling Complex Socio-technical Systems*. Ashgate Publishing.

ICS. (2024). Shipping and World Trade: Driving Prosperity. https://www.ics-shipping.org/shipping-fact/shipping-and-world-trade-driving-prosperity/

Kirwan, B., & Ainsworth, L. K. (1992). *A Guide To Task Analysis: The Task Analysis Working Group*. Taylor & Francis. http://books.google.se/books?id=BfXl8Y8qLz8C

Koester, T., Anderson, M. & Steenberg, C. (2007). Decision Support for Navigation. FORCE Technology, Draft Report. DMI 107-27358.

Lützhöft, M., & Bruno, K. (2009). Talk and Trust Before Technology: First Steps Toward Shore-Based Pilotage. In *RINA HF in Ship Design and Operation Conference* (pp. 25–26).

Militello, L. G., & Hutton, R. J. B. (1998). Applied Cognitive Task Analysis (ACTA): A Practioner's Toolkit for Understanding Cognitive Task Demands. *Ergonomics*, 41(11), 1618–1641.

Orasanu, J., & Connolly, T. (1993). The Reinvention of Decision Making. In G. A. Klein, J. Orasanu, R. Calderwood, & C. E. Zsambok (Eds.), *Decision Making in Action: Models and Methods*. Ablex.

Patton, M. Q. (2002). *Qualitative Research & Evaluation Methods*. Sage Publication.

Sakar, C., & Sokukcu, M. (2023). Dynamic Analysis of Pilot Transfer Accidents. *Ocean Engineering*, 287, 115823.

Schröder-Hinrichs, J. U., Hollnagel, E., & Baldauf, M. (2012). From Titanic to Costa Concordia—A Century of Lessons Not Learned. *WMU Journal of Maritime Affairs*, 11, 151–167.

Sharma, A., Nazir, S., & Ernstsen, J. (2019). Situation Awareness Information Requirements for Maritime Navigation: A Goal Directed Task Analysis. *Safety Science*, 120, 745–752.

Stanton, N. A. (2006). Hierarchical Task Analysis: Developments, Applications, and Extensions. *Applied Ergonomics*, 37(1), 55–79.

Stanton, N. A., Salmon, P. M., Walker, G. H., Baber, C., & Jenkins, D. P. (2005). *Human Factors Methods. A Practical Guide for Engineering and Design*. Ashgate.

Swedish Maritime Administration. (2024). Navigational Shore Assistance SEGOT. https://www.sjofartsverket.se/nsagot Date accessed: s.

UNCTAD. (2021). *Review of Maritime Transport 2021*. UN.

UNCTAD (2023). Review of Maritime Transport 2023. Towards a Green and Just Transition. United Nations publication. Sales No. E.23.II.D.23. Geneva.

Vairo, T., Quagliati, M., Del Giudice, T., Barbucci, A., & Fabiano, B. (2017). From Land-To Water-Use-Planning: A Consequence Based Case-Study Related to Cruise Ship Risk. *Safety Science*, 97, 120–133.

Van Westrenen F. (1999). The Maritime Pilot at Work: Evaluation and Use of a Time–To–Boundary Model of Mental Workload in Human–Machine Systems. Ph.D. thesis, Delft University of Technology, Delft, The Netherlands.

Van Westrenen, F., & Praetorius, G. (2014). Maritime Traffic Management: A Need for Central Coordination? *Cognition, Technology & Work*, 16, 59–70.

10 Fuzzy Assessment of Variability in FRAM Models

Ivenio Teixeira de Souza, Assed Naked Haddad, and Riccardo Patriarca

CONTEXT: THE CONSTRUCTION OF TAILING DAMS

The FRAM was initially developed as an essentially qualitative approach, including linguistic assessment of variability outcomes. Arguably, the FRAM evaluation involves parameters with a high degree of uncertainty due to the vagueness of human knowledge and the inherent variability of operations.

The variability in a single function and among functions in an FRAM model is usually handpicked from a list of values, and this setting has been reported as a factor that may lead to low efficiency and poor thoroughness in the analysis (Duan et al., 2015). Needless to say, the quality of a FRAM analysis directly depends on experts' opinions and on the information they provide for each function (Salehi et al., 2020). As expected, the adoption of the FRAM in earlier studies has shown limitations in coping with this type of opinion (Patriarca et al., 2020).

The Fuzzy Sets Theory (FST) represents one of the strategies used in this sense, as it helps minimizing the arbitrariness or subjectivity arising from the expert's judgements (Hirose and Sawaragi, 2020; Slim and Nadeau, 2020). Combining FRAM modelling with FST is a challenging task, as explored in earlier studies by Bellini et al. (2020), Slim and Nadeau (2020), and De Souza et al. (2022).

This chapter builds on these concepts: we aim to showcase how the FRAM can be used to identify and prioritize critical aspects of socio-technical systems through quantitative analysis, especially when FST is employed to synthesize expert opinions, representing these judgements as fuzzy sets to assess phenotypes of variability.

As a hands-on case, we examine a case study involving an emergency response to the collapse of a mine tailings dam. Tailings dams are earth-filled embankments used to store by-product waste materials from mining operations, which are composed of ground rocks, sand, chemical reagents, and processed water (Adiansyah et al., 2015; Islam and Murakami, 2021). Typically, tailings dams are constructed upstream, downstream, or in the centreline method. Upstream raising is the least expensive but also the least safe.

DOI: 10.1201/9781003518167-13

Domain Expert

This method of construction is associated with the highest number of incidents, representing 76% of the cases; downstream and centreline raised tailings dams represented 15% and 5% of the global cases, respectively (Rico et al., 2008). The storage of tailings from mining operations is particularly prone to emergencies, mainly due to structural collapses resulting from natural phenomena such as slope instability, earthquakes, dam overtopping, seepage, foundation issues, structural issues, and mine subsidence, among others (Rose et al., 2023). Despite their low frequency, dam failures are not rare, and significant mining dam disasters have occurred in recent years. To name a few: the Williamson Mine in Tanzania and the Jagersfontein Mine in South Africa in 2022; the Wenquan Township dam in Shanxi Province, China, in 2022; the Yedikardes village dam in Giresun, Turkey, in 2021; the Tieli dam failure in China, in 2020; the Córrego do Feijão Mine in Brumadinho, Brazil, in 2019; the Cadia Mine in Australia in 2018; and the Fundão Mine in Mariana, Brazil, in 2015.

Among other factors, the rupture of mining dams has been attributed to operational negligence and a lack of storage control inspections (Noraishah Ismail et al., 2021). Are these really the only reasons, or do they represent signs of more sophisticated systemic hazards? Our idea is to tackle emergency management systems as a unit of analysis and see if losses incurred from these accidents could be significantly reduced by engineering more resilient emergency responses.

THE DAM I IN BRUMADINHO

We will iterate the assessment on the tragic case of Dam I at the Corrego do Feijao iron mine in Brumadinho, Brazil. Dam I was a tailings dam that began its construction in 1976 and was operated directly by Vale S.A. since 2001, but the dam stopped receiving tailings in 2016 and had been equipped with remote monitoring instrumentation. It was designed and constructed in the upstream raising method, and the waste generated by the mining operation was periodically accumulated through successive uphill deposition until the dam was eventually decommissioned in 2016. On 25th January 2019, dam suffered a catastrophic failure, releasing a mudflow of approximately 9.7 million cubic meters in a 10-metre wave. The accident killed over 270 people, most of the victims were Vale's employees, and severely contaminated the Paraopeba river basin (Buch et al., 2024).

Domain Expert

Reports indicated that low-amplitude ground vibrations were detected by sensors before the first signals of rupture were perceived on the dam. The rupture started near the central part of the dam, followed by a noticeable leakage at the foot of the dam, and finally the complete failure of the dam, which released the tailings sludge (Rose et al., 2023). The tailings wave then quickly reached the mining facility, destroying the administrative centre, dining hall, maintenance office, railway network, a hotel, several rural properties, a railway bridge, and 132 railway cars (Porsani et al., 2019; Rose et al., 2023).

TABLE 10.1

Profiles of the Experts Involved in the Analysis

Expert	Age	Specialty Area	Education Level	Professional Experience
Expert 1	45	Safety Engineer	Ph.D.	17
Expert 2	36	Environmental Engineer	M.Sc.	12
Expert 3	57	Industrial Engineer	Ph.D.	28

DATA COLLECTION

Data collection was based mainly on document analysis, such as the official accident report by an independent consulting committee (CIAEA, 2020), as well as recent scientific studies on the accident (Buch et al., 2024; Cheng et al., 2021; Porsani et al., 2019; Rose et al., 2023). Such documents were analyzed, aiming to understand the accident dynamics and how an emergency plan is currently structured.

Additional data were obtained through discussions with three experts in safety management systems and emergency response, whose profiles are summarized in Table 10.1. Three rounds of online semi-structured discussions were conducted for the analysis, each lasting ca. 1 hour.

The first round of discussions was arranged to confirm the list of functions involved in the dam collapse emergency response process and revise the FRAM model being developed.

In the second meeting, the emphasis was on discussing the functional variability. Each expert was responsible for judging the variability and dampening capacity (DC) in each identified function and sending the results via email. Each one performed their analysis independently of each other, and they used the same data sources to support their analysis. In turn, the first author of this article was responsible for organizing and computing the gathered data.

The third meeting focused on validating the results obtained after the computation.

METHODOLOGY

This section details the novel integration of FST in FRAM models to refine and systematize the assessment of variability by means of systems experts. However, before accessing the technicalities of FST, we would like to invite you to reflect on a metaphorical representation of the couplings, as seen by the FRAM, which is useful for representing the fuzzy steps we are looking at.

CAN A COUPLING BE INTERPRETED AS A MASS–SPRING–DAMPER SYSTEM?

Modern socio-technical systems present dynamic and complex behaviour arising from non-linear couplings. The notion of "coupling" refers to the phenomenon that two or more elements of a system interact with each other closely in various ways.

From the perspective of the Normal Accident Theory (NAT), as proposed by Perrow (1984), the systems present interactions or couplings that are classified as loose or tight. In systems presenting loosely coupled interactions, events can occur independently of each other. Conversely, tight couplings mean that different elements in the system are highly dependent on each other, and an event in one element of the system can easily propagate throughout the system. Furthermore, interconnected elements are unstable and subject to some disturbing forces, leading the system to vary.

These interactions are hard to model, evaluate, and predict. But that's exactly where the FRAM challenges the *status quo* of existing modelling approaches.

The FRAM is concerned with showing the interactions among functions by modelling nonlinear dependencies, looking at their functional aspects, and defining the system variability. Performance variability usually emerges from these interactions, influenced by technical, human, and organizational aspects. This variability is intrinsic to the system and plays a pivotal role in which people alone or collectively adjust their performance in response to situational demands or disturbances to achieve a system's goal (Hollnagel, 2015).

We argue in this chapter on a novel analogy: assimilating couplings to a harmonic oscillator. Generally, this system consists of three parts, which separately are mass, spring, and damper that store kinetic and potential energy. The springer/damper represents the coupling variability, as relevant among two masses captured by the functions.

Method Geek

As illustrated in Figure 10.1, an FRAM model can be represented by the logical relationship and implications of a mass–spring–damper system, in which the mass refers to the functions i and j, which are activities performed to achieve a certain goal. The spring K represents the capacity of an agent to vary its actions, i.e., the output variability in the function i, whereas the dashpot C represents the capacity of a function j in dampening the same oscillations allowed by the spring K.

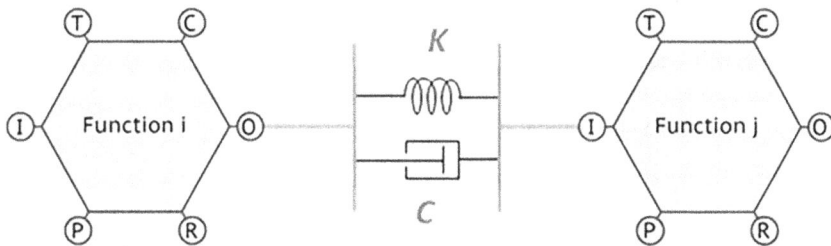

FIGURE 10.1 Metaphorical representation of a FRAM coupling as a mass–spring–damper system.

The metaphor of the mass–spring–damper can also be reflected in the way variability is managed. The variability can be a result of the variability of the function itself (internal or endogenous variability), it can be due to the variability of the working environment (external or exogenous variability), and it can finally be a result of influences from upstream functions (upstream-downstream coupling) (Hollnagel, 2012). Referring to the FRAM building steps, once agreed on a scope (Step 0) and the relevant functions of the system (Step 1) have been described, Step 2 concerns the identification of how variability manifests (the phenotypes). This step can be performed by using either a simple or a more elaborate solution, depending on the number of phenotypes being investigated, theoretically infinite (cf. notion of metadata).

One of the main issues in defining separately the Output variability lies in the complexity of predicting the behaviour throughout the system. For example, a generic upstream function that presents Output variability "too late" may cause a loss of time in the downstream function, amplifying the variability. On the other hand, in terms of precision, the same Output may dampen the variability, if "precise". Therefore, we trust that integrating the phenotypes of variability presents a feasible solution for the analysts. This approach enables a structured and reliable way to assess and address the dynamic changes within the system. Accordingly, we show here an indicator called Magnitude of Variability (MV), which comprises the two phenotypes of the simple solution (i.e., timing and precision), as proposed by Hollnagel (2012), yet it integrates them to define an overall variability level.

But MV alone does not fully capture the actual behaviour of variability. Indeed, as the variability from upstream functions produces a response in upstream–downstream couplings, it should be possible to create an index based on the coupling variability to identify highly connected couplings and then potential criticalities. This perspective reflects the idea that every downstream function presents a DC to deal with the MV from upstream functions. Thus, the Coupling Response Index $\left(\text{CRI}_{ij}\right)$ is proposed as a novel indicator to examine how system performance affects and is affected by the coupling variability. CRI_{ij} expresses the response in pairwise coupling and translates the effect of variability in the upstream i-th function, dampened or amplified by the downstream j-th function. Since system resilience is achieved by controlling variability rather than by constraining it. Therefore, as the CRI is mostly related to how the downstream function deals with the upstream function's variability and the level of variability that the downstream function can handle (Kim and Yoon, 2021), it becomes evident that high dampening is desirable to attain a low chance of propagating variability for other functions, whereas low dampening is desirable for amplifying positive effects, enhancing its occurrence. As pointed out by Hollnagel (2012), the dampening not only serves the purpose of preventing things from going wrong but also contributes to things going right.

In our metaphorical case, MV reflects the spring pulses, while CRI shows the dashpot features, both dynamically represented. The results of this conversion can support identifying the potential threats and opportunities that need more attention in each coupling.

The CRI, such as the MV, is computed by using rule-based knowledge that emulates human assessment, thereby formalizing and structuring it without overly simplifying the process. Unlike methods that rely on closed analytical equations, our proposal rejects oversimplifications, rather offering a non-invasive way to systematically translate and codify human judgements. This effort underpins the development of a fuzzy rule-based FRAM, justified by the objective to accurately capture the nuances and uncertainties inherent in human judgement. In this way, we aim to pair the holistic modelling that can be achieved through FRAM functions with a comprehensive assessment that can be achieved through FST.

Method Geek

In mechanical systems, for example, with only a mass and a spring, there is no mechanism by which the vibrational energy can be dissipated. Thus, a damper element is added to dissipate a portion of that energy, thereby reducing the amplitude of the oscillations, until the system comes to equilibrium. As previously illustrated in Figure 10.1, the damper element is represented, for instance, by a dashpot C, which is typically envisioned as consisting of a cylinder filled with a viscous fluid in which the motion of the mass element is resisted by viscous drag. Such systems present a damping ratio, symbolized as ζ (zeta), which is a percentage of the damping for which the system returns to equilibrium as quickly as possible. Similarly, in FRAM models, the damper plays the role of attenuating undesirable effects resulting from uncontrolled variability, or even amplifying the positive variability. The DC of a function is defined as the capability of this function in a certain context to absorb the magnitude of variability (MV) and somehow provide acceptable outcomes (Bellini et al., 2017). In other words, the DC (later reflected in the value of the CRI) is emulating ζ. As such, it is driven to reduce oscillation amplitudes to maintain the system under control since it is associated with the energy dissipation that occurs inside the downstream function due to resilience properties like, for instance, the buffering capacity, flexibility, margin, tolerance, and cross-scale interactions, as identified by Woods (2006).

Fuzzy Sets Theory (FST)

Fuzzy set theory (FST) (Zadeh, 1965) is a mathematical approach that deals with uncertainty. It is widely used for computing the natural language of humans to resolve the ambiguity of the outputs and to present more comprehensible results. Unlike axioms of classical sets, where elements are either in or out of a set, the theory defines a fuzzy set as a set that allows partial membership of elements. This means an item can somewhat belong to a set, much like being only partly inside a circle and partly inside another circle. This flexibility helps deal with situations where things are not just black or white but have various shades of grey.

It allows us to quantify and work with degrees of truth (called "membership"), rather than forcing every variability assessment into rigid, binary categories. This approach is particularly useful in complex decision-making processes where nuances matter and simple yes/no answers are insufficient.

In more technical terms, given a fuzzy number denoted by $\tilde{N}$ as a fuzzy sub-set of real numbers (R), and its membership function (MF) being defined as $\mu_{\tilde{A}}(x): R \rightarrow [0,1]$, where $\mu_{\tilde{A}}(x)$ is the membership function for fuzzy set A. For example, a triangular fuzzy number $\tilde{N}(l,m,u)$, where l, m, and u were denoted as the smallest membership value, the most possible value, and the greatest membership value, respectively. The membership grade $\mu_{\tilde{A}}(x)$ is defined as follows:

$$\mu_{\tilde{A}}(x) = \begin{cases} \dfrac{x-l}{m-l}, & \text{if } l \leq x \leq m; \\ \dfrac{u-x}{u-m}, & \text{if } m \leq x \leq u; \\ 0, & \text{otherwise} \end{cases} \tag{10.1}$$

AGGREGATING EXPERT OPINIONS

The aggregating process is crucial to the application of FST, and it implies employing a transformation function to unify the multi-granularity of opin-ions. There are several available techniques for aggregating expert opin-ions, such as voting, similarity aggregation method (SAM), game theory, fuzzy Delphi method, max–min Delphi method, and Technique for Order of Preference by Similarity to Ideal Solution (TOPSIS) (Yin et al., 2020). The chosen technique for demonstrative purposes was the SAM (Hsu and Chen, 1996) for aggregating the expert opinions. SAM is a simple method using the index of consensus among experts to aggregate individual fuzzy opinions. Thus, the agreement degree (AD) between expert E_i and expert E_j can be esti-mated through Eq. (10.2).

$$AD = \dfrac{\int_x \left(\min\{\mu_{Ni}(x), \mu_{Nj}(x)\} \right) dx}{\int_x \left(\max\{\mu_{Ni}(x), \mu_{Nj}(x)\} \right) dx} \tag{10.2}$$

Afterwards, an agreement matrix (AM) of pairwise agreement among experts is given by the matrix in Eq. (10.3). The diagonal elements of the matrix are all 1 because it is assumed that no expert influences itself.

$$AM = \begin{bmatrix} 1 & a_{12} & \cdots & a_{1n} \\ a_{21} & 1 & \cdots & a_{2n} \\ \vdots & \vdots & \cdots & \vdots \\ a_{n1} & a_{n2} & \cdots & 1 \end{bmatrix} \tag{10.3}$$

The relative agreement (RA) of an expert E_i ($i = 1, 2, \ldots n$) can be estimated as shown in Eq. 10.4

$$RA_{E_i} = \left(\frac{1}{n-1} \sum_{j=1}^{n} \left(RA_{ij} \right)^2 \right)^{1/2} \tag{10.4}$$

For each linguistic variable, a fuzzy number $\tilde{N}$ can be obtained as in Eq. 10.5.

$$\tilde{N} = \sum_{i=1}^{n} \left(\frac{RA_{E_i}}{RA_{E_i} + \ldots + RA_{E_n}} \right) . \tilde{n}_i \tag{10.5}$$

SETTING A FUZZY INFERENCE SYSTEM (FIS)

A FIS comprises a formulating process and a mapping from a set of inputs to an output by using fuzzy sets. The Mamdani model is usually used to describe complex systems in terms of simple and natural language. The steps on how to construct a fuzzy inference system will not be discussed here in depth; rather, only the aspects concerning the operation of the FIS will be explained. For further details on the design and construction of the complete fuzzy-FRAM architecture, we invite the interested reader to go through the linked manuscript by De Souza et al. (2022).

The Mamdani fuzzy system consists of four stages: fuzzification, knowledge base, inference engine, and defuzzification (Geramian and Abraham, 2021).

First, the fuzzification stage is an interface that converts the values from the real system into fuzzy linguistic terms. Furthermore, at this stage, the variables to be measured and MFs are defined. The degree of membership specifies how closely a value matches the description of that variable. The range of crisp values is converted into fuzzy numbers to compute the FIS.

Second, the construction of fuzzy rules is crucial for computing an FIS, and they can be defined as a conditional statement that characterizes the relationships between the inputs and the outputs in the form IF–THEN, as illustrated below.

$$R = \text{If} \begin{cases} X_{1,1} = & \ldots & \text{AND} & X_{2,1} = & \cdots & \text{AND} & X_{3,1} = & \ldots \\ X_{1,2} = & \ldots & \text{AND} & X_{2,2} = & \ldots & \text{AND} & X_{3,2} = & \ldots \\ \vdots & \vdots & \vdots & \vdots & \vdots & \vdots & \vdots & \vdots \\ X_{1,k} = & \ldots & \text{AND} & X_{2,k} = & \cdots & \text{AND} & X_{3,k} = & \ldots \end{cases}$$

Then Output is O_k

In this example (X_1, X_2, and X_3) are linguistic variables as well as inputs of the FIS. Also, R is the number of rules to compute the inputs to produce an output, whereas O is the eligible linguistic label for the output in the k-th rule of the FIS.

Generally, the number of rules directly depends on the number of input variables (n) and the number of linguistic labels (A_i) that these inputs concern. The number of rules in the inference system is calculated as in Eq. (10.6):

$$R = \prod_{i=1}^{n} A_i \qquad (10.6)$$

Third, in the inference engine, the rules are evaluated through, for example, a Product or Minimum method. Afterwards, the evaluation results for all rules are aggregated, for example, via a Maximum or Sum method.

Fourth, and finally, the defuzzification process is introduced at this stage to convert the fuzzy set result obtained by the inference engine into a crisp value. Among the many methods that have been proposed in the literature in recent years, this study used the centroid technique (Yager and Filev, 1993) because it is maybe the most popular and reflects reality quite accurately, and is given by Eq. (10.7):

$$x^* = \frac{\int \mu_{\tilde{a}}(x).x dx}{\int \mu_{\tilde{a}}(x) dx} \qquad (10.7)$$

This fuzzy-driven assessment extends and specifies the traditional steps of the FRAM, as sketched in Figure 10.2. We aim to clarify that FST specializes in Step 2 and Step 3 for evaluating and prioritizing relevant functions.

STEP 0: RECOGNIZE THE PURPOSE OF THE ANALYSIS

This step aims to define the purpose of the analysis and the boundaries of the systems under study using the FRAM.

STEP 1: IDENTIFY AND DESCRIBE THE SYSTEM'S FUNCTIONS

This step comprises the identification of relevant functions representing the socio-technical system. Information and data are collected to construct a model based on the FRAM principles.

STEP 2: IDENTIFY VARIABILITY

It is at this stage that the integration of FST comes to enhance the traditional FRAM. For this purpose, the MV is proposed as a novel indicator to consistently define the output variability. The MV represents the overall variability from an upstream function i, and it is used to understand the implications of the coupling variabilities of one entire upstream function to the associated downstream function. To make this step more pragmatic, some sub-steps are necessary.

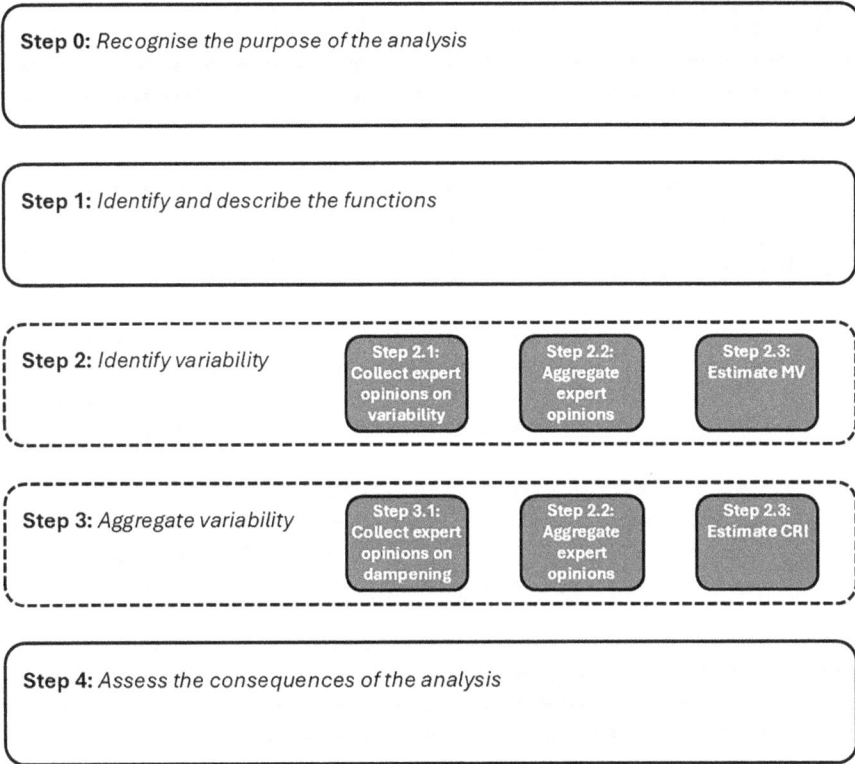

FIGURE 10.2 Framework of the fuzzy FRAM assessment: the FRAM basic steps are represented in rectangles with bolded lines, while the steps specified by FST are depicted in dotted line rectangles.

Method Geek

SUB-STEP 2.1: COLLECT EXPERT OPINIONS ON VARIABILITY

Experts are asked to specify the variability in terms of timing (too late, too early, on time, or not at all) and precision (precise, imprecise, wrong, and acceptable).

SUB-STEP 2.2: AGGREGATE EXPERT OPINIONS ON THE VARIABILITY

Consequently, the opinions arising from experts are aggregated following Eqs. (10.2)–(10.4). To quantify each evaluation, fuzzy scales are associated with each linguistic label, as indicated in Table 10.2. Then, a fuzzy number is obtained by using Eq. (10.5).

SUB-STEP 2.3 – ESTIMATE MV

The MV_i is a tuple of the form $MV_i = \left(V_i^T, V_i^P \right)$ which is obtained by computing the first FIS. At the same time, a set of fuzzy rules should be constructed to compute the FIS accordingly.

TABLE 10.2

Linguistic Labels, Descriptions, and Fuzzy Scales Adopted for the Fuzzy System

	Description	Fuzzy Scale
	Input $\left(V_i^T \right)$	
Not at all (NA)	It represents the possibility that an output is too late to be used for its purposes, or even not produced at all	(0, 0, 1)
Too late (TL)	The function is executed with insufficient speed or presents delayed output	(0, 1, 2)
Too early (TE)	The function is executed faster than required or exhibits premature output	(1, 2, 3)
On time (OT)	The function is executed in an acceptable timeframe	(2, 3, 3)
	Input $\left(V_i^P \right)$	
Wrong (WR)	It requires the downstream function improvisation, which usually amplifies the function variability	(0, 0, 1)
Imprecise (IM)	The output function is incomplete, inaccurate, ambiguous, or in other ways misleading	(0, 1, 2)
Acceptable (AC)	It requires some adjustment in the downstream function	(1, 2, 3)
Precise (PR)	It satisfies entirely the requirements of the downstream function	(2, 3, 3)
	Output $\left(MV_i \right)$	
Very low (VL)	Function's timing and precision have a negligible effect on how downstream functions are performed	(1, 1, 3.25)
Low (LO)	Function's timing and precision have a limited effect on how downstream functions are performed	(1, 3.25, 5.5)
Medium (ME)	Function's timing and precision have a potential effect on how downstream functions are performed	(3.25, 5.5, 7.75)
High (HI)	Function's timing and precision have a serious effect on how downstream functions are performed	(5.5, 7.75, 10)
Very high (VH)	Function's timing and precision have a huge effect on how downstream functions are performed	(7.75, 10, 10)
	Input $\left(DC_j^P, DC_j^T \right)$	
Very low (VL)	The function has a negligible capacity for dampening disturbances from the upstream function	(1, 1, 3.25)
Low (LO)	The function has a limited capacity for dampening disturbances from the upstream function	(1, 3.25, 5.5)

(Continued)

TABLE 10.2 (Continued)
Linguistic Labels, Descriptions, and Fuzzy Scales Adopted for the Fuzzy System

	Description	Fuzzy Scale
Medium (ME)	The function has an occasional capacity for dampening disturbances from the upstream function	(3.25, 5.5, 7.75)
High (HI)	The function dampens most disturbances from the upstream function	(5.5, 7.75, 10)
Very high (VH)	The function neutralizes all disturbances from the upstream function	(7.75, 10, 10)
	Output $\left(\mathrm{CRI}_{ij}\right)$	
High dampening (HD)	The coupling's response is characterized by a slow approach to the equilibrium, and the magnitude of variability decreases over time	(0, 0, 1)
Low dampening (LD)	The coupling's response is characterized by exponential decay of the magnitude of variability	(0, 1, 2)
No-effect (NE)	The coupling's response is neutral, and the effects of variability can be neglected	(1, 2, 3)
Slight amplifying (SA)	The coupling's response is characterized by a slight amplification of the magnitude of variability, but the functional resonance is less pronounced	(2, 3, 4)
High amplifying (HA)	The coupling's response is maximized, and the system can be driven to the functional resonance	(3, 4, 5)

Note: Values can be adapted for the specific case at hand.

STEP 3: AGGREGATE VARIABILITY

Through this step, we can finally evaluate the upstream-downstream coupling effects. First, the DC is evaluated to understand the behaviour of a downstream function when faced with its incoming coupling variability. Then the combination of MV and DC, through specific rules, allows setting the CRI to fully understand the implications of a coupling.

Method Geek

SUB-STEP 3.1: COLLECT EXPERT OPINIONS ON DC

In this sub-step, a collection of data is required to identify the DC of each function under analysis. Accordingly, the experts should evaluate the level of dampening in terms of timing and precision. Specifically, the DC is evaluated based on the five attributes of dampening, as listed in Table 10.2.

SUB-STEP 3.2: AGGREGATE OF EXPERT OPINIONS ON DC

The opinions arising from experts are aggregated following Eqs. (10.2)–(10.4). For each evaluation, fuzzy scales are associated with every linguistic label, as indicated in Table 10.2. Then, a fuzzy number is obtained by using Eq. (10.5).

SUB-STEP 3.3 – ESTIMATE CRI

The CRI_{ij} is a tuple of the form $\text{CRI}_{ij} = \left(\text{MV}_i, \text{DC}_j^P, \text{DC}_j^T \right)$, which is obtained by computing a second FIS. Here, CRI_{ij} represents the coupling response index. DC_j^T represents the dampening capacity of a downstream j-th function in terms of timing, and DC_j^P represents the dampening capacity of a downstream j-th function in terms of precision. The obtained result for CRI_{ij} is thus a metric ranging between 0 and 5, and a set of linguistic labels is defined for the output variable of FIS, as indicated in Table 10.2. At the same time, a set of fuzzy rules should be constructed to compute the FIS accordingly.

STEP 4: ASSESS THE CONSEQUENCES OF THE ANALYSIS

To help decision-makers in suggesting actions for controlling the variability, this approach supports the identification of what is perceived as the most critical functions. For this purpose, a cumulative CRI_i is obtained for each function, aiming to define the priority of intervention. In this sense, a greater CRI_i value indicates higher chances for functional resonance in the system.

Method Geek

The cumulative CRI_i is obtained by considering the number of downstream couplings, where CRI_{ij} is the value of the k-th CRI among two functions F_i and F_j, as indicated in Eq. (10.8).

$$\text{CRI}_i = \sum_{i \neq j}^{k} \text{CRI}_{ij} \qquad (10.8)$$

RESULTS

In this section, we present a walkthrough application to illustrate the proposed methodology. An FRAM model was constructed inspired by the emergency response to dam failure, such as the collapse of Dam I of the Córrego do Feijão iron mine in Brumadinho, Brazil.

The FRAM was constructed using FRAM Model Visualizer (FMV) software. For the inference, the MATLAB Fuzzy Logic Toolbox was used because it provides user-friendly graphical interfaces and reliable environments for structuring and computing fuzzy systems. The details of the implementation are presented onward.

RESULTS OF STEP 0: THE PURPOSE

The scope of the analysis is limited to the response activities in case of emergency, such as those generated by an accident. It does not encompass, therefore, functions related to prevention and recovery activities, which we considered out of scope for brevity. Moreover, the system is described with a small set of macro-level functions aiming to highlight the main couplings for applying the methodology. These assumptions do not affect the validity of the method and help to simply demonstrate the utility of the method.

RESULTS OF STEP 1: THE FUNCTIONS

The first step consists of identifying the relevant functions, describing them, and defining how they interact. To identify the specific functions and build a FRAM model, information and data available in the documents presented in the data collection section were used. In addition, three safety experts were consulted during the identification and description of functions accurately. As a result, a total of 15 essential functions were identified, numbered from F1 to F15, as shown in Figure 10.3). Of these 15 functions that built the FRAM model, 12 are foreground functions, whereas only three are characterized as background functions, including <Provide protocols of emergency> (F1), <Provide instruments for monitoring> (F3); and <Provide training of the response team> (F4). In this case, the background functions are those responsible for supplying the resources and work conditions necessary to monitor and respond to disturbances on dam operation. The foreground functions will be the focus of the analysis and can produce variable outputs. Among the foreground functions, nine functions present variability, and they are analyzed by the experts, supported by the FST to support their judgement for the variability analysis of outputs. These functions are, namely, the following (cf. Figure 10.3):

- <Conduct geotechnical operations> – F2
- <Evaluate dam collapse> – F5
- <Emergency action plan coordination> – F8
- <Operate monitoring center> – F6
- <Execute emergency action plan> – F9
- <Identify and prepare resources> – F13
- <Mobilize operational team> – F14
- <Allocate resources> – F10
- <Activate alert system> – F7

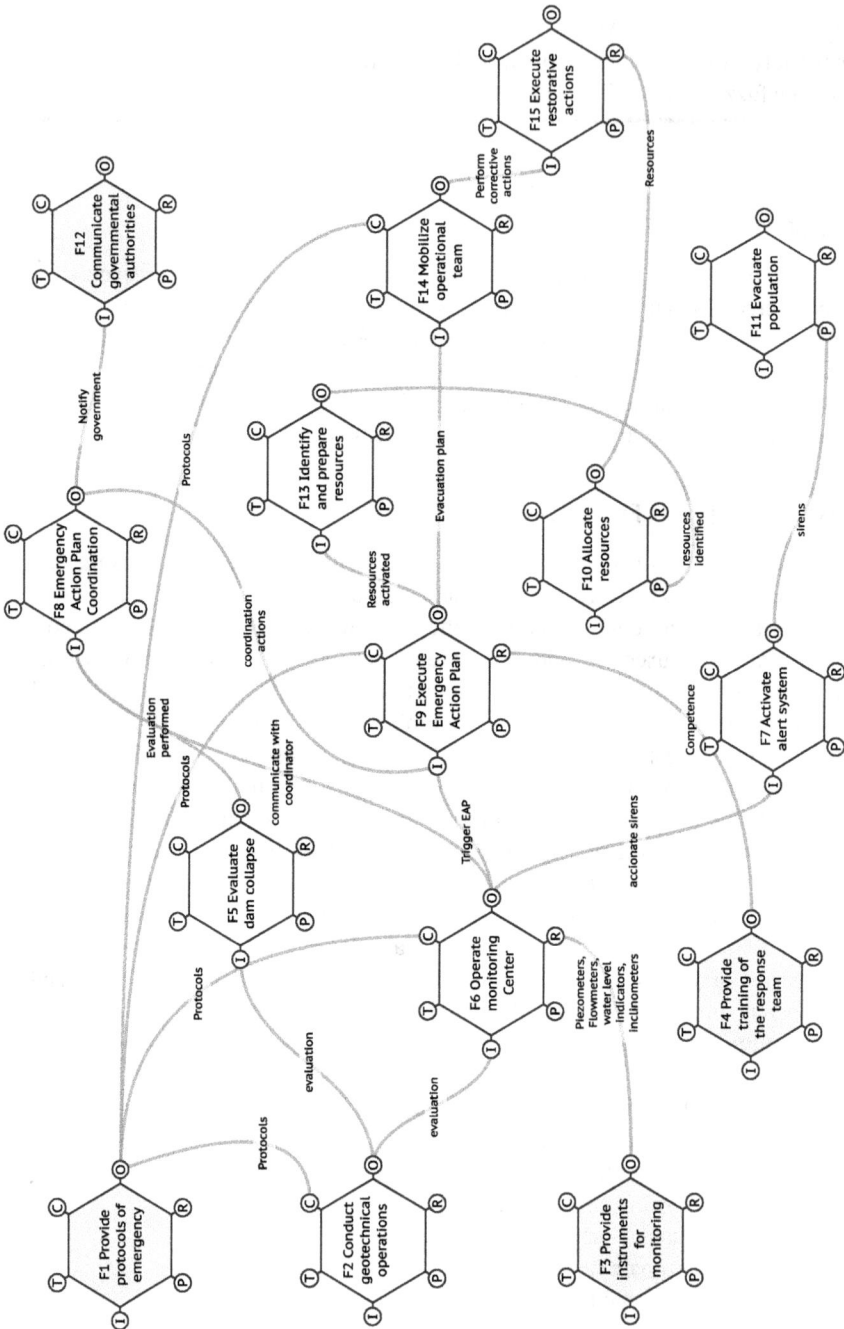

FIGURE 10.3 FRAM model for the emergency response during the rupture of tailings dam.

More specifically, when looking at the model, it is possible to distinguish actual work practices. <Provide protocols of emergency> (F1) comprises a set of instructions gathered in a document that specify relevant actions, roles, and actors to cope with emergency events. In the case of this study, the Emergency Action Plan structure can be regarded as work-as-imagined, which may be slightly different from the work-as-done adopted and performed by companies.

The response phase starts with the function <Evaluate dam collapse> (F5), which involves detecting and reporting the incidents, based on the input from F2, which provides information on hazard identification and risk assessment. Additionally, <Conduct geotechnical operations> (F2) feeds the <Operate monitoring center> (F6) with valuable data to support the team in making decisions. In turn, F6 serves as the input for other functions and activities. For example, the <Activate alert system> (F7), is driven to advise the population and workers to evacuate the area in the incident. Additionally, function F8 represents actions from the emergency coordinator on the incident. <Execute Emergency Action Plan> (F9) is then carried out with input from the plan defined by F8 and information made available by F6. Therefore, F9 triggers the identification and preparation of resources (F13), outlining the resources required to address and tackle the consequences of the event. Subsequently, required resources are allocated (F10) with a precondition provided by F13. This implies, for instance, that imprecise evaluation or even delay in defining necessary resources could negatively impact resource provision in the incident scenario. Consequently, the restorative actions (F15) are affected by these shortcomings. The function <Mobilize operational team> (F14) serves as input for F15 to start the operations for rescuing people affected by the incident.

In this case study, the function <Communicate governmental authorities> was not investigated because it constitutes an external boundary of the system, and therefore, it can be neglected in this demonstrative analysis. In turn, F11 and F15 were evaluated only on the capacity to dampen the magnitude of variability from their respective upstream functions.

RESULTS OF STEP 2: THE FUZZY VARIABILITY

This step firstly aimed at collecting expert opinions for describing functional variability. The 9 functions presenting variability have been qualitatively evaluated in terms of the two phenotypes' timing (on time, too early, too late, or not at all) and precision (precise, acceptable, imprecise, and wrong). A list of functions presenting variability was submitted via email to each expert asking for their opinion on the variability.

Afterwards, the individual responses to each function have been aggregated using Eqs. (10.2)–(10.4), and subsequently an aggregated fuzzy number was generated for each variability by using Eq. (10.5), based on associated fuzzy scales to each term that characterizes the variability, as previously shown in Table 10.2. Next, the data gathered were used to run the first FIS for calculating the MV_i for each function under analysis.

Table 10.3 summarizes the entries received by the experts, as tabulated by the first author, while Table 10.4 shows the results of the assessment.

TABLE 10.3
Qualitative Evaluations Expressed by the Experts for the Variability

Expert	Phenotypes	Functions								
		F2	F5	F6	F7	F8	F9	F10	F13	F14
E_1		OT	OT	TL	NA	TL	TL	OT	OT	TL
E_2	V_i^T	TL	TE	TL	NA	TL	NA	OT	TL	NA
E_3		TL	OT	OT	NA	NA	TL	OT	TL	OT
E_1		IM	IM	IM	IM	WR	IM	AC	AC	IM
E_2	V_i^P	WR	IM	IM	IM	WR	WR	AC	PR	AC
E_3		AC	AC	WR	IM	IM	WR	AC	IM	IM

TABLE 10.4
Results for the Magnitude of Variability (MV$_i$)

Function	Input, $\tilde{N}$		Output
	Timing, V_i^T	Precision, V_i^P	MV_i
F2	(0.522, 1.522, 2.261)	(0.330, 0.997, 1.997)	7.75
F5	(1.733, 2.733, 3.000)	(0.264, 1.264, 2.264)	5.51
F6	(0.522, 1.522, 2.261)	(0, 0.733, 1.733)	7.86
F7	(0, 0, 1)	(0, 0, 1)	9.28
F8	(0, 0.733, 1.733)	(0, 0.267, 1.267)	8.41
F9	(0, 0.733, 1.733)	(0, 0.267, 1.267)	8.41
F10	(2.000, 3.000, 3.000)	(0, 0, 1)	7.75
F13	(0.522, 1.522, 2.261)	(1.003, 2.003, 2.670)	5.49
F14	(0.658, 1.322, 1.993)	(0.264, 1.264, 2.264)	7.05

STEP 3: THE AGGREGATION OF THE VARIABILITY

The FRAM model showed 14 foreground functions that deal somewhat with upstream variability. Firstly, such functions have been qualitatively evaluated in terms of DC, that is, DC_j^T and DC_j^P. All the selected upstream-downstream couplings were listed and then submitted via email to each expert to assign their opinion. Basically, each expert was asked to what extent the function can dampen the variability arising from an upstream function by using labels as indicated in Table 10.2. Moreover, the experts were guided to take account of the associated descriptions to judge the DC in order to facilitate understanding of the meaning of each label. For instance, a downstream function has a very high DC if such a function is capable of effectively controlling all the effects of upstream variability. After that, an online meeting was organized to show the results to all the experts.

Method Geek

By receiving the responses, the first author appropriately tabulated the data, as shown in Table 10.5.

Afterwards, the responses have been aggregated by using Eqs. (10.2)–(10.4), and then an aggregated fuzzy number was generated to DC_j^T and DC_j^P by using Eq. (10.5). Next, data from MV_i, DC_j^T and DC_j^P were used to run the second FIS for calculating the CRI_{ij} for each coupling under analysis whose results are shown in Table 10.6.

TABLE 10.5
Qualitative Evaluations Expressed by the Experts for the Dampening Capacity

Expert	Coupling	F_{2-5}	F_{2-6}	F_{5-8}	F_{8-12}	F_{8-9}	F_{6-8}	F_{6-9}	F_{6-7}	F_{9-13}	F_{9-14}	F_{13-10}	F_{14-15}	F_{10-15}	F_{7-11}
E_1		HI	ME	VH	VH	LO	ME	LO	VL	ME	LO	VL	VL	LO	VL
E_2	DC_j^T	ME	HI	HI	HI	ME	LO	ME	VL	ME	ME	VL	VL	LO	VL
E_3		VH	LO	HI	VH	VL	HI	LO	VL	HI	LO	LO	LO	VL	VL
E_1		HI	HI	VH	HI	ME	ME	VL	VL	LO	LO	LO	VL	LO	VL
E_2	DC_j^P	ME	LO	ME	HI	LO	LO	ME	VL	ME	VL	VL	VL	LO	VL
E_3		HI	HI	HI	VH	LO	LO	LO	VL	ME	LO	VL	LO	VL	VL

TABLE 10.6
Result for the CRI_{ij} for Each Coupling F_{ij}

Coupling	MV_i	DC_j^T	DC_j^P	CRI_{ij}
F_{2-5}	7.75	7.757	7.156	2.00
F_{2-6}	7.75	5.500	6.575	2.12
F_{5-8}	5.51	8.350	7.757	1.47
F_{6-7}	7.86	1.000	1.000	3.5
F_{6-8}	7.86	5.500	3.844	2.55
F_{6-9}	7.86	3.844	3.243	2.78
F_{7-11}	9.28	1.000	1.000	3.68
F_{8-9}	8.41	3.243	3.844	3.02
F_{8-12}	8.41	9.400	8.350	2.11
F_{9-13}	8.41	6.094	4.905	2.52
F_{9-14}	8.41	3.844	2.650	3.02
F_{10-15}	7.75	2.650	2.650	3.29
F_{13-10}	5.49	1.600	1.600	2.96
F_{14-15}	7.05	1.600	1.600	3.20

Amongst 14 couplings represented in Table 10.6, only F_{5-8} presented a CRI value less than 2. This means that the coupling's response is characterized by exponential decay of the magnitude of variability. This response can occur due to the high capacity of dampening of the F8 <Emergency Action Plan coordination>. This phenomenon occurs even when F5 <Evaluate dam collapse> presents MV = 5.51, which is considered high, mainly due to influences from the variability in terms of precision. In this case, two experts judged the output variability in F5 as "imprecise", as noted in Table 10.3. Evaluating a rupture event is a demanding process, and it requires response actions without a complete understanding of the event. Uncertainty elements associated with this function might be related to the human factor in interpreting data, as well as the capacity for coordinating and communicating that is required to interact with different actors. Thus, the response might be instantaneous; however, disturbances such as the incompleteness of information or even noisy communication can lead to imprecision in the output function.

The couplings F_{2-5}, F_{2-6}, F_{8-12}, F_{6-8}, F_{6-9}, F_{9-13}, and F_{13-10} returned CRI values between 2 and 3. These values indicate that the corresponding couplings cannot only rapidly respond to changes in input but also have the ability to allow variability to deal with disturbances and return to equilibrium after being disturbed. This property is beneficial in activities where adjustments are necessary to maintain performance and stability. Couplings exhibiting this setting are crucial for achieving specific performance goals of the system. Understanding their behaviour and properties is essential for managers working in similar fields where the dynamic plays a significant role (e.g., healthcare, air traffic management, among others). As the dynamism of systems usually requires tightly coupled interactions, that is, many interdependencies among elements, they are highly prone to variability. Besides, the working conditions are generally non-linear, which means that small changes in the initial conditions cause huge effects on the outcomes, and hazards are emergent and cannot be easily anticipated. Therefore, enhancing the dampening capacity of the system might be appropriate to allow the system to be robust, not so rigid that it breaks, but also constraining the system to oscillate at extreme amplitudes to the point of resonance.

On the other hand, the couplings F_{8-9}, F_{6-7}, F_{9-14}, F_{14-15}, F_{10-15}, and F_{7-11} returned CRI values of more than 3, making them prone to propagate variability in downstream functions, even though at a lower potential.

For example, as noted in Table 10.5, the coupling F_{7-11} is those presenting the highest CRI value. This is due to the alert sirens having been installed, but none were activated across the entire facility (Rose et al., 2023), which justifies the high score for the MV. Consequently, F11 presented a very low capacity for dampening the unacceptable variability stemming from F7.

Another example is given in the couplings F_{8-9} and F_{9-14}, which both present the same values for the MV and CRI, as shown in Table 10.5. It should be noted that $\left(DC_{14}^{T}\right)$ is slightly higher than $\left(DC_{9}^{T}\right)$, whereas $\left(DC_{14}^{P}\right)$ is reduced if compared to $\left(DC_{9}^{P}\right)$. This means that F14 compensates for its limited capacity of dampening the incoming variability in terms of precision with its improved capacity of dampening in terms of timing. Therefore, even if the MV remains constant, DC has a significant impact on the CRI.

In essence, we refer to DC in downstream functions as the "ability to improvise" for managing unacceptable variability stemming from upstream functions. Improvisation refers to a quick and innovative adaptation in the response phase, which includes preparing new procedures upon realizing that established protocols no longer adequately work (Son et al., 2020).

In some cases, couplings presenting low dampening constitute an undesirable effect that can greatly reduce the performance of a system. It can be especially challenging in systems that require a fast response time, such as emergency response. Couplings with low dampening refer to a situation where a system's response to an external stimulus is too slow due to excessive protocols, bureaucratization, and over-control, resulting in a loss of efficiency and massive consumption of energy. On the other hand, couplings tending to amplify the variability are more flexible and therefore have fast response times, which is beneficial to dynamic and unpredictable systems since it facilitates the occurrence of the variability if it seems to go in the right direction. The downside, of course, is that the faster it is, the less dampening, which allows functions to vary more intensely and reinforce each other, thereby causing the variability of one function to be unusually high. In these cases, the system's designers must carefully balance the trade-off between thoroughness and efficiency (cf. ETTO principle (Hollnagel, 2009)) in operations to provide an adequate response.

Finally, the analysis developed in Step 3 should be used to gain a quick idea of where the critical couplings are in the system and then to delve deeper into understanding why and how this criticality occurs by analyzing the dynamics of each critical function, as outlined in Step 4.

RESULTS OF STEP 4: ASSESSING PRIORITIES FOR MANAGEMENT ACTIONS

Each one of the nine functions has been evaluated in terms of criticality, as obtained by the cumulative CRI, and indicated by Eq. (10.8). In this step, the cumulative CRI is useful to represent the potential of functions to reverberate for different downstream functions. Based on this analysis, we can suggest some actions to enhance such functions in order to improve their capacity to vary without losing control.

Table 10.7 shows the result of the cumulative CRI_i for all functions investigated in this case study.

Note that the top three rankings for critical functions were F6, F9, and F8, while F5 is the function exhibiting the lowest criticality and, consequently, the lowest risk priority. F6 is the most critical function; it presents considerable MV = 7.86 within a 10-range and three downlinks, which is considered another contributing factor to the criticality. Therefore, the higher the CRI and the greater the number of downlinks of one upstream function, the larger the spreading area will be, and consequently, the more critical this function will be. Note that the final ranking orders can certainly be influenced by dampening levels in downstream functions.

The variability spreads the range of unwanted outcomes to be discovered from the couplings among functions gradually, and the variability effect becomes amplified. Therefore, the higher the MV and the greater the number of downstream couplings, the more likely it is that the output of this function will be highly variable and, consequently, more critical.

TABLE 10.7

Ranking of Critical Functions, as Obtained from the Cumulative CRI

Function	CRI Cumulative	Ranking
F6	8.83	1
F9	5.54	2
F8	5.13	3
F2	4.12	4
F7	3.68	5
F10	3.29	6
F14	3.20	7
F13	2.96	8
F5	1.47	9

Domain Expert

For example, comparing the performance of F6 and F8, it is noticed that F6 can strongly influence F7, which in turn has reduced or no capacity for controlling the variability from F6. Meanwhile, F8 exhibits $MV = 8.41$ of potential influence in F12; however, the latter has a very high capacity of dampening the output variability from F8, mainly in terms of timing $\left(DC_{12}^{T}\right) = 9.400$.

Although F7 has the highest MV and CRI, respectively, it only appears in the fifth position of criticality. Following the principle of out-degree, which yields the cumulative CRI, F7 has a single downstream coupling, which makes it less critical than other functions that present multiple downstream couplings, for example, F6 and F9. Moreover, the negative outcome perceived in F7 represents the effect of functional resonance, stemming from the propagation of uncontrolled variability in previous functions.

To adequately manage the system variability, it is essential to update some functions. For example, the findings from the case study emphasized that reduced communication efficiency within F6 introduces delays across downstream functions in the emergency response process, particularly along the coupling of F6–F8. Notably, F8 plays a crucial role in coordinating the Emergency Action Plan, as this function provides instructions and actions for executing the plan promptly. However, performing an effective Emergency Action Plan becomes more challenging when the emergency is unpredictable. Accordingly, the input of the function F8 is served by the output of functions F5 and F6. Once F5 provides information on the dam collapse, an enhancement strategy could be re-characterizing this function to serve as input for the F6. In turn, the coupling F6–F8 would become unnecessary and could be omitted. Concurrently, F6 would benefit from the available information in F5; it would enable the activation of F7 promptly and with acceptable precision. Such actions might reduce the criticality of F6 to a lower level.

Furthermore, given the substantial uncertainty inherent in emergencies, the coordination should not wait for sufficient information to be obtained to implement emergency response actions. Instead, another strategy comprises strengthening the coordination group in F8 with improvisational capacities. This implies the coordination group should be able to develop new plans and procedures upon recognizing that the prescribed ones are inadequate to deal with the current situation. In essence, these abilities can strengthen the DC for managing the variability in F5 since most experts judged their output as imprecise.

The proposed method allows for identifying the most critical functions and thus proposing approximate adjustments to improve the dynamic conduct during a collapse, for example. Moreover, other FRAM-based prospective analysis by using multiple scenario simulations, may be useful to identify other disturbances and variabilities in the system and, consequently other criticalities. Thus, plans and procedures can be updated in order to enhance the response abilities of workers, citizens, and governmental entities.

CONCLUSIONS

The purpose of this chapter was to showcase the integration of the FRAM with fuzzy analytics to identify and prioritize critical functions involved in emergency response activities. At the same time, we argued for the relevance of using FST to cope with the uncertainty and ambiguity of experts' opinions on variability.

We hope we have made an interesting and understandable case on how to integrate variability qualitative assessments via fuzzy rule-based reasoning. To this extent, we have provided detailed guidance on FST integration and introduced, respectively, MV and CRI to identify critical functions.

However, this chapter does not represent the culmination of this line of research. Here, we calculated the criticality of functions using a single index, but the broader concept of functional resonance within the system warrants deeper investigation. This exploration will require blending qualitative logic with expert validation and adopting meta-heuristic quantification approaches. We also acknowledged that our analytical approach presumes equal relevance among experts, suggesting the potential for an improved method that weights experts' contributions according to their significance, particularly in specialized contexts.

A case study of the tailings dam collapse at Córrego do Feijão in Brumadinho, Brazil, served as a practical application of our methodology. The results affirmed the viability of our approach in assessing and ranking the criticality of functions to enhance variability management. In conclusion, the FRAM confirms itself as a thorough analysis method for systemic analyses that incorporate variability and nuances of normal work. The FST, once integrated with the FRAM, offers a flexible framework to address expert opinions and refine the process of assessing and aggregating variability, which comes in handy, especially in the case of large models.

REFERENCES

Adiansyah, J.S., Rosano, M., Vink, S., Keir, G., 2015. A framework for a sustainable approach to mine tailings management: Disposal strategies. *J. Clean. Prod.* 108, 1050–1062. https://doi.org/10.1016/j.jclepro.2015.07.139

Bellini, E., Ceravolo, P., Nesi, P., 2017. Quantify resilience enhancement of UTS through exploiting connected community and Internet of Everything emerging technologies. arXiv 18, 1–34.

Bellini, E., Coconea, L., Nesi, P., 2020. A functional resonance analysis method driven resilience quantification for socio-technical systems. *IEEE Syst. J.* 14, 1234–1244. https://doi.org/10.1109/JSYST.2019.2905713

Buch, A.C., Sims, D.B., Marques, E.D., Silva-Filho, E.V., 2024. Case studies of assessment of human health risks after the dam failures of the Córrego do Feijão Mine and Fundão in Brazil. *Int. J. Mining, Reclam. Environ.*, 1–15. https://doi.org/10.1080/17480930.2024.2330227

Cheng, D., Cui, Y., Li, Z., Iqbal, J., 2021. Watch out for the tailings pond, a sharp edge hanging over our heads: Lessons learned and perceptions from the brumadinho tailings dam failure disaster. *Remote Sens.* 13. https://doi.org/10.3390/rs13091775

De Souza, I., Rosa, A.C., Haddad, A., Patriarca, R., 2022. Soft computing for nonlinear risk assessment of complex socio-technical systems. *Expert Syst. Appl.* 206, 1–12. https://doi.org/10.1016/j.eswa.2022.117828

Duan, G., Tian, J., Wu, J., 2015. Extended FRAM by integrating with model checking to effectively explore hazard evolution. *Math. Probl. Eng.* 2015, 11.

Extraordinary Independent Consulting Committee for Investigation (CIAEA), 2020. *Executive summary of the independent investigation report: Failure of Dam 1 of the Corrego de Feijao Mine*. Brumadinho, MG.

Geramian, A., Abraham, A., 2021. Customer classification: A Mamdani fuzzy inference system standpoint for modifying the failure mode and effect analysis based three dimensional approach. *Expert Syst. Appl.* 186, 115753. https://doi.org/10.1016/j.eswa.2021.115753

Hirose, T., Sawaragi, T., 2020. Extended FRAM model based on cellular automaton to clarify complexity of socio-technical systems and improve their safety. *Saf. Sci.* 123, 104556. https://doi.org/10.1016/j.ssci.2019.104556

Hollnagel, E., 2009. *The ETTO principle: Efficiency-thoroughness trade-off: Why things that go right sometimes go wrong*, Ashgate Publishing.

Hollnagel, E., 2012. *FRAM: The functional resonance analysis method - Modelling the complex. Socio-technical systems*. Ashgate Publishing.

Hollnagel, E., 2015. Why is work-as-imagined different from work-as-done?, in: Wears, R.; Hollnagel, E.; Braithwaite, J. (Ed.), *Resilience of everyday clinical work*. Ashgate Publishing.

Hsu, H., Chen, C., 1996. Aggregation of fuzzy opinions under group decision making. *Fuzzy Sets Syst.* 79, 279–285.

Islam, K., Murakami, S., 2021. Global-scale impact analysis of mine tailings dam failures: 1915–2020. *Glob. Environ. Chang.* 70. https://doi.org/10.1016/j.gloenvcha.2021.102361

Kim, Y.C., Yoon, W.C., 2021. Quantitative representation of the functional resonance analysis method for risk assessment. *Reliab. Eng. Syst. Saf.* 214, 107745. https://doi.org/10.1016/j.ress.2021.107745

Noraishah Ismail, S., Ramli, A., Abdul Aziz, H., 2021. Research trends in mining accidents study: A systematic literature review. *Saf. Sci.* 143, 105438. https://doi.org/10.1016/j.ssci.2021.105438

Patriarca, R., Di Gravio, G., Woltjer, R., Costantino, F., Praetorius, G., Ferreira, P., Hollnagel, E., 2020. Framing the FRAM: A literature review on the functional resonance analysis method. *Saf. Sci.* 129, 104827. https://doi.org/10.1016/j.ssci.2020.104827

Perrow, C., 1984. *Normal Accidents - Living with High-Risk Technologies*. Harper Collins Publishers.

Porsani, J.L., de Jesus, F.A.N., Stangari, M.C., 2019. GPR survey on an iron mining area after the collapse of the tailings Dam I at the Córrego do Feijão mine in Brumadinho-MG, Brazil. *Remote Sens.* 11, 1–13. https://doi.org/10.3390/RS11070860

Rico, M., Benito, G., Salgueiro, A.R., Díez-Herrero, A., Pereira, H.G., 2008. Reported tailings dam failures. A review of the European incidents in the worldwide context. *J. Hazard. Mater.* 152, 846–852. https://doi.org/10.1016/j.jhazmat.2007.07.050

Rose, R.L., Mugi, S.R., Saleh, J.H., 2023. Accident investigation and lessons not learned: AcciMap analysis of successive tailings dam collapses in Brazil. *Reliab. Eng. Syst. Saf.* 236, 109308. https://doi.org/10.1016/j.ress.2023.109308

Salehi, V., Veitch, B., Smith, D., 2020. *Modeling complex socio - technical systems using the FRAM: A literature review*, 1–25. https://doi.org/10.1002/hfm.20874

Slim, H., Nadeau, S., 2020. A mixed rough sets/fuzzy logic approach for modelling systemic performance variability with FRAM. *Sustainability* 12, 1–21. https://doi.org/10.3390/su12051918

Son, C., Sasangohar, F., Neville, T., Peres, S.C., Moon, J., 2020. Investigating resilience in emergency management: An integrative review of literature. *Appl. Ergon.* 87, 103114. https://doi.org/10.1016/j.apergo.2020.103114

Woods, D.D., 2006. *Essential characteristics of resilience, resilience engineering: Concepts and precepts*. In Hollnagel E., Woods D., Leveson N. "Resilience Engineering Concepts and Precepts". CRC Press, ISBN: 9781315605685. https://www.taylorfrancis.com/books/edit/10.1201/9781315605685/resilience-engineering-erik-hollnagel-david-woods

Yager, R.R., Filev, D., 1993. On the issue of defuzzification and selection based on a fuzzy set. *Fuzzy Sets Syst.* 55, 255–271. https://doi.org/10.1016/0165-0114(93)90252-D

Yin, H., Liu, C., Wu, W., Song, K., Liu, D., Dan, Y., 2020. Safety assessment of natural gas storage tank using similarity aggregation method based fuzzy fault tree analysis (SAM-FFTA) approach. *J. Loss Prev. Process Ind.* 66. https://doi.org/10.1016/j.jlp.2020.104159

11 Pathways to Industry 5.0 and Society 5.0

Socio-Technical Systems Modelling through FRAM and Discrete-Event Simulation

Enrique Ruiz Zúñiga, Ehsan Mahmoodi, and Naruki Yasue

CONTEXT: FROM INDUSTRY 4.0 TO SOCIETY 5.0

Industrialized countries frequently encounter challenges in retaining their expanding production sectors domestically. Production facilities, services, and suppliers are commonly established abroad to cope with higher production costs and the shift from mass production to mass customization. To overcome this trend, governments and organizations are putting significant efforts into the paradigms of Industry 4.0/5.0 in Europe/United States and Society 5.0 in Japan. The main objectives of these paradigms are enhancing feasibility, profitability, flexibility, adaptability, and predictivity of the interactions of people, machines, and resources, and more recently, strongly considering a sustainable and human-centric perspective (Hitachi-UTokyo-Laboratory, 2020). This interest increased significantly with the introduction of the notion of Industry 4.0, but it has evolved via the other notions of Industry 5.0 and Society 5.0. This chapter shows something more about each of them, and it attempts to clarify how they demand joint socio-technical system modelling.

INDUSTRY 4.0

Industry 4.0 is considered a key step in the evolution of the industry, following the previous integration of automation and information technology, developments in mass production and electricity, and industrial steam power and mechanization. Industry 4.0 is a paradigm based on digitalization, connectivity, and the emerging technologies of the Internet of Things (IoT) and Cyber-Physical Systems (CPS), aiming to increase efficiency in manufacturing. This industrial next step has seen the integration of IoT to build CPS to allow actuators on an industrial shop floor (such as machines, robots, processors, computers, and workers) to be connected to their surrounding environment (databases, monitoring and control systems, and to the outside

DOI: 10.1201/9781003518167-14

world). This connectivity allows interoperation and cooperation to achieve individual as well as jointly aggregated goals. This can help to accomplish increased flexibility levels required in production to adapt to variable capacity and customized demand.

Domain Expert

Industry 4.0 is a promising paradigm based on the emerging technologies of IoT and CPS, initially supported by the German government and then by the European Commission to maximize industrial system performance. The implementation of these Industry 4.0 technologies can significantly increase production quality, efficiency, flexibility, and security. Nevertheless, while Industry 4.0 has been well-known worldwide for its potential to revolutionize production and improve industrial performance, it has also received criticism for its focus on efficiency and productivity, leaving aside broader societal considerations. One of the primary concerns is the potential negative impact on employment. Mainly due to the escalation of automation and artificial intelligence, there is a belief that jobs in certain sectors may become obsolete, leading to significant unemployment and socioeconomic disparities. The potential of losing human autonomy and devaluing human labour with automated systems is a considerable concern that needs to be addressed. Another criticism focuses on its environmental impact. While Industry 4.0 promises increased energy efficiency and reduced waste in industrial processes, the rapid expansion of smart factories and the proliferation of IoT devices can lead to increased industrialization and consumption rates. Critics also argue that without proper regulations, circularity, and sustainable practices, Industry 4.0 may be overshadowed by negative ecological consequences. Other concerns are privacy and cybersecurity issues, investment risks, missing skills and know-how, and missing collaboration spirit, strategy, and spontaneous decision-making.

INDUSTRY 5.0

The evolution of Industry 4.0 into Industry 5.0 was defined to argue for a more comprehensive approach that combines industrial performance with broader societal development and well-being. It is strongly fuelled by the European Commission, aiming to establish a collaborative ecosystem where advanced technologies and human workers cooperate to enhance productivity, innovation, and job quality within the manufacturing sector (Breque et al., 2021). Industry 5.0 intends to find a balance between the strengths of technology as well as the indispensable abilities and needs of humans in order to drive advancements in industrial processes and in the work of our future industrial society. It advocates strategies that prioritize human-skill enhancement and recognizes the need for sustainability and circularity practices, responsible resource management, and job creation. Moreover, major critics of Industry 4.0 might share the common thought of ensuring that industrial progress does not come at the expense of human livelihoods. Hence, Industry 5.0 can be defined as the evolution of production systems to a more human-centric approach, where human workers collaborate symbiotically to achieve productivity, innovation, well-being, and work–life balance.

The concept of Industry 5.0 recognizes that besides the benefits of automation and digitalization, there are limitations to what digitalization and technology can achieve by themself. It is, therefore, necessary to acknowledge the valuable skills, creativity, and adaptability of human workers and Industry 5.0 to integrate advanced technologies such as robotics, automation, CPS, artificial intelligence, and the IoT to enhance human performance but also well-being and life (Breque et al., 2021).

Domain Expert

One of the key drivers of Industry 5.0 is the concept of "Cyber-Physical-Human Systems" (CPHS), also referred to as Cyber-Socio-Technical Systems (CSTSs), considering advanced technologies to create collaborative and interconnected industrial environments in which humans and machines work together in a coordinated and self-adaptable manner. By combining the strengths of humans and machines, Industry 5.0 aims to enhance human potential while providing a more inclusive and comfortable work environment, instead of viewing technology as a substitute for human labour.

The modelling of socio-technical systems is a key part of pursuing this paradigm.

Socio-technical systems in Industry 5.0 heavily rely on the understanding and management of human and machine equipment performance and relations. Since this paradigm emphasizes the collaboration between humans and machines to optimize manufacturing processes and achieve sustainable performance, an in-depth analysis of socio-technical systems is crucial for implementing Industry 5.0 successfully. Some examples are the consideration and interrelation analysis between technical components, such as automated tools and machines, robots, and data analytic software tools, as well as social aspects, including planning and performance of human workers, physical and psychical characteristics, organizational structures, and cultural factors. The understanding of the information flows of these entities and their interactions can impact work processes, job design, skill requirements, and needs of the organizational structure. By considering socio-technical factors, organizations can design manufacturing systems that balance the strengths of technology with the capabilities and preferences of human workers, leading to improved job quality and overall performance (Breque et al., 2021). Nevertheless, socio-technical systems analysis should also highlight the need for public organizations and policymakers to address issues related to ethics, privacy, inequality, and system resilience in the public and private sectors, ensuring that these paradigms prioritize societal well-being and sustainability. Therefore, there is a need to extend this paradigm of Industry 5.0 to society.

Society 5.0

Societal evolution refers to the continuous development and transformation of social structures, norms, and institutions over time. It encompasses changes in values, technology, economy, and governance that shape the dynamics of societies. The rapid societal evolution being experienced nowadays demands, first, continuous growth and, second, continuous improvement, the former being more questionable in terms of sustainability, while the latter is a requirement for survival. This societal evolution has been widely supported by the development of agriculture and industry at large-scale levels. This would have been impossible without technology. Technology can be considered as applying scientific knowledge for practical purposes, applying scientific methods to solve problems, and providing more efficient solutions. Therefore, a key aspect of the development of societies is the advanced interaction between humans and technology, something typically intrinsic in socio-technical systems.

The shift towards a human-centric approach aims to ensure technology serves as an enabler for the progress of society as a whole, rather than exclusively focusing on the development of the industrial sector. This has been better reflected in what could be considered the Japanese version and extension of Industry 5.0, Society 5.0, joining efforts by governments, institutions, and corporations. Society 5.0 is a holistic vision extending beyond the production sector, first drafted in 2017. It seeks to build a society where technology enhances human well-being across various domains, including healthcare, transportation, energy, education, and governance, with emphasis on utilizing technology to address complex societal challenges and aiming for a sustainable, inclusive, and human-centric society.

There are equivalent governmental strategic plans worldwide to support the development of their industrial sectors, such as Produktion 2030 in Sweden, Industrial Internet of Things in the United States, or Made in China 2025. Nevertheless, the Japanese version also focuses on demographic growth and includes transport, energy, and healthcare sectors, providing the base for a better future smart society. Society 5.0 represents a human-centric society where technology and innovation are used to address societal challenges as well as improve the well-being of individuals; it promotes the coexistence of humans and advanced technologies to create a productive, sustainable, and inclusive society. Slightly different from the mentioned industrial revolutions, the concept of Society 5.0 is based on previous stages of human societal development, leading to significant advancements (Hitachi-UTokyo-Laboratory, 2020):

- Society 1.0 was characterized by agrarian societies
- Society 2.0 marked the Industrial Revolution and the birth of mass production
- Society 3.0 was marked by the age of information and globalization, with the birth of the internet and digitalization
- Society 4.0 brought the integration of CPS, artificial intelligence, and automation

Domain Expert

Society 5.0 appears, therefore, as an innovative response to remedy shortcomings such as social inequality, unemployment, and environmental issues and build a more efficient and sustainable future. These societal drawbacks tend to be widened by the emerging in the last decades of technological developments. Society 5.0, therefore, aims to promote human well-being individually in an efficient growing, sustainable society, a "super smart society". A super-smart society can be defined as a society where digital transformation, driven by emerging technologies, is empowered to address the holistic needs of individuals, communities, and the environment, accompanied by social progress, inclusivity, and sustainable development (Holroyd, 2022).

Similarly to Industry 5.0, Society 5.0 emphasizes that technology must be aligned with social values, well-being, and sustainable development. Nevertheless, it highlights the essence of socio-technical systems by emphasizing the use of advanced technologies in different sectors of society and considering social dynamics. This paradigm acknowledges that the successful implementation of technology-driven solutions requires a comprehensive understanding of the social context, organizational structures, and human behaviour to allow policy-makers, engineers, and designers to better comprehend the functioning of society and the social implications of technology to model and anticipate social behaviours for society improvement (Hitachi-UTokyo-Laboratory, 2020).

A key aspect of Society 5.0 is the convergence of the physical and cyber-physical worlds. Through the integration of automated technologies, data, and connectivity, the emergence of CPS emerges to integrate the physical aspects of life with digital platforms to monitor, control, and predict the behaviours of complex systems. This convergence of managing huge amounts of data and connectivity can enable structural and societal transformative possibilities, allowing for enhanced decision-making, personalized experiences, optimized resource allocation, and improved quality, helpful for most, if not all, members of society (Hitachi-UTokyo-Laboratory, 2020). From the engineering perspective of improving society, following the paradigms of Industry 5.0 and Society 5.0, this synergy of technological tools to enhance the performance of systems as well as the well-being of workers has to be pursued through the use of computerized tools for system design, analysis, and improvement.

SUPPORTING THE 5.0 TRANSITION VIA SOCIO-TECHNICAL SYSTEMS MODELLING

What Industry 5.0 and Society 5.0 have in common is a clear reference to socio-technical systems as a unit of analysis. These latter refer to the interaction between people and technology, aiming to understand the complex relationship between social and technical elements, commonly studied in work environments. Socio-technical systems are joint items designed to achieve specific goal-directed behaviour, usually related to coping with improved environmental complexity, dynamism, new technology, and competition. The understanding, analysis, and improvement of socio-technical systems are necessary to take part in the fourth industrial revolution.

Method Geek

To analyze the socio-technical dimensions, a comparison of the presented paradigms is provided in Table 11.1.

As shown in Table 11.1, several system improvement tools are highlighted for each of the paradigms. Due to a common ground of connectivity and digitalization, these tools or approaches are presented cumulatively, meaning that

TABLE 11.1
Industry 4.0, Industry 5.0, and Society 5.0

Paradigm	Application Area	Motivation	Aim	Common Approaches (Accumulative)
Industry 4.0	Manufacturing and industry	Mass customization, offshoring	Automation, connectivity, efficiency, adaptability, and mass customization	Automated Technology, Cyber-Physical Systems, Internet of Things, Computer Simulation, Optimization, Digital Twins, Value-Stream Mapping, Bid-Data Analytics, Artificial Intelligence, Cloud Computing, Additive Manufacturing, Lean Production, Six Sigma, Failure Mode and Effects Analysis
Industry 5.0	Sustainable and human-centric manufacturing and industry	Lack of human-centricity in industrial development	Sustainable industrial development, human–machine collaboration, human-centric industrial production, mass customization	Socio-Technical Systems, Human Factors Engineering, Human-Centric Design, Collaborative Robotics, Augmented Reality and Virtual Reality, Cognitive Automation
Society 5.0	Smart cities and mobility, healthcare, energy efficiency	Social inequality, unsustainable societal development, unemployment, environmental issues	Human-centric society; efficient and sustainable cities, mobility and energy management; super smart society; technological integration	Physical and Cyber-Physical Worlds' Convergence, Participatory Design, Data-Driven Policy Making, Citizen Engagement Platforms

Source: Bjørn and Boulus-Rødje (2015), Zúñiga et al. (2017), Hitachi-UTokyo-Laboratory (2020), Sony and Naik (2020), Breque et al. (2021).

technology used for the achievement of Industry 4.0 objectives is also commonly used to work in the Industry 5.0 and Society 5.0 direction. Simulation tools have marked the era of digitalization, such as Markov Chain Analysis, DES, System Dynamics, and Monte Carlo Simulation (Bjørn & Boulus-Rødje, 2015; Sony & Naik, 2020).

A common factor to most technologies under the approached column in the previous table is data, digitalization, and computer simulation. Aiming to provide support to these paradigms, this chapter focuses on DES, a well-known computer simulation technique, and the FRAM, a promising system improvement approach from a human factors perspective. A more detailed explanation of a methodology using DES and FRAM as system improvement tools is presented in the next subsection.

DISCRETE EVENTS SIMULATION AND THE FRAM

Developing detailed simulation models of complex systems is essential for thorough analysis and understanding of complex systems. Nowadays, there are diverse techniques available for process improvement. Among them, DES is a powerful and widely recognized technique for approaching the current challenges in complex production systems. DES is a method to model and analyze systems where events occur at distinct points in time, aiming to simulate the behaviour of the system by representing events, state changes, and interactions over time (Banks et al., 2005). When the complexity and/or size of the system to analyze is considerable, DES can provide enormous support for modelling numerous quantitative variables with complex interrelations.

On the other hand, as previously defined in this book, the FRAM is commonly used to analyze complex systems by focusing on the interactions of quantitative or qualitative variables between functions rather than individual components, providing a systemic view of how functions in a system resonate and adapt to changes. While the FRAM commonly supports providing a holistic perspective on system dynamics by focusing on the interactions between functions, DES enables the modelling and analysis of complex systems through the simulation of discrete events measured quantitatively with a simulation clock.

The integration of FRAM and simulation in the context of production systems and socio-technical systems has gained significant attention in recent years. Socio-technical systems are complex systems that consist of interconnected social and technical elements working together to achieve specific goals. These systems involve interactions between people, technology, the environment, and organizational structures to accomplish tasks effectively.

This chapter is intended to serve as an introduction to the terms and applications of FRAM and DES in the context of Industry 5.0 and Society 5.0.

DATA COLLECTION

The data collection strategy applies mixed qualitative methods, involving observations, document reviews, and semi-structured interviews with operators, line or process responsible persons, managers, and stakeholders. Observations included several factory visits following the flows of materials and resources.

Method Geek

The gathered information was summarized in conceptual models or flowcharts, and the WDA was used to identify necessary, existing, and missing data. Existing and necessary data included manufacturing process data, assembly, customization, painting, testing, and packaging. This data was mostly obtained by document reviews and observations. Nevertheless, a common challenge usually faced in socio-technical systems is the lack of available digital qualitative data, as well as limited methods for collecting data considering safety aspects and human well-being. This difficulty can be common even in large-scale manufacturing systems, commonly still lacking data collection and management systems for human-centric qualitative data. In this case, the missing data defined the questions of the semi-structured interviews, focusing on processes, missing performance information, and human-centric aspects such as level of stress, interruptions, delays, and human failures. The interviews were conducted using an open-ended format, allowing the generation and collection of tacit knowledge of novice and expert operators, as well as process or line managers.

In cases where systems are not newly designed or adequately adapted for the digital transformation of human interactions in Industry 5.0 or Society 5.0, they may fail to provide digitized performance data for resources, processes, and products, even with the assistance of simulation and optimization tools. To address these challenges, wearable devices, sensors, and cameras that are connected to artificial intelligence and CPS systems can be utilized. In this case, the information collected by security cameras was not allowed to be used for performance measures. Nevertheless, logged times of automated transports associated with some of the processes facilitated some processing times and statistics about failures and interruptions more related to human factors. Implementing these technologies can improve the availability and collection of data. However, it is important to consider sensitive issues such as the privacy of operators and workers, as well as the concerns of labour unions. Additionally, the quality or reliability of the data being collected also has to be considered, which may be compromised due to human-related behaviours, unexpected situations, or changes in the environment.

METHODOLOGY

While DES is centred on quantitative modelling and simulation, the FRAM traditionally focuses on a qualitative approach grounded in systems thinking. This complementarity allows each approach to offer unique insights into system behaviour and interventions.

The qualitative analysis of FRAM can provide a foundation to identify and understand the complex relationships within a system, highlighting potential vulnerabilities or key aspects, variables, functions, or interrelations of the system. This qualitative understanding can then be incorporated into DES models that represent, in a detailed and quantitative manner, system dynamics, including quantitative variables of specific operations to analyze specific KPIs (Key Performance Indicators), outcomes, or performance. Therefore, a more comprehensive and accurate representation of the system can be achieved, providing enhanced knowledge and decision-making support, especially in complex and large-scale systems. This section presents the methodology for combining DES and the FRAM, going through two practical case studies in the manufacturing sector:

- Case study A, "An integrated discrete-event simulation with functional resonance analysis", is based on a published paper in the *Decision Analytics Journal* (Zúñiga et al., 2023)
- Case study B, "Analyzing Resilient Performance of Human Workers with Multiple Disturbances in Production Systems", is based on a paper published in the *Journal of Applied Ergonomics* (Yasue et al., 2025).

Case study A aims to address complex manufacturing systems by considering human–machine interactions and identifying potential solutions from both quantitative and qualitative perspectives. It focuses on data collection, simulation, bottleneck analysis, and qualitative approaches to enhance the performance of a complex manufacturing system within the yacht construction industry. By integrating quantitative data analysis through DES with the qualitative approaches of WDA and FRAM, the methodology offers a comprehensive framework for identifying system weaknesses, bottlenecks, and interrelations.

On the other hand, case study B investigates the impact of multiple disturbances on human work performance in an automotive engine manufacturing plant and explores how skilled workers contribute to resilient work performance under these disturbances. The research aims to address the knowledge gap in understanding the cumulative effect of various aspects, such as skills, fatigue, and circadian rhythms, on human workers in manufacturing settings. By examining the interaction between workers' skills and other factors within the production system, the study seeks to provide insights into maintaining resilient production performance in factories subject to a variety of disturbances. The study employs the FRAM to model the work of novice and expert workers in a case study focusing on the test and repair process of an engine assembly line simulated with DES.

In both case studies, the methodology followed is structured into four different phases, starting with problem setting, and progressing towards a DES model first, and a joint WDA-FRAM model later, iterating these last two steps where necessary (see Figure 11.1).

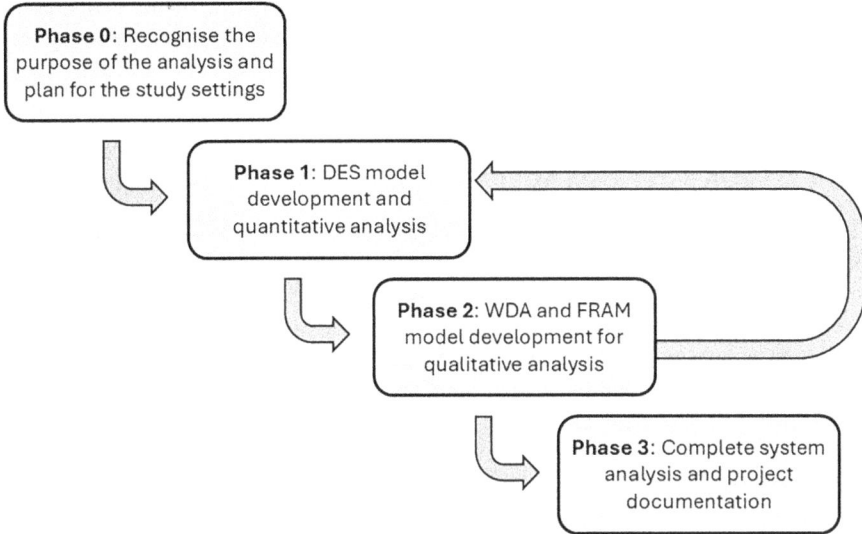

FIGURE 11.1   Proposed methodology with DES, WDA, and the FRAM.

PHASE 0: RECOGNIZE THE PURPOSE OF THE ANALYSIS AND PLAN FOR THE STUDY SETTINGS

The preliminary phase of this methodology represents the identification of the problem and the definition of the objectives of the project, key to understanding the size and complexity of the project and being able to define the project team and project plan, involving everyone from managers to operators to understand the whole picture of the system and to visualize its possible target conditions or potential improvements. A flexible project plan is usually required in this type of exercise, accepting variability in the execution of functions and continuous back and forth among functions and properties of the system. This phase usually stops with a high-level conceptual representation of the system to narrow down the scope of the project and identify the need for data to be collected. This phase refers to the traditional Step 0 of the usual FRAM model development process.

PHASE 1: DES MODEL DEVELOPMENT AND QUANTITATIVE ANALYSIS

Once the data has been collected, the conceptual representation of the system can then be converted into an initial DES simulation model. One recommended practices to achieve this conversion is the DES methodology proposed by Banks et al. (2005).

The obtained simulation model has to be verified and validated, double-checking all the collected data and processes modelled with the help of subject matter experts and domain experts in charge of the system and different processes. Thereafter, a bottleneck study and some initial experiments are recommended to analyze the performance of the current state of the system and define and test some "what-if" scenarios to improve the system. This can also be guided by the mentioned simulation methodology of Banks or equivalent. This step is somehow unrelated to the traditional building steps of the FRAM, yet it allows increasing system knowledge to an extent that facilitates the subsequent functional modelling.

PHASE 2: WDA AND FRAM MODEL DEVELOPMENT FOR QUALITATIVE ANALYSIS

The DES model usually does not cover all inherent features of the system at hand, especially with respect to qualitative variables and processes, as frequently happens in socio-technical systems. These systems present challenges if interpreted with purely quantitative approaches due to the lack of data or measurements and due to abstract performance measures, qualities, and objectives. Once the primary weaknesses or bottlenecks within the system have been identified through simulation, a more in-depth analysis of these processes can be conducted using the WDA for the identification of quantitative factors or variables associated with the identified bottlenecks. Data collection methods, including interviews, documentation analysis, and Gemba walks, are employed to gather insights, with particular attention to refined details often omitted from manuals or formal instructions that may influence system performance. Factors to consider include the diverse performance levels of individuals engaged in the same process, such as their skill sets, qualifications, experience, age, and personality traits, alongside interpersonal relationships among employees, equipment malfunctions, delays in material delivery, and operational interruptions. Therefore, the WDA is developed by adhering to a systematic procedure: establishing the purpose and intended use of the WDA, defining abstraction levels (abstraction-decomposition space) and boundaries, identifying constraints along with potential information sources, iteratively refining the abstraction-decomposition space through multiple iterations, and finally validating the WDA. The various abstraction levels fulfil distinct roles; they provide answers to "how" when moving down through the WDA from higher abstraction levels and address "why" when ascending from lower levels. Typically, the upper section of the WDA is designated as the "Functional Purpose", which aligns with the overall objectives of the study, such as the reduction of quality defects, while the lower section, "Physical Function", delineates the specific steps that facilitate the process or its enhancement.

After defining the WDA, the focus shifted towards creating FRAM representations. The WDA was analyzed from the perspective of the problem descriptions specific to each case study. Subsequently, the aims and objectives were re-evaluated, and the primary functions – typically aligned with the corresponding hierarchical levels of the WDA – were selected for inclusion in the FRAM model. The authors of this chapter believe that implementing WDA and then the FRAM can be more enriching in obtaining hidden interrelations of epistemic actions. With respect to the FRAM implementation, its design process followed the traditional Steps 1, 2, and 3 for function identification and then variability identification and aggregation.

PHASE 3: COMPLETE SYSTEM ANALYSIS AND PROJECT DOCUMENTATION

An iterative process combining WDA and FRAM can be successful when the size and complexity of the system require analysis at several levels, such as strategic, tactical, or operational, to analyze qualitative factors of the identified limiting process by DES.

Thus, an additional iterative step can be considered to enrich the previously defined DES model with the knowledge obtained by WDA and FRAM, aiming to maximize system performance. In this case, the additionally quantified qualitative aspects are reintroduced in the previous DES model, focusing on the identified bottlenecks and considering the key processes or variables found with the WDA and FRAM. The WDA, as well as the DES and FRAM models, were constructed based on collected data from the industrial partners and verified through workshops and group discussions with company experts and the participant group of operators. The results were then documented to ensure proper variability management once they fully understood what the process bottlenecks are and their relations to key qualitative variables identified by the WDA and FRAM. This can then be translated into recommendations to decision-makers to define performance risks or safety issues, as well as procedures or recommendations of the key identified processes having relevant significance for the performance of the system. A useful tool here can be the utilization of an A3 report (typical of lean production) to summarize the objectives, methodology, models, and results of the system improvement project.

RESULTS

The findings of Case Study A show that working with FRAM and DES in the context of Industry 5.0 and Society 5.0 can provide valuable insights to stakeholders about system dynamics, where and how to drive improvements, and how to enhance overall system performance within complex and dynamic socio-technical improvements. Several key manufacturing processes were identified, and their socio-technical systems were analyzed from a human-centric perspective to increase the performance of the system, adapting to production demand changes. The methodology served as a guideline for managers and stakeholders and provided a structured approach for system improvement. The involvement of a diverse project team was crucial for the understanding of the system, conceptual modelling definition, and model verification and validation. In a similar manner, the availability of digital and organized data, such as skills of human resources, processing times, and process instructions and requirements, was also crucial for the construction of simulation models and WDA and FRAM approaches. Future work includes further data collection, case studies, and potential applications in different departments.

For Case Study B, diverse simulations have been conducted, considering the operator's skill level (expert or novice) and the types and timings of variabilities occurring, to investigate the combined effects of disturbances on work performance. Key findings reveal a significant correlation between increasing variability and the performance outcomes for both novice and expert operators. The DES model identified the bottleneck of the system in the engine test operations, showing that multiple disturbances make it challenging to maintain work performance, with more than eight compounded variabilities significantly hindering recovery. The study identified critical variables that have a substantial impact on work performance, such as human–machine interaction and operational support, organizational factors, and availability of resources.

This emphasizes the necessity for proactive management strategies aimed at mitigating disturbances to sustain operator performance and overall system efficiency, prioritizing the monitoring and assessment of these disturbances as well as implementing effective strategies to control and prevent their interplay. Targeted suggested interventions addressing these were designing intuitive equipment, enhancing organizational processes, and synchronizing work schedules with circadian rhythms to boost operational efficiency, minimize risks, and improve overall performance for operators. It was learned that managers should foster a culture that recognizes and rewards adaptive responses to disruptions, leading to a more efficient and resilient production system. This can involve expanding traditional training programmes to include the development of soft skills such as adaptability, critical thinking, and situation awareness. These competencies can enable workers to anticipate issues, adapt to evolving circumstances, and make informed decisions swiftly.

By analyzing socio-technical systems with WDA, FRAM, and DES to pursue the goals of Industry 5.0 and Society 5.0, the authors of this chapter proposed a methodological solution to ensure that technological advancements are aligned with societal needs, getting closer to merging cyberspace with physical space to obtain a knowledge-intensive society and facilitate archiving a data-driven society, respecting human values, and contributing to a more inclusive, sustainable, and human-centric future.

Method Geek

Nevertheless, the problem definition, system conceptualization, data collection, model verification and validation, and asking the right questions for the design of "what-if" scenarios are not straightforward. To successfully obtain significant positive outcomes in this kind of system improvement project with computerized system improvement tools, it is necessary to consider expertise in the use of system improvement tools, digital data availability, involvement of management and stakeholders, coping with reaction to change, credibility, and generational breach, as well as overcoming potential previous negative experiences.th

CONCLUSIONS

Production systems around the world are intensively competing to follow the change in trend from mass production to mass customization. Well-consolidated organizations with decades of expansions and adaptations are also struggling to cope with the increased complexity of socio-technical systems required to move forward in society and industry paradigms. Society 5.0 and Industry 5.0 are two interrelated concepts that foster the evolution of systems towards a more advanced and inclusive future. While both initiatives share the goal of harnessing the potential of technology to benefit society, they approach it from different perspectives, focusing on the complexity of societal development and the transformation of production systems, respectively.

Complexity emerges as organizations seek to optimize operations while fostering a collaborative and adaptive organizational ecosystem. Achieving suitable interactions between humans and machines requires a deep understanding of human-centric design, training, and system performance analysis, design, and improvement. Understanding and managing this complexity requires a holistic and interdisciplinary approach that considers the interdependencies, variability, emergent behaviours, and feedback loops within the interconnected systems. The combination of WDA, FRAM, and DES can bring together the strengths of qualitative analysis and quantitative modelling. By applying these tools effectively, organizations can enhance efficiency, performance, and resilience while adapting to potential changes and achieving sustainable improvements.

Managing complexity in both Industry 5.0 and Society 5.0 paradigms necessitates continuous learning, digital and available data, adaptability, and collaboration among various systems and stakeholders. Furthermore, by prioritizing the management of critical disturbances, cultivating a resilient workforce, and fostering a culture of continuous improvement, organizations can better handle the complexities of modern production environments and societies and build long-term operational resilience.

REFERENCES

Banks, J., et al. (2005). *Discrete-Event System Simulation.* Pearson Prentice.

Bjørn, P. and N. Boulus-Rødje (2015). "Studying technologies in practice: 'Bounding practices' when investigating socially embedded technologies." In V. Wulf, K. Schmidt, & D. Randall (Eds.), *Designing Socially Embedded Technologies in the Real-World,* 341–356. Springer. https://doi.org/10.1007/978-1-4471-6720-4_14

Breque, M., et al. (2021). *Industry 5.0 – Towards a sustainable, human-centric and resilient European industry.* E. C. D.-G. f. R. Innovation. Publications Office of the European Union.

Hitachi-UTokyo-Laboratory (2020). *Society 5.0: A People-Centric Super-Smart Society.* Springer Nature.

Holroyd, C. (2022). "Technological innovation and building a 'super smart' society: Japan's vision of society 5.0." *Journal of Asian Public Policy,* 15(1), 18–31.

Sony, M. and Naik, S. (2020). "Industry 4.0 integration with socio-technical systems theory: A systematic review and proposed theoretical model." *Technology in society* 61: 101248.

Yasue, N., et al. (2025). "Analyzing resilient performance of human workers with multiple disturbances in production systems." *Journal of Applied Ergonomics,* 122, 104391.

Zúñiga, E. R., et al. (2017). The internet of things, factory of things and industry 4.0 in manufacturing: current and future implementations. *15th International Conference on Manufacturing Research ICMR 2017. Incorporating the 32nd National Conference on Manufacturing Research.* University of Greenwich, London, September 5–7, 2017.

Zúñiga, E. R., et al. (2023). "An integrated discrete-event simulation with functional resonance analysis and work domain analysis methods for industry 4.0 implementation." *Decision Analytics Journal,* 9, 100323.

Section III

Step 3
Aggregate Variability, Actual, and Potential

12 The Milky Way to FRAM
Comparing Instantiations via Network Metrics in a Dairy Production Plant

Andrea Falegnami and Andrea Tomassi

CONTEXT: THE DAIRY PRODUCTION

At the end of 2022, we were asked to provide advice on the implementation of safety procedures in the dairy industry. Apart from the considerable financial reward (but only for scientific reasons), we decided to accept the assignment in order to make sure our competences could make some impact in a real and dynamic context. Our team consisted of two complementary professionals: a researcher in socio-technical systems and industrial safety management, with experience in the FRAM, and a trainee in operations management and industrial safety, eager to see theoretical knowledge applied in practice. Our work was carried out in close collaboration with dairy industry personnel, trying to understand their specific needs and concerns. The project was successfully completed, meeting the customer's requirements, that is, increasing staff awareness of actual work practices and sustaining operations by safer recommended practices (an ambitious outcome for the traditional Step 0 of the FRAM!).

Since the management was not interested in the specific method but in the results, we were left free to select the method we considered most appropriate to study operations. Eventually, we decided to opt for the FRAM, as soon as the preliminary data gathering emphasized the inherent variability of operations and limited visibility of it, at least at that time.

What follows is a faithful account of the case study and the methods used for analysis.

In production plants, engineering resilience is crucial due to the complex and interdependent processes involved. These facilities must maintain high standards and coordinate various procedural and technological elements to ensure continuous and safe production. Resilience Engineering (RE) focuses on both qualitative and quantitative ways to make this possible, as demonstrated by the works estimating resilience in production lines by showing initial capacity loss and subsequent recovery paths (Caputo et al., 2023) or several attempts gathered from literature (Patriarca et al., 2018). The need for an accompanying resilience-based approach to the manufacturing workforce has also been demonstrated by the need to identify a list of essential skills and competencies to enhance organizational flexibility (Borella & Borella, 2021).

DOI: 10.1201/9781003518167-16

The management of a diverse range of products in dairy plants, each with specific production requirements, benefits from resilient performance management. Food production plants require continuous communication and coordination among operators, quality control experts, and laboratory technicians are vital. Consider microbiological interactions and contamination risks: several studies emphasize the complexity of identifying the most effective measures to reduce production diseases in organic dairy farms. It notes that the systemic roles of variables are highly diverse across farms, making it crucial to adopt farm-specific approaches. The complexity arises from the need to consider multiple interrelated factors and the specific context of each farm to improve health management effectively (Edwards, 1980; Krieger et al., 2017; Simões et al., 2015).

Domain Expert

Indeed, dairy production facilities represent vital contexts for RE as they constitute complex systems prone to criticalities both in terms of personnel and procedural/technological aspects. The staff must comply with a large number of hygiene and health regulations and procedures to ensure food safety. A lengthy and intricate series of checks and inspections involves the personnel operating on the production lines as well as those involved in transportation and quality control, who must continuously liaise with laboratory analysis technicians. Even excluding the possibility of contaminating dairy products when handling them, the concurrent issues of productivity and workplace safety remain significant, especially with the complicated machinery used. From a technological standpoint, dairy plants must address challenges related to both obsolescence and machinery maintenance. Sudden breakdowns or equipment malfunctions can cause disruptions in production and potentially jeopardize the quality of the final product. Concerning obsolescence, it should be emphasized that the dairy industry operates with highly specific and costly machinery custom-made by specialized companies. Consequently, addressing impending obsolescence through direct machine replacement on-site is unlikely. Instead, periodic equipment "revamping" practices are more probable, often involving gradual automation of production lines. This means that, from an organizational perspective, issues related to obsolescence can be traced back to maintenance-related problems, both of which configure as conditions for partial or total disruption of a production line. The complexity of the machinery requires specialized personnel for its management and maintenance, and ongoing training is essential to ensure efficient and safe plant operation. There is always the possibility that some operators may be uninformed about appropriate safety practices, especially in the case of some asynchrony between the training they are offered and the pace at which standards or procedures may change at the blunt end. For example, simply disconnecting the power supply to the machinery is insufficient for it to enter a safe state – it could still possess a potential energy component (e.g., electrostatic, thermal, chemical, pneumatic, and gravitational) capable of causing significant harm to both objects and people. In addition to the criticalities described, there is also the challenge of adequately managing shift changes to prevent fatigue-induced errors.

USE CASE: THE LOTO (LOCK OUT – TAG OUT) PROCEDURES

The interplay between human factors and technology introduces vulnerabilities that standard procedures alone cannot address. To this extent, resilience engineering enhances the system's ability to anticipate, absorb and recover from disruptions – both technological and biological – ensuring that dairy production facilities maintain operational integrity, adapt to changing conditions and maintain high standards of food safety and quality.

Consequently, a resilience engineering approach seemed the natural choice when a European player in the dairy industry requested a prospective assessment of the operating conditions of its packaging plants: How would the socio-technical system of a packaging line adapt if LOTO (Lock Out – Tag Out) procedures were implemented from scratch?

Domain Expert

The LOTO (Lock Out – Tag Out) procedures represent a fundamental cornerstone in the management of workplace safety, especially in highly complex industrial settings such as milk and cream packaging plants. These procedures are designed to protect workers during the maintenance or repair of machinery and equipment, effectively preventing the risk of accidents related to hazardous energy. In summary, the LOTO process involves the physical blocking of a machine's energy mechanism through the application of locking and closure devices (lockout) and the application of tags (tagout) to clearly indicate that the machine or equipment is out of service and must not be operated. Before carrying out any maintenance intervention, personnel are required to strictly follow the LOTO procedures, which include verifying the deactivation of the energy supply, applying locking devices, and attaching visible warning labels. These procedures are designed to significantly reduce the risk of energy-related injuries during maintenance activities, ensuring a safer working environment compliant with health and safety regulations.

The dairy farm examined in this case study faced the need to draw up and then implement LOTO procedures to ensure the safety of its operators when maintaining machinery. Developing such procedures is not an activity that fits all possible industrial systems; each company must develop appropriate LOTO procedures, adapting the general regulations to its own context. For this reason, instead of proceeding directly to the drafting of such procedures, in our case, the company opted for a preliminary and, as far as possible, in-depth analysis of two different hypothetical implementation situations, comparing them with a basic situation corresponding to daily production under normal conditions. During maintenance or even improvement operations on a production line, the overall productivity of the line is reduced (or entirely eroded) because the machines concerned temporarily cease to participate in the production process, and the company is interested in assessing the possible systemic effects in these use cases.

As already mentioned, part of the LOTO procedures is derived from rather strict regulations. The rest are implemented by the company according to internal standards. LOTO procedures are detailed protocols, even down to the colour and shape of the locks and the operators to whom they are assigned. As such, when effectively implemented, they should represent both a prescribed variety of work (Work-As-Prescribed, WAP) and a normative variety of work (Work-As-Normative, WAN) (Moppett & Shorrock, 2018). In our case, however, the LOTOs were not at such an advanced stage of design, as they remained at the ideal stage of a partially under-specified concept. In this respect, the blunt-end operators being interviewed could only provide a vague (and very meagre) description, especially in terms of the impact on packaging activities. They limited themselves to indicating the personnel affected by the application of LOTOs, as well as the line sections that would be cut off from the production process as a result of the application of padlocks to the machinery under maintenance.

The blunt-end operators and the sharp-end operators represent, respectively, the operators furthest and closest to the examined production process: blunt-end operators are responsible for supervision and management activities at a strategic level, focusing their attention on organizational aspects, production planning, and quality control. These operators are often involved in defining production processes, drafting standard operating procedures, and managing human and technical resources within the dairy company. On the other hand, sharp-end operators are more directly involved in the daily packaging and production operations. These operators work in the field, directly operating packaging machines, monitoring production processes, and ensuring that quality and safety standards are met. They are responsible for executing operational activities, quality control of products, and resolving any operational issues that may arise during the production process. For example, in the case at hand, the sharp-end operators are the workers directly operating on the packaging machinery, and the blunt-end operators represent the management as a collective agent: the client who asked for the analysis.

The client's need was twofold: on the one hand, the company had to comply as quickly as possible with the regulations on LOTO procedures, which, at the time of the consultation, were still non-existent; on the other hand, it wanted, as far as its discretion allowed, to define optimal LOTO procedures in terms of productivity as well as safety. In fact, the interest of the blunt-end operators was mainly directed to the comparison of alternative scenarios, in consideration of several factors: impacts on machinery, personnel, procedures and, more generally, on production. In addition to the identification of activities associated with greater potential instability of the entire production process, certain performance indicators were required to allow monitoring and possible timely responses (e.g., leading and lagging indicators). Leading indicators are proposed metrics that provide insights into the future performance trend. In contrast, lagging indicators are retrospective metrics that indicate past or current performance. While leading indicators are useful for anticipating changes and implementing preventive measures, lagging indicators offer an assessment of past performance and can be used to evaluate the effectiveness of already implemented strategies. In such cases, top management often expects numerical quantifications, although these, especially when the underlying assumptions are ignored, can be prone to instill feelings of false confidence and control over the system.

Scenario 0 – Baseline (The Everyday Production Process)

Domain Expert

In the process of transforming raw milk into finished and semi-finished products in a dairy company, milk is delivered daily by designated supplier transporters. After the recognition and documentary control of the tanker trucks at the weighbridge, sampling and laboratory analysis are performed. Once the delivery is accepted, the tanker truck is unloaded and weighed again. The raw milk is then stored in separate tanks and mixed if necessary. After analytical and operational checks, the milk can be sent for technological treatment. During this process, it is preheated, cleaned, skimmed, pasteurized, and cooled before being sent to storage tanks. The semi-finished products undergo further checks before being sent for packaging, where they are packed into cardboard boxes or PET bottles according to the daily production schedule. Products with different characteristics are designed for different market segments and therefore require different packaging. Cartons and PET bottles are then palletized. Before distribution, the products undergo further quality control and sample analysis. The packaging department consists of two lines for carton packaging and three lines for PET bottle packaging. The two carton packaging lines are similar and use identical machinery, while the PET packaging lines have some differences, with one line being more sophisticated than the other two in terms of the machinery present. The identical lines cover most of the PET production. The more advanced line covers not only the peak demand but also the rest of the production requirements, allowing a wider range of end products to be packaged.

In the search for a good ETTO compromise (Hollnagel, 2009), it was decided to limit the study to three of the five packaging lines: the two carton lines and the more versatile PET packaging line, on the assumption that the model obtained would be sufficiently representative for the company's purposes. The main carton packaging machines form, fill, disinfect and track the cartons. On the other hand, PET bottles are formed from preforms, sterilized, blown, and filled before being labelled, inspected, and packaged. For the most part, operators follow similar procedures for both types of packaging, carrying out tests, loading materials, and monitoring the process, the difference being that PET packaging takes place virtually continuously, without interruption.

Scenario 1 (Operator 1 in Support)

In this instantiation, LOTO procedures affect one or more machines on one of the two carton packaging lines. Therefore, under these rather general LOTO implementation assumptions, the interruption of an entire packaging line was considered. There is a loss of efficiency for the plant due to personnel becoming unemployed for accidental reasons. A direct economic loss, because the unemployed personnel burn money for nothing, but also an opportunity cost, because they could be used to redistribute the workload. Therefore, in the present scenario, the production manager finds it necessary to reallocate the worker who is unemployed due to the machine breakdown on the PET line.

Domain Expert

The decision to redeploy the workers to a line of a completely different type from the original line was taken in order to increase the complexity of the modelling. The underlying assumption is that increasing the number of workers accustomed to working on the same machines is more likely to lead to a more efficient outcome. Conversely, adding more operators to a different line, especially if they are not sufficiently familiar with it, does not necessarily increase overall efficiency. In addition, it was considered more likely that the PET packaging line would need to be rebalanced due to the wider product range.

Depending on the expected workload, the production manager instructs the operator on the procedures and activities to be carried out on the new line. The activity begins with the inspection of the machines and packaging necessary for the packaging of PET bottles to ensure that everything is in order and ready for the start of production. Once the line is ready, the operator starts the packaging process, following the instructions given and using the machines involved, such as the PET bottle blowing machine, the filler for filling the bottles with milk and the labeller for applying the labels. During the packaging process, the operator is responsible for supervising the machine, periodically checking the quality of the product and replenishing used packaging. It is essential that the new operator coordinates with the personnel already on the PET line to ensure a smooth workflow and a seamless transition. At the end of the shift, the operator reports any problems encountered during the shift to the production manager and provides a summary of the activities performed. This scenario involves the packaging operator, the production manager and the personnel already on the PET line, as well as the machines used in the PET bottle packaging process.

Scenario 2 (Operator 1 Removed)

As in this scenario, one operator is assumed not to be present on the line, given that it may not be beneficial to increase the number of workers on a line, although economic evaluations such as the one carried out in the previous section claim otherwise. Most of the time, system resilience manifests itself precisely in the face of these situations in which human beings quickly adapt to changing circumstances. However, it is worth modelling a scenario of the application of LOTO procedures in which, instead of being reallocated to the PET line, the operator is simply removed for the downtime.

In this instantiation, the operator is completely removed from the milk carton line to allow for maintenance or repair. The production manager coordinates maintenance activities with specialized technical personnel and ensures that all safety procedures are carefully followed. The technical staff performs necessary maintenance operations on the carton milk packaging line, which may include replacing damaged components, lubricating moving parts, or resolving electrical or mechanical faults. During this process, it is crucial for the technical staff to strictly adhere to LOTO procedures to ensure the safety of the entire maintenance operation. Once maintenance and repair operations are completed, the technical staff verifies that the line is fully operational and safe for production restart. The operator then receives clearance from the production manager to return to the line and resume carton milk packaging activities. In this instantiation, the packaging operator, the production manager, specialized technical personnel, and support staff are involved, along with the machines used in the carton milk packaging line and the tools used for maintenance operations.

WHY THE FRAM?

We chose the FRAM, mainly because of its undisputed ability to describe socio-technical systems. It should be noted, however, that the adoption of the FRAM met some initial resistance. To date (2024), outside of high-tech sectors (e.g., aerospace) or those where systemic safety is the primary objective (e.g., healthcare), the FRAM is still relatively rare to find, especially in contexts such as food production. This is an aspect that deserves to be highlighted, in part because of the specificity of social systems that has been noted by disciplines other than resilience engineering and perhaps should not be overlooked (Thaler & Sunstein, 2009). Complementing the FRAM with a semi-quantified approach based on Monte Carlo simulation and multi-layer networks has facilitated its promotion and the fulfilment of client requirements.

The Monte Carlo method is a simulation technique used to solve complex problems by employing probability distributions capable of describing a particular random phenomenon. This approach is particularly useful when obtaining an exact analytical solution is not feasible or when assessing the behaviour of a system in the presence of numerous variabilities and/or uncertainties, as in the case of descriptions of socio-technical phenomena. Readers interested in the Monte Carlo method applied to the FRAM used in this approach can refer to the seminal work of Patriarca et al. (2017), which inspired the present research. In addition to comparing different instantiations, adopting a network perspective makes it possible to identify the most "central" activities within each potential instantiation in which LOTOs are applied to the packaging line. Once these activities have been identified, we will aim to define performance indicators related both to specific activities (crucial to disseminating information and/or influencing other activities) and to the topology of the whole network for the instantiation under consideration (Sujan et al., 2021). This assessment will only be performed on one instantiation at a time.

DATA COLLECTION

The perspective adopted in the consultancy work reflects the conceptual framework of WAx, which places particular emphasis on the position of the analyst as an integral part of the socio-technical system to be modelled.

Domain Expert

In the world of operational dynamics, the WAx framework delves into the intricacies of how work unfolds at both sharp and blunt ends of operations. At the forefront is the concept of WAD_{SO} (Work-As-Done by the Sharp-End Operator), where the sharp-end operator's actual execution diverges from what was initially envisioned, termed as WAI_{SO}. This gap between imagined and actual work is partially bridged by $WADI_{SO}$, where the operator discloses the reality of their actions. Contrastingly, the blunt-end operator prescribes work methods, captured as WAP_{BO} (Work-As-Prescribed by the Blunt-end Operator), which may not always align with actual practices. WAN encapsulates the normative aspects of work, while WAI_{BO} captures the blunt-end operator's idealized perception. WAO_{BO} offers insights into the observed reality from the blunt-end perspective, while $WADI_{BO}$ represents the disclosed reality by the blunt-end operator. Further layers involve analysts' interpretations: $WAI_{Analyst-of-WAI}$ and $WADI_{Analyst-of-WAI}$ delve into tacit and explicit knowledge, respectively, regarding the imagined work by the analyst. Conversely, $WAO_{Analyst}$ portrays the work as observed by the analyst. Finally, $WAI_{Analyst-of-WAD}$ and $WADI_{Analyst-of-WAD}$ explore the analyst's understanding of the actual work executed by the sharp-end operator, both tacitly and explicitly disclosed. The interested reader may find additional information in the paper entitled "WAx: An Integrated conceptual framework for the analysis of Cyber-Socio-Technical Systems" (Patriarca et al., 2021).

Method Geek

From a theoretical perspective, WAD and WAI are both social constructs that cannot be accessed directly, as extensively discussed by Patriarca et al. (2021). Therefore, the analysts must content themselves with explicit constructions of their own knowledge as the best proximal objects to consider ($WADI_{Analyst}$-of-WAx). The WAx framework traces the knowledge transfers theoretically necessary to move from one variety of work to another. The WAD and the various WAI are transferred to explicit counterparts constructed by the analyst (i.e., $WADI_{Analyst}$-of-WAx). Referring to WAx, the analyst can therefore, in principle, consider any biases and/or information losses associated with the modelling choices made and/or the analyzed context.

The consultants aimed to construct three different instantiations, respectively corresponding to the $WADI_{Analyst}$-of-WAD (i.e., the baseline scenario of routine performance under standard conditions), the $\text{WADI}_{\text{Analyst}}\text{-of-WAI}_1$ (i.e., the ideal scenario where the operator on the carton milk line under maintenance is transferred to support the PET line), and the $\text{WADI}_{\text{Analyst}}\text{-of-WAI}_2$ (i.e., the ideal scenario where the operator on the maintenance line is simply removed).

It was necessary to pass through intermediate varieties of work such as the $\text{WAI}_{\text{Analyst-of-WAD}}$ and the $\text{WAI}_{\text{Analyst-of-WAI}}$, starting from naturalistic observations ($\text{WAO}_{\text{Analyst}}$), meetings with the blunt-end operators, and norms analysis (WAP_{BO} e WAN), as well as documentary analyses (the various WAP) and interviews (WADI).

Figure 12.1 illustrates the conceptual process adopted throughout the study. The process is represented as mostly sequential and separated into distinct subphases; however, it was actually an iterative process, only theoretically sequential, which underwent continuous reassessment. The direction of the arrows in the graphical representation partially suggests this aspect.

Typically, the construction of a proximal WAD representation in the FRAM goes through several stages, the first of which is a preliminary documentary analysis. The assumption is that the knowledge embedded in the regulatory and prescriptive documentation represents a z variant of the intentions of the legislator and the socio-technical system designer (WAI). This is undoubtedly a partial view that bears little resemblance to reality, but it is nonetheless fundamental for an initial sketch of the system under analysis, especially in terms of macro-activities. What emerges is typically used as a skeleton on which functions not explicitly mentioned in the documents but observed in the field can be based, as well as inconsistencies with what is stated. Nevertheless, it is, one might say, the basic step in any standard FRAM document collection, from which all variants can be obtained by supplementation.

Congruent with the paragraph above, the FRAM instantiations related to the three scenarios were modelled following an initial data collection phase based on documentary studies (WAN + WAP). Field observation sessions (WAO) and semi-structured interviews (WADI), both with blunt-/sharp-end operators, made it possible to define most of the FRAM functions that would be modelled in the scenarios ($\text{WAI}_{\text{Analyst-of-WAD}}$, $\text{WAI}_{\text{Analyst-of-WAO1}}$, $\text{WAI}_{\text{Analyst-of-WAO2}}$). This concludes the preliminary data collection stage.

To achieve the objective of the analysis, which involves comparing different scenarios and developing Key Performance Indicators (KPIs) at the system level, it is imperative to model the corresponding FRAM instantiations. Once the relevant functions and their placement within the resulting stream derived from the instantiation process are identified, the basic topology of the corresponding networks is attained. This modelling is essential because equivalent networks are required to obtain centrality values, which in turn are required to obtain KPIs. More in detail, the baseline scenario was modelled first. The documents provided an initial sketch of the instantiation,

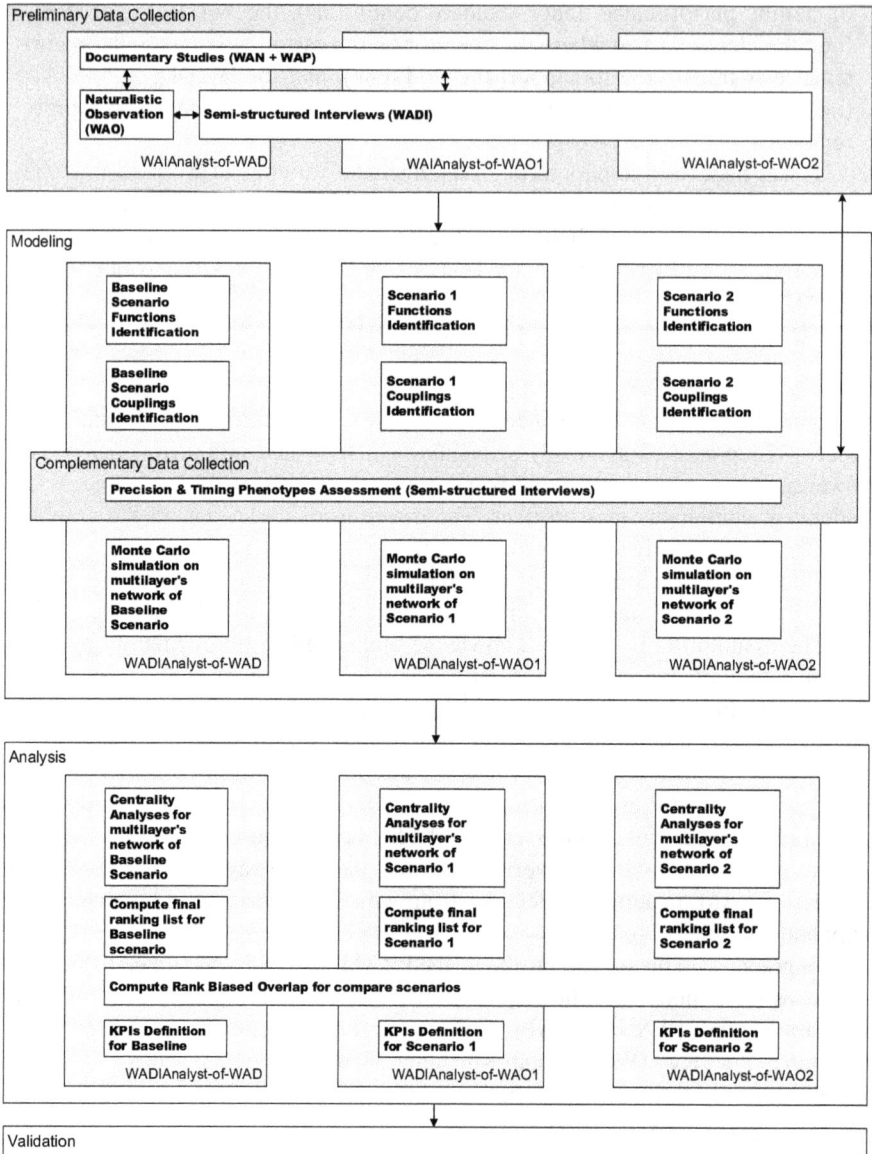

FIGURE 12.1   The process followed throughout the study.

later refined by directly observing operators working on the packing line. The result-ing instantiation, shown in Figure 12.2, has 44 functions and 181 couplings.

Once the set of activities to focus on was established, the consultants proceeded to outline the remaining two instantiations: extensive document research was necessary before repeated discussions with supervisors and maintenance personnel to design these settings in which LOTO procedures were imagined to be applied.

Scenario 1 consists of 38 functions and 149 couplings (Figure 12.3), while Scenario 2 consists of 32 functions and 121 couplings (Figure 12.4). Many of the

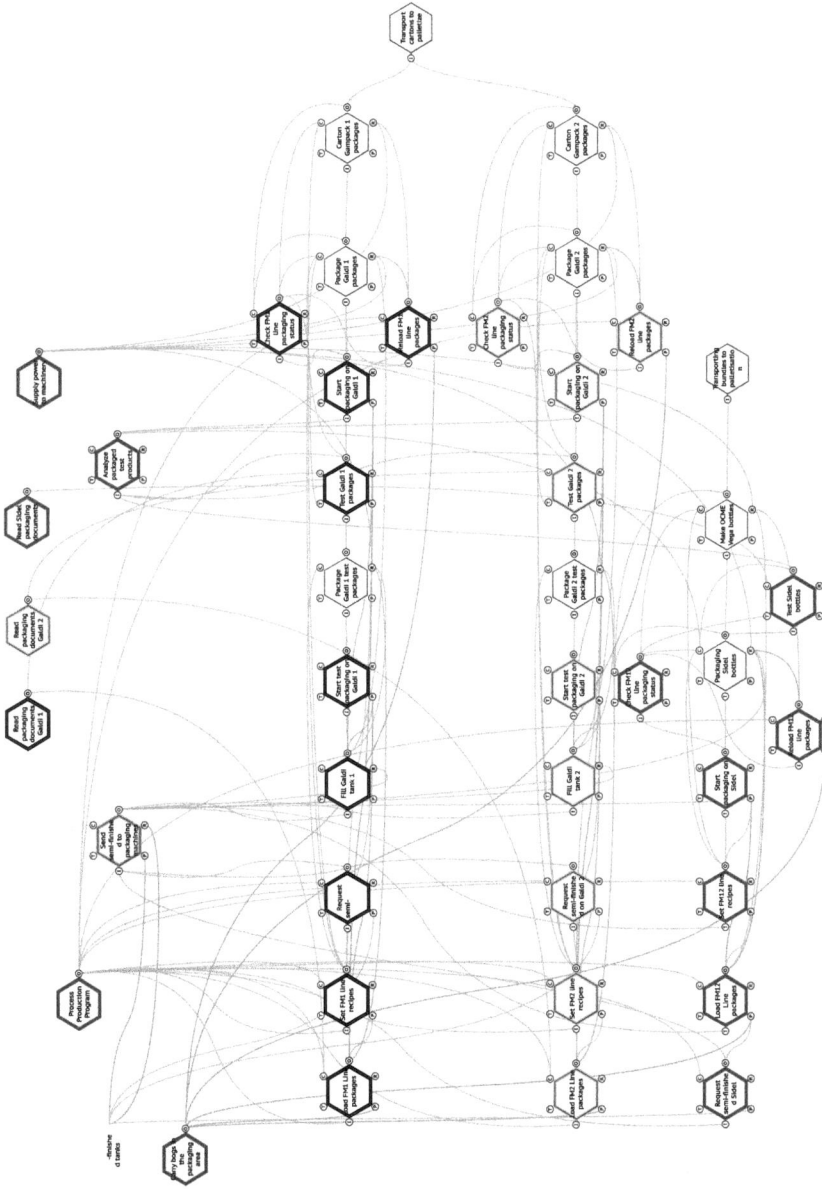

FIGURE 12.2 The FRAM visual representation of Scenario 0. The LOTO procedure in such a scenario is not put in place; therefore, all the functions performed by the maintenance officer are missing.

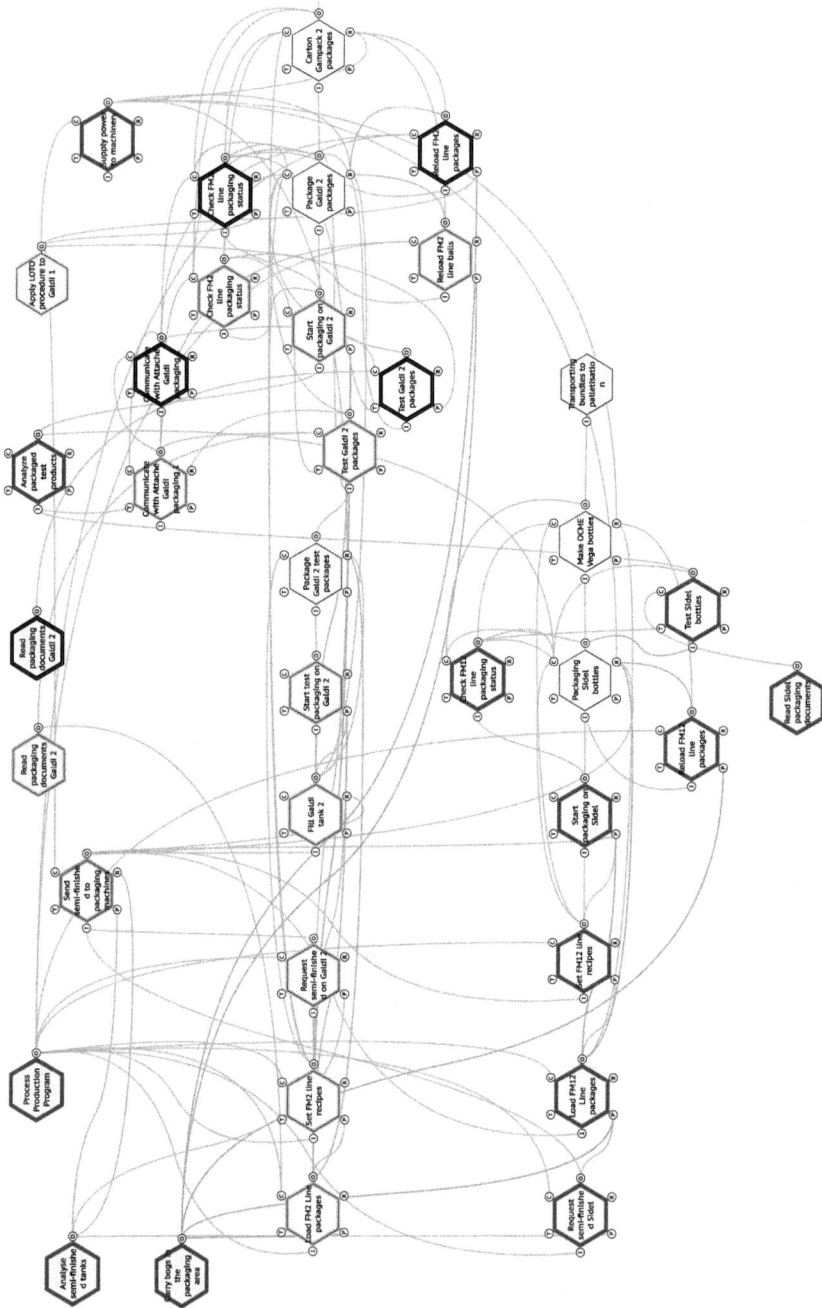

FIGURE 12.3 The FRAM visual representations of Scenario 1 (a). Different from Scenario 0, here the maintenance officer is responsible for LOTO-related tasks. The Appendix summarizes all scenario functions, indicating whether the function belongs to a given scenario, its socio-technical owner, and its type.

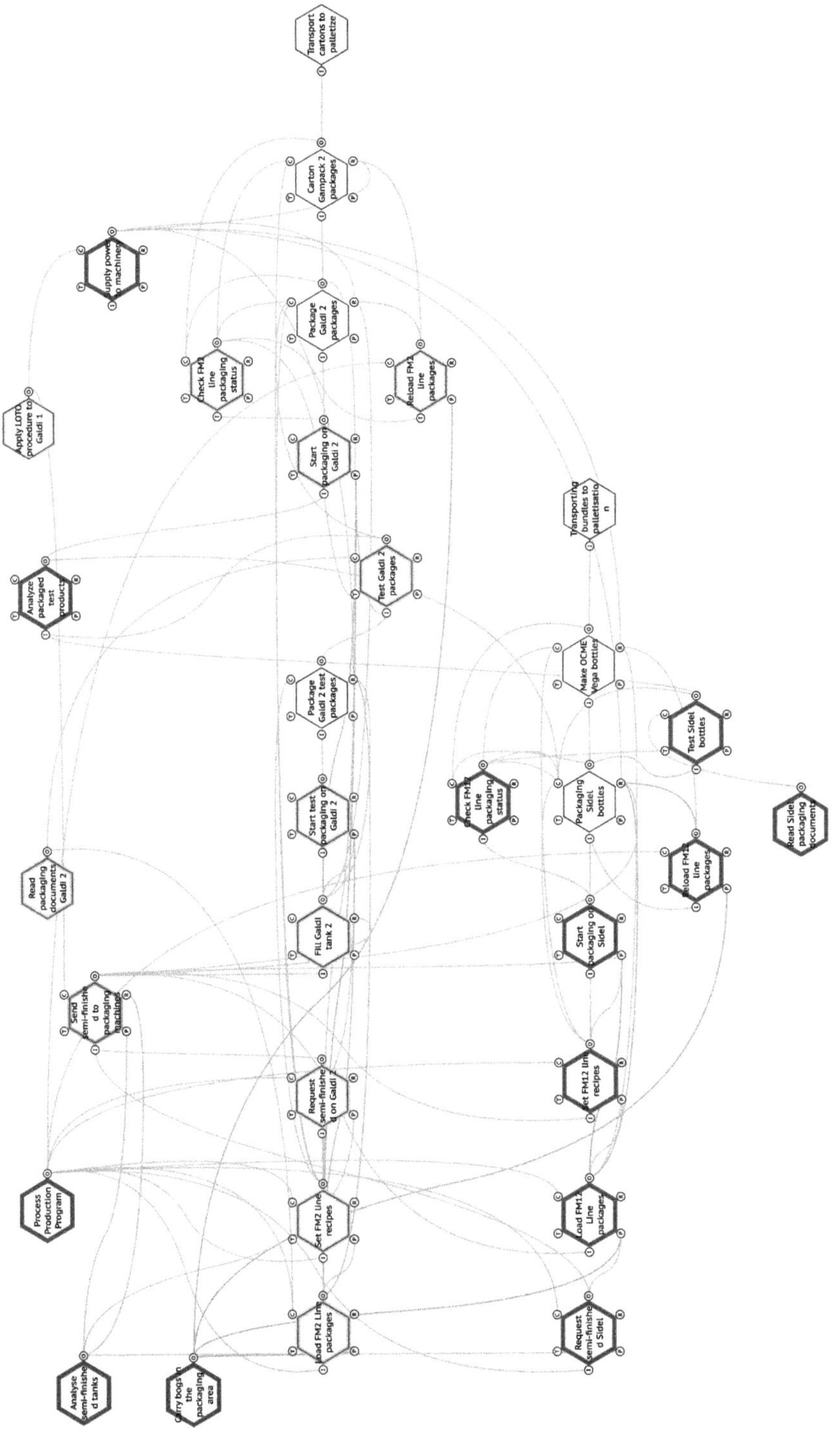

FIGURE 12.4 The FRAM visual representations of Scenario 2. Different from Scenario 0, here the Maintenance Officer is responsible for LOTO-related tasks. Note that Scenario 2 manifests fewer functions even if they are locally more connected due to the concurrent activity of the redeployed operator. The Appendix summarizes all scenario functions, indicating whether the function belongs to a given scenario, its socio-technical owner, and its type.

functions belonging to the baseline scenario represent the core of the other two scenarios since these are "reduced" variants of the former.

The process for assessing variability (see next section) largely follows the logic adopted in Patriarca et al. (2017). The assessment of the distribution of Timing and Precision phenotypes is required for the semi-quantification analysis. Therefore, a second round of data collection was necessary after having modelled the functional stream (complementary data collection, see Figure 12.1). It should be noted that, from a consultancy point of view, the whole analysis can be seen as a kind of what-if analysis, involving two hypothetical situations whose actual operating conditions have never been observed. Assessing Timing and Precision phenotype distributions, as in the above-cited work led by Patriarca, required retrieving information from several types of sources. The primary sources included documents, such as data logs of packaging machinery, customer orders, and maintenance records. In parallel, following an ethnographic approach, direct observation of workers during the packaging process was conducted, along with semi-structured interviews with technicians on topics such as packaging process analysis, machine repair (frequency of malfunctions and machinery unavailability, downtime, etc.), and safety (incident frequency, minor process variations, shift changes, etc.).

Domain Expert

In detail, three sessions of semi-structured interviews were necessary. The use of this combination of data collection techniques made it possible to capture diverse perspectives of the same phenomena, all mapped within the WAx framework. The participants included 17 individuals (e.g., 8 packaging operators, 2 maintenance technicians, 1 safety manager, 1 production manager, 1 quality control officer, 2 laboratory technicians, and 2 warehouse workers) across 7 sociotechnical roles (e.g., Packaging Operator Cardboard machine 1; Packaging Operator Cardboard machine 2; Packaging Operator PET machine; Plant Operator; Maintenance Operator; Packaging Machinery; and Organizational Agent, that is, the collective agent representing the entire dairy company).

Overall, the research involved two months of on-site work with the aim of understanding normal functioning during packaging activities. The consultants were allowed to observe every packaging, maintenance, and warehouse activity, interacting directly with the personnel at work, except in cases of obvious danger. The familiarity gained with the staff partially helped overcome some social resistance between the analysts and the operators.

The result of this extensive data collection has been a set of discrete distributions for the phenotypes with the associated scores. For some coupling, it was available more than a distribution (e.g., when two or more subject expert matters provide different assessments). In such cases, the final distribution was made after consensus had been reached. Table 12.1 shows an excerpt of one couple (Timing and Precision) of distributions for a given coupling.

TABLE 12.1

Excerpt from the Phenotype Distributions for the Couplings from the Function <Analyze Packaged test products> to <Start packaging on Cardboard Machine 1>

Upstream Function	Output	Downstream Function	Function Type	Aspect	D_T_On Time	D_T_Too Early	D_T_Too Late	D_T_Not at All	D_P_Precise	D_P_Acceptable	D_P_Imprecise	Coupling
Analyze packaged test products	Positive response on the analysis of test packages Cardboard Machine 1	Start packaging on Cardboard Machine 1	Human	Input	0.7	0	0.25	0.05	0.5	0.5	0	C30
					V_T_On Time	V_T_Too Early	V_T_Too Late	V_T_Not at All	V_P_Precise	V_P_Acceptable	V_P_Imprecise	
					0.5	1	1	2	1	1	2	

Note: The values for the Timing phenotype (V_T) are reported below the corresponding likelihood of event (D_T). The grey shading is proportional to the numerical value. The same information is given for the Precision phenotype.

METHODOLOGY

It is worth resuming the diagram proposed in Figure 12.1 since, for the purpose of the consultancy, FRAM modelling was aimed at defining KPIs and comparing alternatives. The possibility to identify the most central functions (according to different ideas of network centrality) for each scenario is provided by the theory of multilayer networks. The approach adopted hereinafter is an adaptation of the work of Falegnami et al. (2020) based on the translation of FRAM instantiations into multilayer networks.

Method Geek

A multilayer network is a mathematical model that makes it possible, in its most extensive formulation, to represent the co-presence of a number of simultaneous relations in principle without limits on the number of entities and the number of relations. These relations can be of any nature, e.g., of precedence, association, information flow, matter, time, or money. In the general case, this mathematical object can be represented by means of tensors, the manipulation of which requires advanced mathematical skills not necessary in the case study addressed here. In the cited article, the authors prove the transferability of a FRAM model into a tensor without loss of information. The interested reader is warmly invited to get lost in the reference list of this chapter.

The FRAM allows for the examination of variability and interaction dynamics among system functions because it is based on the premise that each function within a system possesses intrinsic variability that can be amplified or dampened by interactions with other functions. These interactions, or couplings, are often nonlinear and highly dependent on specific operational conditions. Utilizing Monte Carlo simulation, it is possible to explore a wide range of system instantiations which can be synthesized into the Variability Priority Number (VPN). The VPN is a numerical index that quantifies the variability of a system function based on the combination of timing and precision variations of its outputs, and the influences of operational conditions (Patriarca et al., 2017).

Method Geek

The VPN can be calculated in five steps:

- **Definition of Output Variability**: Each Output of a function varies in terms of Timing (e.g., too early, on time, too late, or never) and Precision (e.g., precise, acceptable, imprecise, or erroneous). A numerical score is assigned to each state of these variables. Table 12.2 reports an excerpt of these scores as values of a discrete distribution whose values are indicated correspondingly as V_T and V_P, and whose corresponding likelihoods of event are indicated as D_T and D_P.

TABLE 12.2

Summary of Key Centrality Metrics Used in Complex Network Analysis, Providing a Brief Explanation, the Corresponding Formula, and Their Specific Applications within Any FRAM Instantiation

Centrality Metric	Brief Explanation	Formula	Application in FRAM
Degree Centrality	Measures the importance of a node by counting its direct connections (local index).	$CD(v) = \deg(v)$	Evaluates the prominence of a function based on the number of direct connections it has with other functions. Useful for identifying functions with high direct interaction, akin to what has been done in Lundberg & Woltjer (2013).
Eigenvector Centrality	Assigns a centrality score to each node based on the centrality of its connections within the context of the whole network (systemic index).	$CE(v) = \dfrac{1}{\lambda} \displaystyle\sum_{u \in adj(v)} C_E(u)$	Identifies functions connected to important functions, useful for assessing the global influence potential of a function.
PageRank Centrality	A variant of Eigenvector centrality evaluates the importance of a node based on the quality and quantity of incoming links (systemic index).	$PR(v) = \dfrac{1-d}{N} +$ $d \displaystyle\sum_{u \in adj(v)} \dfrac{PR(u)}{outdeg(u)} PR(v)$	Helps identify functions that receive influence from authoritative functions, useful for understanding the importance of a function within the entire network.
Betweenness Centrality	Indicates how much a node lies on the shortest paths between all other nodes (systemic index).	$CB(v) = \displaystyle\sum_{s \neq v \neq t} \dfrac{\sigma_{st}(v)}{\sigma_{st}}$	Identifies functions that act as brokers between different components, useful for detecting key functions in variability control.

(Continued)

TABLE 12.2 (Continued)
Summary of Key Centrality Metrics Used in Complex Network Analysis, Providing a Brief Explanation, the Corresponding Formula, and Their Specific Applications within Any FRAM Instantiation

Centrality Metric	Brief Explanation	Formula	Application in FRAM
Closeness Centrality	Measures how close a node is to all other nodes in the network (systemic index).	$$CC(v) = \frac{1}{\sum_{u \in V} d(v,u)} CC(v)$$	Evaluates the accessibility of a function relative to all others, useful for identifying functions that can quickly influence or be influenced by others.
Hubs and Authorities	Hubs are important nodes that point to important Authorities; Authorities are important nodes that are pointed to by important Hubs (systemic index).	$$Hub(v) = \sum_{u \in adj(v)} Auth(u)$$ $$Auth(v) = \sum_{u \in adj(v)} Hub(u)$$	Identifies upstream Hubs that can anticipate and propagate resonances, and downstream Authorities that can accumulate resonance signals from the whole network.

Note: These metrics help in evaluating the importance and influence of functions (nodes) in a network, facilitating the identification of critical functions that can affect variability control and resonance propagation in a defined instantiation.

- **Calculation of Output Variability (OV):** The OV is the product of Timing and Precision scores (V_T and V_P) for a specific Output. This represents the intrinsic variability of the Output before any interaction with other functions. Since the scores are expressed as random variables at each run of a Monte Carlo simulation, their product will produce a different result.
- **Effects of Coupling:** Couplings between functions can amplify or dampen this variability. For each coupling, amplification or damping factors are defined for both Timing and Precision. These factors modify the OV based on the nature of the interaction between functions. Such an amplifying (dampening) effect can be described with a random variable as well. However, in our work, this was not the case. It has been chosen to be expressed as deterministic values associated with the aspects involved in the coupling and with the specific functions considered.

- **Influence of Operational Conditions**: Specific operational conditions (e.g., high workload and adverse weather conditions) can further modify the variability of the output. An index of Scenario Performance Condition Impact (SPCI) quantifies this effect for each combination of function and operational condition. This is often a nuance intended to lessen the effects of too simplistic modelling. Since the case at hand was for the most unknown, the consulting group decided to avoid useless, overly complicated calculations.
- **Calculation of VPN**: The final VPN for each coupling in a specific scenario is calculated by combining the OV modified by coupling factors and the effects of operational conditions. This index, therefore, reflects the total expected variability for a function under a given set of operational conditions.

Higher VPN values indicate greater variability and therefore greater uncertainty and potential for negative impacts on the system, making that function a priority target for mitigation interventions or further analysis.

In the semi-quantified FRAM through Monte Carlo methodology, each function's timing and precision variability are represented by discrete probability distributions, of which an example is given in Table 12.2. The simulation aggregates these distributions to calculate the VPN for each function coupling, considering amplification or damping effects in various operational scenarios. In the cited research study by Falegnami et al. ((2020), VPNs are used as weights for the associated multilayer network. In this way, the analyst can identify the most central nodes (i.e., functions). Although this method provides a useful insight, in this research it may not fully capture the range of possible system behaviours, especially under stress or unusual conditions, since the consultants did not possess enough information to model either the dynamic effects of amplifying (dampening) of the couplings nor the influence of the operating conditions. In fact, VPNs used as weights of the multilayer network are a summary of all Monte Carlo simulation runs. For this reason, since each simulation run offers a unique perspective on potential variability configurations, the consultants decided that rather than running 1000 simulation runs to obtain the weights of a network, it would be worthwhile to obtain 1000 networks – one for each run (as per the notion of a FRAM instantiation). The weights, resulting from sampling the distributions defined for Coupling Variability (CV_{ij}), define as many networks as there are simulation runs, each of which is potentially different from the others. The Coupling Variability (CV_{ij}) of a particular coupling between two functions, i and j, is an index that measures how much the variability of an output of one function can influence another function to which it is connected.

For each of these networks, centrality analyses were conducted separately, allowing the identification of functions relevant for comparative analyses and for defining the KPIs, such as the most central "emerging" ones.

In the realm of complex network analysis, the concept of centrality plays a fundamental role in understanding the importance and influence of nodes within a network (Borgatti & Everett, 2006; Newman, 2010). Some reflections are listed below:

Degree centrality, for example, measures the importance of a node by counting the number of direct connections it has with other nodes in the network. Nodes with a high value of degree centrality are those with a large number of direct connections. While often representing a good criterion for evaluating the prominence of a node, this is a criterion of local centrality, that is, not capable of fully accounting for the effects of distant nodes.

- Eigenvector centrality assigns a centrality score to each node based on the centrality of the nodes it is connected to. In other words, nodes with many connections to important nodes have a higher centrality score.
- PageRank centrality is a variant of Eigenvector which evaluates the importance of a node based on the quantity and quality of its incoming links. It forms the basis of how search engines assess the relevance of web pages based on their popularity and the authority of the sites linking to them.
- Betweenness centrality indicates the extent to which a node lies on the shortest paths (i.e., geodesics) between all other nodes in the network. Nodes with high betweenness centrality act as brokers between different components (macro-functional groups) of the instantiation.

All these "centralities" no longer refer solely to local effects but adopt a "global" perspective of the network. Beyond theoretical differences, each of these methods assigns a higher value to nodes with many connections to other important nodes, and often the choice of one criterion over another depends on the type of network available, how edges are defined, or the need for numerical convergence. In the context of FRAM, therefore, an activity that possesses high betweenness could represent a preferred function in terms of variability control. It is also important to define Hubs and Authorities, which share a circular definition:

- Hubs are nodes that are more important the more they depart arcs towards important Authorities;
- Authorities are nodes that are more important the more they receive incoming arcs from important Hubs.

These centralities are mainly used to assess the importance of upstream or downstream nodes in directed networks. Important upstream hubs could, in the FRAM context, be used to identify nodes potentially capable of anticipating the initiation and propagation of resonances. Authorities among downstream nodes could be prone to accumulating resonance signals from the instantiation as a whole (see Table 12.2).

As a result, centralities will vary from network to network, that is, from instantiation to instantiation. Functions that are more central in one simulation run may be less central in another. For this reason, rather than crystallizing variability into a static configuration, it is worth considering the potential variability represented by the ensemble of all these networks. It is from the simultaneous evaluation of the entire cloud of networks resulting from the Monte Carlo simulation that we may obtain more meaningful centrality values.

Analogous to the probability cloud of an electron's position around the nucleus of an atom, we called the cloud of resulting networks from the application of the Monte Carlo simulation the "variability orbitals". Introducing the idea of variability orbitals proposes a vision where functions can oscillate within certain limits, much like electrons in an atomic orbital. This metaphor helps visualize variability not as a single value or a fixed range but as a probabilistic space region that dynamically changes with the system's conditions. Each function can thus have different "states" of variability that interact in complex and often unpredictable ways with other states of variability. Below is the description of how the FRAM analysis process was structured, incorporating this analogical concept of variability orbitals.

STEP 0: RECOGNIZE THE PURPOSE OF THE ANALYSIS

The initial step in conducting an effective FRAM analysis is to clearly recognize and define its purpose within the context of the specific operational environment. In our case, the primary objective was prospective since the aim was to improve safety and resilience within the dairy production plant by implementing a robust LOTO procedure. This purpose required not just technical adjustments but a broader alignment with the specific needs and dynamics of a socio-technical system. For our team, understanding the unique operational complexities and goals of the client organization was essential. Management was primarily interested in practical results, specifically in improving operator awareness and safety rather than in the technical specifics of the chosen analysis method.

STEP 1: IDENTIFY AND DESCRIBE THE SYSTEM'S FUNCTIONS

The first step was to identify the system functions participating in each scenario (see Table 12.1 and Figure 12.1), as specified in the Data Collection section.

STEP 2: IDENTIFY VARIABILITY

For each scenario, data regarding the Timing and Precision phenotype distributions were gathered. The functions' relative position within the stream from the previous step is consolidated during the modelling activity. Afterwards, the analysts defined the arrangement of functions for each of the three instantiations. This means that the analyst knows the final stream of an instantiation. In terms of networks, this implies that for each of the networks corresponding to the three scenarios, all the starting nodes and all the ending nodes, that is, the edges, are defined. However, the weights of these edges have not yet been evaluated. To do so, it is necessary to identify the potential variabilities of each function within the instantiation. In terms of networks, this means describing the FRAM instantiation as a stochastic network.

The variabilities, akin to "probability orbitals", represent the range of possible ways in which each function can behave depending on the surrounding conditions, as outlined above. These "variability orbitals" reproduce the potential modulation of the reverberation of variability in the instantiation.

STEP 3: AGGREGATE VARIABILITY

The third step of the FRAM method involves aggregating variabilities through couplings. In this enriched model, couplings are considered an expression of the variability orbital, and the aggregation of variability is partially deferred to the phase of evaluating central nodes.

Using Monte Carlo simulations, various configurations can be explored through variability orbitals. Each simulation run modifies the weights of the couplings, leading to different system response configurations in that particular scenario. In practice, at this stage, there are as many networks as the number of runs defined for each instantiation (e.g., 1000 in our case).

Based on the different centrality scores, it is possible to identify which functions are worth paying attention to. Additionally, it is also feasible to define performance indicators to monitor and thus manage the system in that specific scenario to help maintain its resilient performance.

In this application, a Monte Carlo simulation is used to generate the complex of weights (the orbital) and thus the networks associated with the three investigated scenarios. For each of the networks, centrality scores are calculated, which are relative measures assimilated into lists ranking the functions. The phase of aggregating variabilities concludes by aggregating all the ranking lists into a single definitive ranking list for that scenario. This is done for all three scenarios. To obtain the final ranking list, an aggregation method described in a paper entitled "A minimum violations ranking method" is employed (Pedings et al., 2012).

Method Geek

The minimum violations ranking method utilizes linear programming to optimize a ranking aggregation function defined by conformity between input rankings, ensuring that the final ranking is as consistent as possible with the input lists.

The key steps are:

i. **Formation of Conformity Matrix (C):** The matrix C is formed based on a certain definition of conformity. In our case, each element c_{ij} is defined as the difference between the number of lists that rank element i above element j and the number that rank j above i.

$$c_{ij} = n_{ij} - n_{ji}$$

where n_{ij} is the number of lists that rank article i above article j, and n_{ji} is the number of those that do the opposite.

ii. **Optimization via Linear Programming (LP)**: Solving the LP optimization problem where the objective is to maximize the sum of the products of the elements of matrix C and a matrix of decision variables X, subject to constraints ensuring anti-symmetry, transitivity, and non-negativity of rankings.

More in detail:

Objective Function: The objective is to maximize the sum of the products of the matrix elements C and a decision variable matrix X. Essentially, this means each element c_{ij} of the conformity matrix C is multiplied by a corresponding decision variable x_{ij} in matrix X, and the goal is to maximize the sum of all these products. The matrix X effectively determines the ranking of items relative to each other.

$$Maximize \sum_{i,j} c_{ij} x_{ij}$$

Matrix X: This is a variable decision matrix where each element x_{ij} can be 0 or 1, corresponding to when item i is ranked lower than item j, or i is ranked higher than item j.

iii. **Derivation of Final Ranking**: The optimal ranking is determined by ordering the columns of the solution matrix X in ascending order under specific constraints:

- Antisymmetry, which ensures that if item i is ranked higher than item j, then the opposite is not considered possible, and it can be stated as

$$x_{ij} + x_{ji} = 1 \text{ for all } i \neq j$$

- Transitivity prevents circular reasoning in rankings. If I is higher than j, and j is higher than k, then I should also be higher than k:

$$x_{ij} + x_{jk} + x_{ki} \leq 2 \text{ for all distinct } i, j, k$$

- Non-Negativity, which ensures that the decision variables cannot be negative, thus:

$$x_{ij} \geq 0$$

Implementing these constraints ensures that the derived rankings are consistent, logical, and as aligned as possible with the aggregated preferences indicated by matrix C.

At this stage, the possible alternative scenarios in the application of LOTO procedures can be compared. For this purpose, Rank-Biased Overlap (RBO) was adopted as the key tool for quantifying the similarity between centrality rankings in different operational scenarios.

RBO, developed by Webber et al. (2010), is particularly useful in contexts where ranking lists may differ in length or be incomplete (as in our case, given the different number of functions belonging to each scenario) and where items at the top of the lists deserve predominant consideration. The importance of this analytical tool stems from its ability to assign decreasing weights to items based on their position in the list, reflecting the importance of higher positions in operational contexts where higher centrality functions play a critical role in determining system resilience.

During the analysis, the parameter p, which determines the number of positions considered at the top of the rankings, was set to 0.9, indicating a strong emphasis on top-listed items and underscoring the importance of centrality functions in evaluating system operational effectiveness. In parallel, parameter k, representing the depth of evaluation for extrapolation, was configured by default to half the maximum length between the two compared lists, thus balancing the accuracy of the calculation with the practicality of execution when comparing lists of variable length.

The calculation of RBO returns a value ranging from 0 (no similarity) to 1 (perfect similarity). This quantitative value provides a direct measure of the correlation between two analyzed scenarios or configurations.

The assessment of the orbital requires the Monte Carlo method to generate a certain number of networks associated with a given instantiation. Then, centrality associated with each of these networks is calculated to obtain the final ranking list. Finally, by calculating RBO, it is possible to compare the relative ranking lists and thus the corresponding scenarios. The authors used the R programming language for all these tasks, which is particularly suitable because it implements libraries for sampling from various types of probability distributions for the analysis of complex networks, as well as for linear algebra and optimization.

Below is the pseudocode outlining the implementation logic.

///////Function Definitions:

i. Function sample_value (probabilities, values):
  - Takes probabilities and corresponding values as input
  - Returns a value randomly sampled based on the given probabilities
ii. Function run_monte_carlo_simulation (couplings_path, num_runs = Max_num_runs):
  - Reads couplings data from table files (e.g., .csv | .xlsx) specified by couplings_path
  - Extracts timing and precision probabilities

- Initializes a weights matrix for the Monte Carlo simulation
- Runs a Monte Carlo simulation for a specified number of runs
- For each run:
- Samples timing and precision values based on their respective probabilities
  - Calculates alphaT and alphaP based on sampled values.
  - Calculates weights based on sampled values and calculated alphas.

/// The entire weights matrix expresses the probability of the orbital connections between the functions

- Returns the weights matrix.

iii. Function create_graphs_for_all_runs(weights, function_list_path):
- Takes weights as input
- Reads functions data from tables files (e.g., .csv | .xlsx) specified by function_list_path
- For k 1 to Max_num_runs:
- Constructs a graph based on functions specified in function_list_path and weights(:,k)
- Returns graph_list

//////Execution:

i. Define file paths: couplings_path_1, couplings_path_2, couplings_path_3, function_list_path

ii. Run a Monte Carlo simulation for each data set:
- Call run_monte_carlo_simulation for each couplings_path
- Store weights in weights_1, weights_2, and weights_3

iii. Create graph lists for each data set:
- Call create_graphs_for_all_runs for each couplings_path and corresponding weights
- Store graph lists in graph_list_1, graph_list_2, and graph_list_3

iv. Calculate centrality measures for each graph list:
- Call calculate_centrality_for_graph_list for each graph list
- Store centrality measures in centrality_list_1, centrality_list_2, and centrality_list_3

v. Compute conformity matrices for each centrality measure:
- Call compute_conformity_matrices for each centrality list and centrality measures
- Store conformity matrices in all_conformity_matrices1, all_conformity_matrices2, and all_conformity_matrices3

vi. Solve linear programming for each conformity matrix:
- Call process_conformity_matrices for each set of conformity matrices
- Store optimized matrices in optimized_matrices1, optimized_matrices2, and optimized_matrices3

    vii. Derive final rankings for each centrality measure:
- Call derive_final_ranking for each set of optimized matrices
- Store final rankings in final_rankings1, final_rankings2, and final_rankings3

    viii. Write final rankings to a file:
- Call write_rankings_to_file (e.g., .xlsx) with all final rankings

    ix. RBO calculations
- Call calculate_rbo for the final_ranking.xlsx

STEP 4: ASSESS THE CONSEQUENCES OF THE ANALYSIS

Based on the insights obtained from the simulation and analysis, it is possible to develop KPIs to monitor and thus manage and/or mitigate the risks associated with variabilities and their potential resonances. For example, this could involve strengthening specific couplings or adjusting operational practices. More evidence is proposed in the Results section of this chapter.

RESULTS

The RBO analysis summarized in Table 12.3 highlights a fundamental dissimilarity between the two maintenance scenarios, suggesting that they are not equivalent in terms of the performance of key functions. When identifying a scenario for the application of LOTO procedures that preserves the behaviour of the normal scenario as much as possible, the option should be for the one in which the operator associated with the maintenance line is removed (i.e., Scenario 2).

Since PageRank informs about the functions most prone to exhibit resonance considering the whole network (i.e., as much systemic information as possible), Scenario 2 appears to be generally preferable to Scenario 1, particularly for preserving more of the PageRank and Authority metrics structure. This counterintuitive result suggests that reallocating the momentarily idle operator to the PET machine line in that operational condition may not be the optimal choice. It is important to note that even if the result was validated ex post during a meeting, it remains purely theoretical,

---

TABLE 12.3
RBO Results

	Baseline vs Scenario1	Baseline vs Scenario2	Scenario1 vs Scenario2
Authority	0.42	0.53	0.44
Betweenness	0.45	0.41	0.36
Hub	0.45	0.44	0.39
PageRank	0.41	0.43	0.34

Note: A value close to 0 means that the ranking lists tend to differ more. A value of 1 means completely identical ranking lists for the considered centrality index. The LOTO scenarios possess a quite different ranking structure. Scenario 2 resembles more the Baseline scenario than Scenario 1.

based on the judgements of the interviewed experts and not fully supported by empirical evidence, also because at the time this chapter is written, the dairy company has yet to implement LOTO procedures. Furthermore, while RBO expresses similarity between rankings, it is still possible that performance in Scenario 1 might be better.

The application of the RBO provides an immediate assessment of the similarity between operational scenarios in which LOTO are implemented and the baseline configuration, highlighting how each of them affects the stability and potential vulnerabilities of the system under different operational conditions. Regarding centrality metrics, the research group has identified Betweenness, PageRank, and pairs of Authority and Hub as key indicators for monitoring and managing critical system functions (Appendix B). Comparing the values in Appendix B, Betweenness reveals functions that act as vital intermediaries in the flow of processes, suggesting a KPI to monitor the number of times interruptions or inefficiencies are mediated by these functions. For instance, the technological function <Packaging PET machine bottles>, operated by the machinery, appears particularly suitable for this purpose (especially because it is present in all scenarios). This function operates as a brokerage nexus in the three different FRAM instantiations. Defining a KPI, such as the rate of semi-finished products boxed compared to a reference value, can inform the process engineer of behaviour within normal limits.

Similarly, the PageRank parameter provides functions that are more relevant in terms of variability and connections to equally variable functions, giving us a systemic picture of variability and indicating functions more prone to functional resonance phenomena. In this regard, as confirmed by RBO, it can be noted how the structure of functions most prone to resonance remains unchanged between the Baseline scenario and Scenario 2. This leads us to define performance metrics for the functions <Carry bogs in the packaging area> and <Process Production Program>, for example, the number of references moved from the warehouse to the number requested by the line, as well as the measure of quantitative deviations for each reference produced by the respective production program anticipated. These metrics are advantageous because they also allow comparison between the two situations.

Lastly, the combined analysis of Authority and Hub enables the identification of functions that not only receive significant inputs from other parts of the system but also broadly influence other functions. We are interested in those Authorities that also have a high Hub value. For example, <Load FM2 Line Balls> allows the definition of an index that can act as a leading indicator; for example, measuring any delays in loading the FM2 line can provide an early understanding of how variations in these functions can reverberate in the system, acting as a predictor of any anomalous behaviours.

The results were presented at a two-hour meeting attended by both management and staff working on the packaging lines. Participants included the safety manager, the production manager and two packaging operators. The results of the analysis were considered interesting and plausible, particularly by the packaging operators. The production manager and the safety manager were more cautious in accepting the results. However, they all agreed that the FRAM analysis had significantly increased their knowledge of potential critical issues in their working environment. The visual representation, together with the "ranking of activities to watch out for", stimulated the critical spirit of all employees and helped to foster a shared corporate culture. The ability to define performance indicators in a systematic way was also highly valued.

CONCLUSIONS

The approach based on orbitals obtained through Monte Carlo simulations allows obtaining a much richer and dynamic view of the process. Each simulation can reveal a different configuration of central nodes and connections, offering a deeper understanding of interdependencies and potential cascading effects in different scenarios.

This approach not only identifies which functions are essential under "standard" conditions, but it also reveals how their importance can change significantly in exceptional scenarios. This way of analyzing socio-technical systems potentially offers an effective framework for developing effective mitigation strategies and designing systems that are both resilient and robust in the face of variability and uncertainties. With such an understanding, real progress can be made towards proactive management of safety and performance, fully leveraging the capabilities of the FRAM to navigate the complexity of increasingly interconnected systems dependent on a wide range of operational variables. However, the methodology used presents some limitations that should be considered in future developments.

This case study was limited to considering exclusively centrality measures, which may not be sufficient to summarize the various characteristics of the socio-technical system under examination. Indeed, the approach based on RBO and centrality, while offering a detailed view of network dynamics, may not fully capture nonlinear interactions. Further exploration of the interaction between different centrality metrics could offer a more holistic and multilevel understanding of network dynamics. These methodological improvements would not only refine the ability to monitor and manage productive operations but also design more resilient and adaptive systems capable of maintaining high performance. However, the proposed methodology proved to be effective in defining process KPIs even in scenarios that had not yet been implemented, as a preliminary low-fidelity simulation.

APPENDIX 12.A

Table 12.A.1 represents a summary for the functions across the different scenarios (e.g., S0 = Scenario 0 - baseline, S1 = Scenario 1, S2 = Scenario 2). The presence of the function in a given scenario is highlighted by the X mark on the corresponding column. For each function is also specified the type (e.g., O = Organizational, H = Human, T = Technological) and the owner (Agent).

APPENDIX 12.B

Table 12.B.1 shows Betweenness (BC), PageRank (PC), Hubs (HC), and Authorities (AC) centrality indices for the different scenarios. The darker the tone, the higher the value. Only the first seven positions are reported for conciseness. It is worth recalling that Betweenness centrality identifies functions that act as key intermediaries in process flows, PageRank highlights functions that are central to variability and interconnectedness, Hub centrality pinpoints functions that influence many others, and Authority centrality marks functions that are significantly impacted by others in the system.

TABLE 12.A.1
List of Functions and Main Properties for Each Scenario

ID	Function Name	Description	S0	S1	S2	Type	Agent
1	Analyze packaged test products	The laboratory shall carry out chemical and cryoscopic analyses of packaged test products.	X	X	X	O	Organizational Agent
2	Analyse semi-finished tanks	Semi-finished milk and cream which have been stored are analyzed by the laboratory before they can be sent to the packaging department.	X	X	X	O	Organizational Agent
3	Apply LOTO procedure to Cardboard machine 1	The operator applies the LOTO procedure		X	X	H	Maintenance Officer
4	Start test packaging on Cardboard machine 1	The cardboard packaging operator shall start packaging the test packages, which shall then be tested and sent to the laboratory for analysis.	X			H	Packaging Machine Operator Cardboard machine 1
5	Start test packaging on Cardboard machine 2	The cardboard packaging operator shall start packaging the test packages, which shall then be tested and sent to the laboratory for analysis.	X	X	X	H	Packaging Machine Operator Cardboard machine 2
6	Start packaging on Cardboard machine 1	On receipt of the positive analytical response, the cardboard packer shall start producing the required reference on the machine.	X			H	Packaging Machine Operator Cardboard machine 1
7	Start packaging on Cardboard machine 2	On receipt of the positive analytical response, the cardboard packer shall start producing the required reference on the machine.	X	X	X	H	Packaging Machine Operator Cardboard machine 2
8	Start packaging on PET machine	The PET packaging officer shall start production of the required reference on the machine.	X	X	X	H	Packaging Machine Operator PET machine

(Continued)

TABLE12.A.1 (Continued)
List of Functions and Main Properties for Each Scenario

ID	Function Name	Description	S0	S1	S2	Type	Agent
9	Load FM1 Line Balls	The cardboard packaging officer shall, depending on the reference to be made, load the cardstock of the packaging machine, the hopper of the carpets, and the cardboard store of the packaging machine.	X			H	Packaging Machine Operator Cardboard machine 1
10	Load FM12 Line Balls	The PET packaging officer shall, depending on the reference to be made, charge the hopper of the preforms, the hopper of the plugs, the coils of the label and the spools of the flask.	X	X	X	H	Packaging Machine Operator PET machine
11	Load FM2 Line Balls	The cardboard packaging officer shall, depending on the reference to be made, load the cardstock of the packaging machine, the hopper of the carpets, and the cardboard store of the packaging machine.	X	X	X	H	Packaging Machine Operator Cardboard machine 2
12	Communicate with Attaché Cardboard machine packaging 1	The packaging worker Cardboard machine 2 communicates with the packaging worker Cardboard machine 1 after the latter has been moved to Cardboard machine 2 due to the application of the LOTO procedure. The two employees must communicate with each other to coordinate their activities.		X		H	Packaging Machine Operator Cardboard machine 2
13	Packaging PET machine bottles	The packaging machine packaging the semi-finished product according to the recipe set by the PET packaging machine.	X	X	X	T	Packaging Machinery
14	Package Cardboard machine 1 test packages	The machine packages a precise number of test and pre-production analysis packages.	X			T	Packaging Machinery

(Continued)

TABLE 12.A.1 (Continued)
List of Functions and Main Properties for Each Scenario

ID	Function Name	Description	S0	S1	S2	Type	Agent
15	Package Cardboard machine 2 test packages	The machine packages a precise number of test and pre-production analysis packages.	X	X	X	T	Packaging Machinery
16	Package Cardboard machine 1 packages	The packaging machine packs the semi-finished product according to the recipe set by the cardboard packaging worker.	X			T	Packaging Machinery
17	Package Cardboard machine 2 packages	The packaging machine packs the semi-finished product according to the recipe set by the cardboard packaging worker.	X	X	X	T	Packaging Machinery
18	Check FM1 line packaging status	When the packaging is started, the cardboard packaging operator begins, in parallel with the packaging work carried out by the machine, to check that the process is being carried out correctly, supervising the packaging and paying attention to possible alerts or alerts from the machine.	X			H	Packaging Machine Operator Cardboard machine 1
19	Check FM12 line packaging status	When the PET packaging is launched, the PET packaging worker begins, in parallel with the packaging work of the machine, to check that the process is properly carried out, supervising the packaging and taking care of possible alerts or alerts from the machine.	X	X	X	H	Packaging Machine Operator PET machine
20	Check FM2 line packaging status	When the packaging is started, the cardboard packaging operator begins, in parallel with the packaging work carried out by the machine, to check that the process is being carried out correctly, supervising the packaging and paying attention to possible alerts or alerts from the machine.	X	X	X	H	Packaging Machine Operator Cardboard machine 2

(Continued)

TABLE 12.A.1 (Continued)
List of Functions and Main Properties for Each Scenario

ID	Function Name	Description	S0	S1	S2	Type	Agent
21	Process Production Program	The production forecast to be made during the following week is drawn up weekly and the quantity to be produced each day for each reference.	X	X	X	O	Organizational Agent
22	Test PET machine bottles	The PET packaging officer shall perform three tests on the bottles to verify the correct packaging: weight, marking, and absence of microleakage.	X	X	X	H	Packaging Machine Operator PET machine
23	Test Cardboard machine 1 packages	The cardboard packaging officer shall perform three packet tests to verify the correct packaging: weight, marking, and absence of microleakage.	X			H	Packaging Machine Operator Cardboard machine 1
24	Test Cardboard machine 2 packages	The cardboard packaging officer shall perform three packet tests to verify the correct packaging: weight, marking, and absence of microleakage.	X	X	X	H	Packaging Machine Operator Cardboard machine 2
25	Make OCME Vega bottles	The machine, positioned at the end of the PET packaging line, protects the bottles with a secondary packaging, wrapping them with a shrink film.	X	X	X	T	Packaging Machinery
26	Supply power to machinery	Machinery, in order to perform its function, needs to be powered by different types of primary energy (electricity, compressed air, etc.).	X	X	X	O	Organizational Agent
27	Set FM1 line recipes	The recipes are pre-inserted inside the machine and the operator selects the reference to be packaged.	X			H	Packaging Machine Operator Cardboard machine 1

(Continued)

TABLE 12.A.1 (Continued)
List of Functions and Main Properties for Each Scenario

ID	Function Name	Description	S0	S1	S2	Type	Agent
28	Set FM12 line recipes	According to the production programme, the PET packaging operator sets the recipe on both the packaging machine and the rendering machine. Depending on the reference, the type of cap, the type of preform, the amount of semi-finished product in the bottle, and the size of the flasks may vary.	X	X	X	H	Packaging Machine Operator PET machine
29	Set FM2 line recipes	The recipes are pre-inserted inside the machine and the operator selects the reference to be packaged.	X	X	X	H	Packaging Machine Operator Cardboard machine 2
30	Carton Gampack 1 packages	The packaging machine carries out its work in series at the packaging machine, dealing with the secondary packaging of the product. Using photocells, the machine is able to start the cardboard when a precise number of packaging is at the entrance so as not to slow the process down.	X			T	Packaging Machinery
31	Carton Gampack 2 packages	The packaging machine carries out its work in series at the packaging machine, dealing with the secondary packaging of the product. Using photocells, the machine is able to start the cardboard when a precise number of packaging is at the entrance so as not to slow the process down.	X	X	X	T	Packaging Machinery
32	Send semi-finished to packaging machines	The process plant operator shall enable the packaging line and packaging machine to be filled with the required semi-finished product.	X	X	X	H	Plant Operator

(Continued)

TABLE 12.A.1 (Continued)
List of Functions and Main Properties for Each Scenario

ID	Function Name	Description	S0	S1	S2	Type	Agent
33	Read packaging documents Cardboard machine 1	The cardboard packer must know the documents governing the actions to be taken during the packaging process.	X			H	Packaging Machine Operator Cardboard machine 1
34	Read packaging documents Cardboard machine 2	The cardboard packer must know the documents governing the actions to be taken during the packaging process.	X	X	X	H	Packaging Machine Operator Cardboard machine 2
35	Read PET machine packaging documents	The PET packaging officer must know the documents governing the actions to be taken during the packaging process.	X	X	X	H	Packaging Machine Operator PET machine
36	Reload FM1 line balls	During packaging, the cardboard packaging officer is responsible for reloading the packaging machine and the packaging machine with the necessary packaging.	X			H	Packaging Machine Operator Cardboard machine 1
37	Reload FM12 line balls	During packaging, the PET packaging clerk is responsible for reloading the packaging machine and the rendering machine with the necessary packaging.	X	X	X	H	Packaging Machine Operator PET machine
38	Reload FM2 line balls	During packaging, the cardboard packaging officer is responsible for reloading the packaging machine and the packaging machine with the necessary packaging.	X	X	X	H	Packaging Machine Operator Cardboard machine 1 / Packaging Machine Operator Cardboard machine 2
39	Request semi-finished on Cardboard machine 1	The cardboard packer displays the tanks available and contacts the process plant operator by telephone to send the semi-finished product to the machine.	X			H	Packaging Machine Operator Cardboard machine 1

(Continued)

TABLE 12.A.1 (Continued)
List of Functions and Main Properties for Each Scenario

ID	Function Name	Description	S0	S1	S2	Type	Agent
40	Request semi-finished on Cardboard machine 2	The cardboard packer displays the tanks available and contacts the process plant operator by telephone to send the semi-finished product to the machine.	X	X	X	H	Packaging Machine Operator Cardboard machine 2
41	Request semi-finished PET machine	The packaging worker contacts the plant operator to request the machine to be fed with the new semi-finished product.	X	X	X	H	Packaging Machine Operator PET machine
42	Fill Cardboard machine tank 1	The cardboard packaging officer shall enable the packaging tank to be filled with the semi-finished product sent by the plant operator. When filled, the control panel indicates that the machine is ready to switch to production.	X			H	Packaging Machine Operator Cardboard machine 1
43	Fill Cardboard machine tank 2	The cardboard packaging officer shall enable the packaging tank to be filled with the semi-finished product sent by the plant operator. When filled, the control panel indicates that the machine is ready to switch to production.	X	X	X	H	Packaging Machine Operator Cardboard machine 2
44	Transport cartons to palletize	The cartons are sent via conveyor belts to the fridge cell, where they will be palletized and shipped.	X	X	X	T	Packaging Machinery
45	Carry bogs in the packaging area	The references to be produced during the day are transported from the warehouse to the dedicated area on the line so that the packaging worker can have them more quickly during production.	X	X	X	O	Organizational Agent

TABLE B.1

Top-Rated Seven Functions Respectively According to BC, AC, HC, PC

Scenario 0

ID	BC	Function Name
5	44	Analyze packaged test products
36	43	Packaging PET machine bottles
3	42	Send semi-finished to packaging machines
4	41	Test Cardboard machine 1 packages
26	40	Test Cardboard machine 2 packages
6	39	Start packaging on Cardboard machine 1
27	38	Start packaging on Cardboard machine 2

ID	AC	Function Name
9	44	Process Production Program
16	43	Carry bogs in the packaging area
13	42	Analyse semi-finished tanks
3	41	Send semi-finished to packaging machines
20	40	Load FM2 Line Balls
14	39	Load FM1 Line Balls
34	38	Request semi-finished PET machine

Scenario 1

ID	BC	Function Name
5	34	Analyze packaged test products
36	33	Packaging PET machine bottles
26	32	Test Cardboard machine 2 packages
3	31	Send semi-finished to packaging machines
30	30	Check FM2 line packaging status
37	29	Test PET machine bottles
27	28	Start packaging on Cardboard machine 2

ID	AC	Function Name
16	34	Carry bogs in the packaging area
9	33	Process Production Program
30	32	Check FM2 line packaging status
45	31	Apply LOTO procedure to Cardboard machine 1
47	30	Communicate with Attaché Cardboard machine packaging 2
46	29	Communicate with Attaché Cardboard machine packaging 1
42	28	Read packaging documents Cardboard machine 2

Scenario 2

ID	BC	Function Name
5	32	Analyze packaged test products
36	31	Packaging PET machine bottles
26	30	Test Cardboard machine 2 packages
3	29	Send semi-finished to packaging machines
37	28	Test PET machine bottles
27	27	Start packaging on Cardboard machine 2
29	26	Package Cardboard machine 2 packages

ID	AC	Function Name
9	32	Process Production Program
16	31	Carry bogs in the packaging area
3	30	Send semi-finished to packaging machines
32	29	Load FM12 Line Balls
39	28	Check FM12 line packaging status
20	27	Load FM2 Line Balls
30	26	Check FM2 line packaging status

Scenario 0

ID	HC	Function Name
20	44	Load FM2 Line Balls
14	43	Load FM1 Line Balls
32	42	Load FM12 Line Balls
40	41	Reload FM12 line balls
28	40	Reload FM2 line balls
11	39	Reload FM1 line balls
34	38	Request semi-finished PET machine

ID	PC	Function Name
9	44	Process Production Program
16	43	Carry bogs in the packaging area
13	42	Analyse semi-finished tanks
3	41	Send semi-finished to packaging machines
20	40	Load FM2 Line Balls
14	39	Load FM1 Line Balls
44	38	Supply power to machinery

Scenario 1

ID	HC	Function Name
28	34	Reload FM2 line balls
20	33	Load FM2 Line Balls
32	32	Load FM12 Line Balls
40	31	Reload FM12 line balls
34	30	Request semi-finished PET machine
21	29	Set FM2 line recipes
26	28	Test Cardboard machine 2 packages

ID	PC	Function Name
47	34	Communicate with Attaché Cardboard machine packaging 2
46	33	Communicate with Attaché Cardboard machine packaging 1
9	32	Process Production Program
16	31	Carry bogs in the packaging area
13	30	Analyse semi-finished tanks
3	29	Send semi-finished to packaging machines
45	28	Apply LOTO procedure to Cardboard machine 1

Scenario 2

ID	HC	Function Name
20	32	Load FM2 Line Balls
32	31	Load FM12 Line Balls
40	30	Reload FM12 line balls
28	29	Reload FM2 line balls
21	28	Set FM2 line recipes
34	27	Request semi-finished PET machine
33	26	Set FM12 line recipes

ID	PC	Function Name
9	32	Process Production Program
16	31	Carry bogs in the packaging area
13	30	Analyse semi-finished tanks
3	29	Send semi-finished to packaging machines
45	28	Apply LOTO procedure to Cardboard machine 1
20	27	Load FM2 Line Balls
32	26	Load FM12 Line Balls

Note: Grey shading indicates decreasing values: darker shades represent higher numbers.

REFERENCES

Borella, L. de C., Borella, de C. M. R. (2021). Skills and competencies for resilience in manufacturing systems: A systematic literature review. *International Journal of Advanced Engineering Research and Science*, 8(8), 374–383.

Borgatti, S. P., & Everett, M. G. (2006). A GRAPH-theoretic perspective on centrality. *Social Networks*, 28(4). https://doi.org/10.1016/j.socnet.2005.11.005

Caputo, A. C., Donati, L., & Salini, P. (2023). Estimating resilience of manufacturing plants to physical disruptions: Model and application. *International Journal of Production Economics*, 266, 109037. https://doi.org/10.1016/j.ijpe.2023.109037

Edwards, C. J. W. (1980). Complexity and change in farm production systems: A Somerset case study. *Transactions, Institute of British Geographers*, 5(1), 45–52.

Falegnami, A., Costantino, F., Di Gravio, G., et al. (2020). Unveil key functions in socio-technical systems: Mapping FRAM into a multilayer network. *Cognition, Technology & Work*, 22, 877–899. https://doi.org/10.1007/s10111-019-00612-0

Hollnagel, E. (2009). *The ETTO principle: Efficiency-thoroughness trade-off: Why things that go right sometimes go wrong*. Ashgate Publishing

Krieger, M., Hoischen-Taubner, S., Emanuelson, U., Blanco-Penedo, I., de Joybert, M., Duval, J. E., Sjöström, K., Jones, P. J., & Sundrum, A. (2017). Capturing systemic interrelationships by an impact analysis to help reduce production diseases in dairy farms. *Agricultural Systems*, 153, 43–52. https://doi.org/10.1016/j.agsy.2017.01.022

Lundberg, J., & Woltjer, R. (2013). The resilience analysis matrix (RAM): Visualizing functional dependencies in complex socio-technical systems. *Proceedings of the 5th Resilience Engineering Association Symposium*, Soesterberg (NL), 25–27 June 2013.

Moppett, I. K., & Shorrock, S. T. (2018). Working out wrong-side blocks. *Anaesthesia*, 73(4), 407–420. https://doi.org/10.1111/anae.14165

Newman, M. (2010). *Networks: An introduction*. Oxford University Press. https://doi.org/10.1093/acprof:oso/9780199206650.001.0001

Patriarca, R., Bergström, J., Di Gravio, G., & Costantino, F. (2018). Resilience engineering: Current status of the research and future challenges. *Safety Science*, 102, 79–100. https://doi.org/10.1016/j.ssci.2017.10.005

Patriarca, R., Di Gravio, G., & Costantino, F. (2017). A Monte Carlo evolution of the functional resonance analysis method (FRAM) to assess performance variability in complex systems. *Safety Science*, 91, 49–60. https://doi.org/10.1016/j.ssci.2016.07.016

Patriarca, R., Falegnami, A., Costantino, F., Di Gravio, G., De Nicola, A., & Villani, M. L. (2021). WAx: An integrated conceptual framework for the analysis of cyber-socio-technical systems. *Safety Science*, 136. https://doi.org/10.1016/j.ssci.2020.105142

Pedings, K. E., Langville, A. N., & Yamamoto, Y. (2012). A minimum violations ranking method. *Optimization and Engineering*, 13(2), 349–370. https://doi.org/10.1007/s11081-011-9135-5

Simões, G. H., Dos Santos Pozza, M. S., Zambom, M. A., Lange, M. J., & Neumann, M. E. (2015). Dairy production system type and critical points of contamination. *Semina: Ciencias Agrarias*, 36(6), 3923–3934. https://doi.org/10.5433/1679-0359.2015v36n6p3923

Sujan, M., Watt, J., De Nicola, A., Villani, M. L., Costantino, F., Di Gravio, G., Falegnami, A., & Patriarca, R. (2021). D4-2: Definition of H(CS)2I indicators—Task 4 – FRAM analysis: Definition of H(CS)2 I indicators (Project Deliverable 4.2; SAF€RA's 2018 Fourth Joint Call: New Technologies, New Trends and Monitoring Safety Performance.).

Thaler, R. H., & Sunstein, C. R. (2009). *Nudge: Improving decisions about health, wealth, and happiness*. Penguin.

Webber, W., Moffat, A., & Zobel, J. (2010). A similarity measure for indefinite rankings. *ACM Transactions on Information Systems*, 28(4). https://doi.org/10.1145/1852102.1852106

13 Understanding Patterns in Mixed Road Traffic

Comparing Functional Critical Paths between Drivers and Automated Vehicles

Niklas Grabbe

CONTEXT: AUTOMATED DRIVING

Automating the driving process, leading to self-driving cars, has remained a prominent human aspiration. In fact, this goal is currently achievable: substantial advancements have been achieved in realizing the necessary vehicle technologies, promising an increase in traffic safety, efficiency, and driving comfort.

In general, automation refers to the full or partial replacement of a function by a machine previously carried out by a human operator (Parasuraman et al., 2000). In terms of automated driving (AD), it is crucial to thoroughly analyze the driving task itself. Geiser (1985) outlines three distinct levels of the driving task: primary, secondary, and tertiary. The primary task involves maintaining the vehicle's course at a designated speed. This task encompasses three hierarchical levels: navigation (strategic route selection), guidance (tactical manoeuvre and trajectory selection), and stabilization (operational control of acceleration and steering). Additionally, secondary driving tasks support the primary task and are contingent upon traffic and environmental conditions. For instance, drivers signal their intentions to other road users using indicators or activate windshield wipers in response to weather conditions. Furthermore, drivers may engage in tertiary driving tasks that enhance comfort or provide information, communication, and entertainment.

The Society of Automotive Engineers (SAE) introduced six levels of driving automation (LoDA) to define and distinguish the roles and responsibilities of both the driver and the automation system in relation to the primary driving task. These levels, outlined in SAE J3016 (2021), are widely accepted and utilized within the

DOI: 10.1201/9781003518167-17

automotive industry. In general, as the level increases, automation takes over more tasks, and at the same time, the driver is increasingly taken out of the loop.

Domain Expert

Specifically, at LoDA 0, the driver executes steering and acceleration of the vehicle, monitors the environment, and is the fallback solution for all driving modes. As automation levels increase, the system gradually assumes more of the driver's tasks across various driving modes. Systems up to LoDA 2 always support the driver in longitudinal and lateral control but do not replace her/him since the driver ultimately monitors the technical system and the traffic, serving as a fallback level. The decisive difference starts at LoDA 3, where the automation monitors the environment itself without drivers as supervisors but as the last fallback level with a corresponding takeover request in case of system limits or malfunctions. The next significant advancement emerges at LoDA 4, where the machine itself becomes the fallback, and the driver is entirely relieved of the driving task in certain limited driving modes. The distinction between LoDA 4 and 5 lies solely in the number of driving modes involved.

Introducing automation into other domains, such as aviation, has shown that anticipated benefits may also be accompanied by unexpected, adverse effects, causing as many problems as solved (e.g., Wiener & Curry, 1980). These adverse effects probably occur due to safety blind spots (Noy et al., 2018), mainly accompanied by the ironies and pitfalls of automation (e.g., Bainbridge, 1983). As the driver is still needed, for example, for supervision and as a fallback, technological changes lead to dynamics and adaptations by the driver collaborating with automation. However, this can lead to adverse outcomes, including risk adaptation in the form of risk or task difficulty homeostasis (Wilde, 1982), automation surprises (e.g., Sarter et al., 1997), or the out-of-the-loop performance problem (e.g., Endsley & Kiris, 1995). These phenomena result in deskilling due to a lack of practice, loss of situational awareness, mode confusion, complacency, over-trust, and mistrust, leading to misuse, abuse, or disuse of automation. Additionally, vigilance problems and boredom may result in distracted driving behaviours. After 40 years, these outdated phenomena remain widespread as the widespread use of automation is increasing. However, due to the introduction of artificial intelligence (AI) relying on opaque machine learning algorithms, the problems may even be exaggerated or at least exhibit a changing characteristic because people have trouble understanding and trusting the AI's behaviour and capabilities (Endsley, 2023). This is particularly relevant with regard to AD, where vehicles are equipped with different LoDAs and possible short-term and frequent transitions.

In addition, a road accident is a rare, Poisson-distributed, multi-causal, and emergent event. For example, drawing from data indicating that drivers experience a fatal accident every 90 million kilometres, conduct approximately 125 observations, and make an average of 12 decisions per kilometre driven,

Fastenmeier (2015) concluded that a wrong driver decision resulting in a fatal accident would likely occur after approximately 10 billion observations and 1 billion decisions. Hence, it is essential to remember that a driver's involvement in a road traffic accident is just one of many interacting elements, including other road users, road layout, environmental conditions, and vehicle components. The driver may not have prevented the accident at the last moment. Removing the driver would eliminate both the negative and positive impacts they contribute to traffic, leading to a differentiated assessment of accident development, particularly in terms of accident prevention processes. All in all, the driver is both an active and a passive participant in an accident and an accident avoidance and compensation element in the same system (Bengler et al., 2017).

Undoubtedly, automation systems capable of rapid, consistent, and precise computation have the potential to eliminate accidents related to typical driver errors, such as speeding, misjudgment of distances, or distraction. However, the current characteristic of drivers to adapt to changing system conditions to compensate for adverse road behaviours and conditions and to prevent accidents constitutes the more challenging task for automation. The numbers above provide insight into the impressive capabilities of humans in completing driving tasks, highlighting the considerable challenge for automation to match or surpass human driving performance.

As we have previously seen, the expected safety benefits of automated vehicles are highly questionable. In fact, systems of LoDA 3 and above have not yet been successfully tested and do not exist as serial vehicles on public roads for a broad operational design domain (ODD), except for narrowed ODDs such as parking, Waymo's robotaxi service in a limited urban area, or the limited functions for use in traffic jams on motorways up to speeds of 60 km/h. In particular, LoDA 4 and 5 cause great difficulties regarding safety. These will therefore be referred to as highly automated driving (HAD). The evidence of the safety of HAD is still lacking due to the so-called "approval trap" (Winner, 2016). The approval trap arises due to the impossible application of the current test concepts on HAD.

The current test concepts follow a track distance and statistical approach covering a required test distance under representative conditions in real traffic without accidents (cf. Wachenfeld & Winner, 2016). For instance, the release of a current driver assistance system requires up to two million test kilometres. However, if we still retain this procedure for HAD, Wachenfeld & Winner (2016) concluded that, for example, for a motorway pilot without reducing the test cases, 6.62 billion test kilometres without an accident occurrence would have to be completed on the public motorway to provide statistical proof of safety through real driving. This is economically and practically not feasible for HAD. Thus, the release becomes a great challenge, or the approval trap, for HAD because the release of serial production of prototypes cannot take place (Wachenfeld & Winner, 2016).

The test dilemma can only be overcome by significantly shortening the required driving distance when still using the current approach or using a completely new approach and perspective. Thus, new test methods and approaches have to be developed. These approaches must consider that automated vehicles face mixed traffic, representing at least a long transition phase during which automated vehicles with varying degrees of automation and manually driven vehicles share the road. Thus, the complex and intractable adaptations in the sociotechnical system (STS) that traffic represents must be acknowledged. Grabbe et al. (2020) argue to broaden the focus more on a systemic level and view automated vehicles in an integrated STS (Noy et al., 2018). In this context, the automated vehicle is just one element interacting with various elements in the entire system, influencing overall performance. For example, research by Preuk et al. (2016) and Ma & Zhang (2022) implies that in mixed traffic, manual drivers adapt their driving behaviour when encountering automated vehicles because those vehicles do not behave like regular drivers. Safety is thus an emergent and complex system property.

A shift to address road safety issues by a systems thinking approach is needed (e.g., Larsson et al., 2010) to understand that supposed causes like driver errors, in fact, mostly represent the effects of system-wide issues as symptoms, rather than the primary cause of accidents (Read et al., 2017). In essence, adopting a systemic perspective enables the identification of potential conflicts within interaction flows and the analysis of trade-offs to uncover unforeseen adverse outcomes, ultimately facilitating the optimization of the overall system design.

In line with this observation, a significant perspective, that is, a complexity-oriented holistic approach based on resilience engineering (RE) (Hollnagel et al., 2012), which considers interactions, processes, and patterns within a complex system that form the adaptive capacity to be resilient, is currently lacking and inevitable as a fundamental basis for the safety assessment of AD. This potentially helps to reveal hidden risks or safety blind spots of AD concerning the overall traffic system performance. Overall, the processes in the road system leading to accident development and accident avoidance have to be differentiated, including the interdependencies between each element in the system. According to Rasmussen (1997), this belongs to the comprehension of the mechanisms characterized by adaptive capacity in the system, which shapes the behaviour to create resilience. However, as the word "mechanism" could be misleading as implying a Newtonian, mechanistic reasoning, which represents a paradox to previously shown safety argumentation, we propose here to replace it with the word "pattern". Patterns represent how sharp-end, as well as blunt-end agents, adapt their behaviour to cope with the complexity of the work (Eurocontrol, 2021a, 2021b; Hollnagel & Woods, 1983), that is, how activities or functions in a system are carried out. If we recognize and understand those patterns, we can use them to understand what happens and anticipate what may happen in the future (Hollnagel, 2016). In particular, among other things, RE relies on the pattern approach, aiming to identify how adaptations in a system work and what drives these adaptation processes to develop empirical patterns of adaptive behaviour (cf. Woods & Branlat, 2017).

Given that systemic analysis methods appear most appropriate for recognizing safety patterns in road traffic concerning the context of AD, Grabbe et al. (2020) conducted a comprehensive comparison of prominent systemic methodologies, exploring their advantages and drawbacks. The authors advocate for initiating the product development cycle with the Functional Resonance Analysis Method (FRAM), ultimately concluding that the FRAM stands out as the most suitable approach for revealing safety patterns in road traffic.

This work uses the FRAM as a method for an integrated qualitative and quantitative proactive risk assessment. The scope of the analysis and the degree of resolution have to be described to set the scene and system boundary for the four steps that follow.

DATA COLLECTION

A scenario-based analysis was conducted to compare critical paths, that is, patterns, among drivers and automated vehicles leading to accident development and avoidance. Then, their potential effects are evaluated to improve the system design. Specifically, the FRAM is applied to an overtaking manoeuvre on a rural road (cf. Grabbe et al., 2022b), comparing five different instantiations (i.e., scenarios) in mixed traffic. Here, the automation of different agents will be altered. Each instantiation includes four agents: (i) ego vehicle (EV), (ii) lead vehicle (LV), (iii) rear vehicle (RV), and (iv) oncoming vehicle (OV). The EV follows LV, and since the EV is under time pressure and wants to reach its destination quickly, LV is travelling below the speed limit; it starts an overtaking manoeuvre. RV follows EV, and OV drives free on the oncoming lane, representing a platoon of OVs. Additionally, the scenario can be divided into five stages according to the behaviour of the EV: follow, swerve, pass, merge, and get-in-lane. Table 13.1 provides an overview of the five scenarios. In the first scenario, only manual drivers exist. In the second and third scenarios, EV or OV are replaced by HAD, respectively. The fourth scenario combines scenarios two and three, where HAD replaces both EV and OV. In the fifth scenario, the EV is removed by shared and traded control between the driver and automation.

TABLE 13.1
Overview of the Five Scenarios and Their Involved Agents with Altered Automation

	Agents			
Scenarios	EV	LV	RV	OV
1	Driver	Driver	Driver	Driver
2	HAD	Driver	Driver	Driver
3	Driver	Driver	Driver	HAD
4	HAD	Driver	Driver	HAD
5	Shared and traded control	Driver	Driver	Driver

Domain Expert

The assumptions are as follows:

- The system remains the same, independent from automation levels, which is why the functional structure of the FRAM model remains the same; hence, only performance variability values will change. The reason is that an FRAM model should treat humans and automation systems as equivalent producers of functions to compare the joint performance of both systems as the net result of the functional resonances.
- No vehicle-to-x (V2X) communication is implemented, and the capabilities of HAD represent the current state-of-the-art.
- Mixed traffic is set as a condition, meaning a mix of manual drivers and HAVs, but no other levels of automation and no complete penetration by HAD.
- The HAV drives compliant but has problems with sensor range and is driving obtusely with little adjustment and compensation, resulting in late reactions and no proactive behaviour.
- The driver tends to drive less compliantly but adapts and compensates better, leading to usually early and on-time reactions and proactive acting in terms of anticipation.

The overtaking scenario, selected due to its status as an accident black spot, involves numerous complex traffic tasks like swerving, speed adjustment, merging, and interactions among various road users. Thus, this scenario serves as a valuable opportunity to illustrate the interplay and complexity within road traffic, showcasing systemic dependencies among different road agents and environmental factors. The specific reasons for the five listed scenarios are the following: scenario one is the baseline, and it is interesting to compare the systemic effects when changing the EV and/or OV by automation, as these two agents are more critical for the successful outcome to overtake safely than LV and RV. Furthermore, it is interesting to analyze the system behaviour in the case of the shared and traded control concept as recommended in Grabbe et al. (2022b).

The following contents are essentially based on the research work of Grabbe et al. (2020, 2022b) and Grabbe (2024), representing a succinct overview of the application of the FRAM in the context of AD. The goals are twofold: first, the application of the FRAM to road traffic is highlighted from a practitioner-oriented perspective by providing insights about the FRAM's practical benefits and challenges concerning a safety assessment of AD; second, the focus is on the methodological advancements of the FRAM by using quantitative metrics concerning complexity and interaction intertwined with performance variability to identify critical patterns for system stability in order to facilitate the system's adaptive capacity.

The FRAM, traditionally a qualitative research method, poses challenges for traditional statistical procedures used in quantitative research due to differences in

TABLE 13.2

Assignment of the Quality Criteria in Quantitative and Qualitative Terms as Well as Their Verification Procedures Based on Anfara et al. (2002) and Creswell & Miller (2000), According to Grabbe et al. (2022b)

Quantitative Term	Qualitative Term	Verification Strategies
Internal validity	Credibility	Prolonged engagement in the field; Use of peer debriefing; Triangulation; Member checks; Time sampling; Persistent observation; Clarifying researcher bias
External validity	Transferability	Provide a thick description; Purposive sampling
Reliability	Dependability	Create an audit trail; Code-recode strategy; Triangulation; Peer examination; Stepwise replication
Objectivity	Confirmability	Triangulation; Practice reflexivity

assessing quality criteria such as internal and external validity, reliability, and objectivity. Anfara et al. (2002) introduced a translation of these criteria into qualitative terms: credibility, transferability, dependability, and confirmability. This translation aims to enhance the evaluation of research quality and rigour, thereby improving trustworthiness. Additionally, Creswell & Miller (2000) outlined various verification strategies to meet these qualitative terms. The assignment of the quality criteria in quantitative and qualitative terms, as well as their verification procedures, can be taken from Table 13.2. Here, the verification strategies underlined boldly are implemented in this research to fulfil the four qualitative terms (cf. Grabbe et al., 2022b).

METHODOLOGY

The four building FRAM steps and the aforementioned quality criteria and verification strategies are intertwined, which is explained below.

Step 0 refers to the analysis of an overtaking scenario on a rural road, used to compare five different instantiations, involving four different agents.

For Step 1, a Work-as-Imagined (WAI) Model and Work-as-Done (WAD) model were created and combined to develop an overall model, as a mere understanding of the assumed overtaking processes is insufficient. This aims to refine and augment the WAI model into a more realistic overall model in iterative steps. Finally, the overall model was formally checked for completeness and consistency and calibrated and validated through a focus group within a peer review workshop, providing face validity to ensure objective, reliable, and valid analysis results based on the FRAM model.

In terms of Step 2, a mix of objective and self-reported data was used to determine performance variability. Objective data was gathered in a driving simulator study, whereas interviews and a survey served as self-reported data.

Concerning Step 3, the aggregation of variability to determine potential functional resonance was achieved using a semi-quantitative approach (cf. Grabbe et al., 2022b) to handle the complex scenario, which would become overwhelming using mere qualitative modelling by the FRAM.

Lastly, in Step 4, the performance variability of the entire system was managed by deriving system design recommendations through a well-reasoned function allocation, that is, the design decision of which system functions are to be performed by humans and which should be automated and to what extent to improve system safety. To achieve the well-reasoned function allocation, the performance variability of the entire system is analyzed by comparing the contributions between driver and automation to road safety based on systemic patterns on both an abstract global level to compare the system performance between different instantiations and a fine-grain level regarding the comparison of individual functions. Quantitative evaluations on the global and functional level enable a systematic and structured analysis even for largely complex FRAM models, ensuring comprehensive results, which then have to be qualitatively reflected in the model to enable the patterns to be fully understood.

Step 1: Identify and Describe the System's Functions

Develop the WAI Model. The WAI model is based on a comprehensive and detailed hierarchical task analysis of driving developed by Walker & Stanton (2017). Here, the tasks and plans are constructed using logical operators such as "And, Or, If, Then, Else, and While". The list of tasks and plans, which are essential for the overtaking scenario, was translated into functions where the logical operators were used to define couplings between each function through their aspects.

Develop the WAD Model. The WAD model used observations and interviews conducted within a static driving simulator as information elicitation. Ten participants with a valid driving license were involved in the study, during which driving data, audio recordings, and driver behaviour by video recordings were captured for analysis. The responses in the interviews, as well as the audio track during the experiment, were collected, categorized, and assigned frequencies. From this processed interview data, as well as objective data streams such as the longitudinal and lateral driving behaviour in response to scenario objects or the behaviour of other drivers, activities for driving tasks were identified and subsequently translated into functions.

Method Geek

The experiment spanned 30 minutes, with each participant navigating the scenario from the perspective of the four agents. Each participant underwent three passes for each perspective: the first pass served for familiarization, the second involved thinking aloud and explaining actions, and the third pass included five simulation stops corresponding to the scenario's stages, where subjects detailed the functions they would perform in the subsequent seconds, encompassing perception, cognition, and action levels. Afterwards, a semi-structured interview was conducted, delving into aspects of the overtaking process from the viewpoint of all four agents. The interview comprised ten questions: the initial six focused on executing the overtaking manoeuvre across its five stages, addressing factors such as decision-making criteria, execution, risk assessment, reactions to risks, and successful completion. The remaining four questions explored broader themes such as environmental perception, time pressure impacts, and factors triggering critical situations.

Develop the overall model. Initially, two researchers independently compared their individually created WAI and WAD models, aiming to integrate them into a comprehensive model. The approach involved leveraging the WAI model as a foundational framework, with additional functions, aspects, and couplings identified by the WAD model incorporated. Subsequently, the two researchers merged their independently generated overall models through collaborative comparison and discussion.

Method Geek

For this activity, a structured process was used where the models were walked through step by step per agent and temporal stage of the scenario, starting with EV in the Follow stage. First, the functions and their granularity level were compared to consider whether functions should be split into more detailed functions or aggregated into more abstract functions. Second, the aspects of each function and the resulting couplings were compared. In particular, it was discussed whether the aspects are assigned to the proper type of aspect (e.g., input or precondition) and which aspects are essential or optional to produce a desired output for each respective function. Lastly, the assignment of the agent and temporal stage to each function was compared.

In the final phase, the researchers iteratively refined the comprehensive model by conducting an in-depth cognitive walkthrough and systematically reviewing the model to identify potential omissions, erroneous couplings, and missing functions. The overall model, as well as the WAI and WAD models, was produced using the software FRAM Model Visualizer (FMV).

Validate the overall model. The overall model was verified for completeness and consistency and calibrated and validated through a focus group within a peer review workshop, providing face validity to ensure objective, reliable, and valid analysis results based on the FRAM model. The verification was implemented after the peer review workshop, having agreed that the overall model accurately reflects the essential processes of the overtaking scenario. Here, the model has been checked and adjusted for consistency and completeness using the FRAM Model Interpreter (FMI), which is a stepwise automatic interpretation of the syntactical and logical correctness of the overall model.

The peer review workshop involved seven experts with strong knowledge and broad experience of either human factors or functional development of automation in the automotive area in industry or academia. The profiles represent a mix of human factors specialists, engineers, and senior researchers with experience ranging from five to thirty years.

Method Geek

The workshop was managed as follows. First, the experts were educated about the FRAM model and its creation process one week before the workshop through a 90-minute recorded video. In addition, general background information about FRAM was given to familiarize the peers with the method, and participants were divided into three groups to provide comments on the specific agents. In the workshop, the overall model was then discussed step by step for each agent. The model was explained and discussed in detail for the respective three groups, and the experts gave their feedback, and the models were iteratively adapted. At a follow-up meeting, the overall model was finally iteratively calibrated and fine-tuned again with all seven peers in a joint two-hour session. To validate the overall model, the peer group reflected on their personal experience and human factors knowledge of driving a car, including manual driving as well as AD. This contained additions, modifications, or deletions regarding functions, aspects, and their couplings, as well as the assignment of agents and temporal stages.

Additionally, Grabbe et al. (2022a) provide a more formal approach to achieving and demonstrating the reliability and validity of a FRAM model. In particular, different criteria (i.e., reliability, verification, and validity) and methods for their evaluation are presented. For example, in terms of validity, construct-, content-, and predictive validity should be distinguished, which can be addressed by face validation, theory-based validation, and a mix of function- and outcome-based validation. For the latter one, Grabbe et al. (2022a) used the signal detection theory in a driving simulator study to evaluate the predictive validity of the FRAM model.

STEP 2: IDENTIFY VARIABILITY

A mix of objective and self-reported data was used to determine the performance variability. Objective data, including driving dynamics and driver and glance behaviour, were gathered in the driving simulator, whereas interviews and a survey served as self-reported data. The data is used to define the performance variability in terms of the different characteristics of the phenotypes of timing and precision for each function, as explained in more detail below. The performance variability is shown via frequency distribution to create a realistic representation of actual everyday performance.

Driving simulator study. A second driving simulator study was conducted where the simulator environment and the setting were the same as mentioned above. Overall, 30 subjects with a valid driving license took part in the study, considering balanced age and gender distribution. It should be emphasized that the sample represents heterogeneous drivers based on mileage and driving styles.

Method Geek

Before the actual test drive, the subjects took a 15-minute drive on a rural road for familiarization. According to the Wiener driving test (Risser & Brandstätter, 1985), an observation period of about 15 min is necessary before drivers show their everyday normal driving behaviour and fall into their regular habits, which should ensure a valid investigation of everyday performance variability. Besides recording driving data, audio tracks, and the driver's behaviour, the glance behaviour was tracked with a head-mounted eye-tracking system focusing on scanning patterns and fixated objects. This ensured insights, especially into the drivers' perceptual behaviour, in addition to executive activities and to record cognitive processes. The participants drove the four agents' perspectives three

TABLE 13.3

Definition of the Timing and Precision Characteristics Using the Lane-Keeping Function as an Example

Phenotype	Characteristic	Definition
Timing	Too early	If the driver already countersteers although the vehicle is driving in the middle of the lane.
	On time	If the driver countersteers in time (the vehicle is approaching the left or right of the lane boundary) to keep the vehicle in the lane.
	Too late	If the driver countersteers too late (the vehicle has already left the lane) to keep the vehicle in the lane.
	Not at all	If the driver does not countersteer at all to keep the vehicle in the lane.
Precision	Precise	If the car always drives perfectly along the centre line between the left and right of the lane boundary.
	Acceptable	If the car always drives between the left and right of the lane boundary.
	Imprecise	If the car crosses the left or right boundary of the lane.

Source: Grabbe et al. (2022b).

times in permutated order, leading to three runs for each agent per subject. The subjects were instructed to reproduce their everyday driving behaviour and complete overtaking manoeuvres and driving tasks as quickly as possible, but as safely as necessary. To determine performance variability, the driving data and glance behaviour were evaluated for each run, which led to a total of 90 data sets per agent and function. Each run was assigned to the different characteristics of the timing and precision phenotypes based on previously established definitions of the characteristics of the phenotypes per function. Here,

Table 13.3 exemplifies this for the lane-keeping function. Finally, this resulted in a frequency distribution of performance variability for each function, for example, for timing, 90% on time and 10% too late, and for precision, 20% precise and 80% acceptable. For the percentage calculation, the numerator is represented by the observed number of runs specified as a respective characteristic of timing or precision, and the denominator is represented by the total number of runs.

Interviews and surveys. Unfortunately, only a few functions' performance variabilities for the driver, including mainly functions referring to actions, could be reliably determined by observation in the driving simulator. However, evaluation of much of the perceptual and cognitive processes remained elusive through this method. Consequently, a subsequent phase involved conducting large-scale structured interviews along with a survey. In essence, the guiding principle for determining performance variability per function was as follows: if the variability of a function could be objectively measured in the simulator study, those values were utilized; otherwise, values from the interviews and survey were employed. Given that subjective assessment through interviews predominantly captured functional variability, the drivers' self-assessment played a pivotal role. Overall, 30 subjects with a valid driving license took part in the study.

Method Geek

The survey covered over 100 functions and was gone through step by step in an interview so that queries could be clarified. The driving tasks were always queried according to the stages of the scenario, and the subjects were informed about the stage at which the driving task was performed. For each driving task, the name of the driving task, the performing agent, a description of the task of the function, and the output of the same were given. This was followed by the evaluation of variability in timing and precision. Here, the subjects expressed in per cent how often or how they perform a driving task in everyday life, concerning timing or precision phenotypes, respectively. The defined value ranges included never (0%), rarely (1–25%), sometimes (26–50%), often (51–75%), usually (76–99%), and always (100%). The cumulative responses for each participant had to total 100% in every instance. Lastly, the performance variability distribution ratings for each function were averaged across all participants for each characteristic.

In light of the limited availability of public data regarding AD performance and driving behaviour, structured interviews along with a survey were conducted to gauge performance variability for automation as a general concept, drawing on the current state-of-the-art of automation systems and near-future advancements. Twelve experts participated in these interviews, bringing diverse perspectives from their roles in AD function development, alongside substantial practical and theoretical insights into the performance of existing series and prototype functions. The interview procedure and questionnaire structure mirrored those used for the driver case, outlined earlier.

Method Geek

However, the individual questions differed slightly; instead of providing frequency distributions for timing and precision characteristics, experts were tasked with selecting the most probable characteristic per phenotype (single choice) within the analyzed scenario for AD, considering the backdrop of short-term automation advancements. Subsequently, all ratings from each expert were aggregated into a frequency distribution of performance variability for each function.

STEP 3: AGGREGATE VARIABILITY

To introduce the semi-quantitative approach, first, a contextualization of the Abstraction/Agency framework by Patriarca et al. (2017a) and the Time-dependent framework by Steen et al. (2021) into a Space-Time/Agency framework (cf. Grabbe et al., 2022b) was developed. Hereby, the functions of the FRAM model are assigned to the four different agents (EV, LV, RV, and OV) and five temporal stages during the scenario (Follow, Swerve, Pass, Merge, and Get in lane). This should ensure enhanced knowledge representation combined with a multi-dimensional approach that has two dimensions: the temporal-spatial levels and the agency levels. Since it is not effective to analyze an STS according to only one level (Patriarca et al., 2017a), this approach simplifies dealing with complexity that requires a system to be structured following different levels of analysis with different resolutions and perspectives. This is evidenced by the interactions® occurring within an agent and between different agents across varying temporal and spatial contexts.

The semi-quantitative approach was implemented with the help of the software myFRAM 1.0.4 (Patriarca et al., 2017b), facilitating the conversion of the FRAM model into a matrix format, thereby enabling quantitative or numerical calculations by assigning numerical scores to each performance variability characteristic. In general, the metrics can be divided into three categories: functional variability, system resonance, and system propagational variability. The functional variability represents the variability that a function directly receives and transfers without considering its interaction and effect in the system sufficiently. Therefore, the system resonance tries to reflect the interaction and complexity of a function in the system, incorporating the non-linearity, emergence, and dynamics of the system. It is a kind of weighting of the impact and affectedness of a function to evaluate the effect of a function variability system-wide. Combining functional variability and system resonance results in system propagational variability, which shows the systemwide impact and affectedness of each function's variability up to a global system variability level.

Method Geek

In the following, the main analysis metrics are briefly described. Also, the basic formulas are given. The definition and calculation of each metric were implemented with myFRAM and MATLAB® 2020. MATLAB was used to calculate centrality metrics and feedback loops as explained further below. Here, the nodes represent the respective metrics, and the structure, that is, which metrics are composed of how, is marked by arrows and their direction from right to left. However, the complete formulas and more details, including a graphical representation of the process, can be found in Grabbe et al. (2022b). It should be noted that these assessments are valid for an instantiation of the model, not the full model, as they rely on its conversion into a graph.

The functional variability is based on the coupling variability (CV) of one foreground function by its downlinks (DL) and uplinks (UL), resulting in two metrics: downlink functional coupling variability (DLFCV) and uplink functional coupling variability (ULFCV). The CV is the product of the upstream output variability in terms of timing and precision, and the associated propagation factors. The DLFCV (Eq. 13.1) is used to understand the implications of the CVs of one entire upstream function j to associated downstream functions i, and the ULFCV (Eq. 13.2) is used to comprehend the impact of the variability of a downstream function i through its incoming CVs of upstream functions j:

$$DLFCV_j = \sum_{i=1}^{n} CV_{ij}$$ (13.1)

$$ULFCV_i = \sum_{j=1}^{n} CV_{ij}$$ (13.2)

However, the metrics mentioned earlier failed to adequately capture the inter-twining functions, as they were solely assessed in isolation, lacking consideration for interactions except for variability propagation factors. Hence, several metrics delineating system resonance were defined and categorized into dimensions of interaction and complexity. On the one hand, the interaction of functions was determined with the following metrics to calculate the degree to which a function interacts with other functions or agents in the system represented as an instantiation:

- Number of downlinks (i.e., outdegree) and uplinks (i.e., indegree) showing how many functions a function can directly influence, and how many functions it is directly influenced by;
- Interrelatedness expressing how many functions a function is linked to within an agent and within the same stage or in different stages;
- Interrelatedness presenting how many functions of other agents a function is linked to and weighting it with the number of different agents;
- Feedback loop factor reflecting the extent to which a function's output can influence its input through direct and indirect feedback loops.

On the other hand, centrality measures from graph theory were utilized to approximately depict parts of the system's complexity. This selection stems from the effectiveness of graph theory in exploring emergent nonlinear characteristics of systems, demonstrating success in elucidating various features of complexity (Falegnami et al., 2020). Also, the translation of a FRAM instantiation into a network using graph theory has been previously implemented by Bellini et al. (2017) and Falegnami et al. (2020), showcasing the overall compatibility of these approaches. Finally, the following three different centrality indices and one self-defined metric were chosen, assuming the best way to represent this complexity:

- Katz-centrality depicting the relative degree of influence of a function within the system to show the extent of indirect impact.
- Incloseness- and Outcloseness-centrality assessing the centrality of a function within a system, which determines its proximity to other functions.

Consequently, a function's high centrality indicates its potential for functional resonance.

- Betweenness-centrality demonstrating the extent to which a function connects with other functions, which underscores its critical role in ensuring the system's success.
- Clustered Variability showing the accumulation of variability both upstream and downstream of a function, which reveals clusters of functions directly linked with high variabilities.

These metrics result in two main indicators of system resonance: the Weight as Upstream (WaU) and Weight as Downstream (WaD). The WaU and WaD reflect the system effect of a function as an upstream and downstream function, respectively. This should simulate the interaction and fit between functions and their inherent complex interdependencies. The WaU and WaD of one function (f) are determined via (Eq. 13.3) and (Eq. 13.4), where the respective metrics are included in the calculation by assigned weighting factors ß:

$$
\begin{aligned}
WaU_f = \ &\beta_1 * N_{DL}^{relative}{}_f + \beta_2 * Intrarelatedness_f^{relative} \\
&+ \beta_3 * Interrelatedness_f^{relative} \\
&+ \beta_4 * FeedackLoopFactor_f^{relative} + \beta_5 * CTV_f^{relative} \\
&+ \beta_6 * Katz-centrality_f^{relative} \\
&+ \beta_7 * Outcloseness-centrality_f^{relative} \\
&+ \beta_8 * Betweenness-centrality_f^{relative}
\end{aligned} \tag{13.3}
$$

$$
\begin{aligned}
WaD_f = \ &\beta_1 * N_{UL}^{relative}{}_f + \beta_2 * Intrarelatedness_f^{relative} \\
&+ \beta_3 * Interrelatedness_f^{relative} + \beta_4 * FeedackLoopFactor_f^{relative} \\
&+ \beta_5 * CTV_f^{relative} + \beta_6 * Katz-centrality_f^{relative} \\
&+ \beta_7 * Incloseness-centrality_f^{relative} \\
&+ \beta_8 * Betweenness-centrality_f^{relative}
\end{aligned} \tag{13.4}
$$

The assignment of the weighting factors was subjective and follows the logic that some metrics weigh more heavily than others. For example, interrelatedness weighs more heavily than interrelatedness since this considers that influencing other agents has a higher system effect than only influencing one's own agent.

In terms of system propagational variability, the WaU and WaD are offset against the CV values of each function, resulting in a relative DLFCV (Eq. 13.5) and relative ULFCV (Eq. 13.6):

$$
DLFCV_j^{relative} = \sum_{i=1}^{n} CV_{ij} * WaU_j * WaD_i \tag{13.5}
$$

$$
ULFCV_i^{relative} = \sum_{j=1}^{n} CV_{ij} * WaU_j * WaD_i \tag{13.6}
$$

Finally, this leads to the overall functional coupling variability (OFCV) (Eq. 13.7), which identifies critical functions with high potential for functional resonance, offering functional prioritization of their impact on the system:

$$OFCV_f = ULFCV_i^{relative} + DLFCV_j^{relative} \qquad (13.7)$$

For example, a high value means that the function has a large systemic effect and/or is largely systemically affected. In the last step, a global system variability (GSV) (Eq. 13.8) could be calculated as the sum of the OFCVs of n functions within the whole system:

$$GSV = \sum_{f=1}^{n} OFCV_f \qquad (13.8)$$

GSV expressed as the Global System Variability shows the accumulated variability of all functions and their interactions with the whole system for one specific condition. This enables, for example, a comparison of system performance between a system where purely human drivers operate and one where an automated system operates with human drivers inside other vehicles.

STEP 4: ASSESS THE CONSEQUENCES OF THE ANALYSIS

The performance variability of the entire system was managed by deriving system design recommendations through a well-reasoned allocation of functions. To achieve this, the performance variability of the entire system is analyzed by comparing the contributions of both the driver and automation to road safety. This analysis is based on three levels:

- First, an abstract global level to compare the entire system performance between different instantiations.
- Second, a fine-grain level regarding the comparison of individual functions.
- Third, based on the evaluations on the functional level, risk functions and their critical paths could be identified, which represent patterns occurring repeatedly over several scenarios. It should be pointed out that these patterns represent leverage points within the system that emerge as the most efficient means to enhance the system's resilience in overtaking scenarios.

Method Geek

On the abstract level, the GSV is evaluated for the overall model or per stage, agent, and function type to get a differentiated view of abstract spots of high destabilization in the system between the scenarios. It has to be noted that the GSV is a relative rather than absolute value, meaning that, for example, the stagewise GSV shows the GSV in relation to the number of functions within a respective stage.

On the functional level, due to the large complexity of the FRAM model, it is impossible to compare all functions. Therefore, a selection has to be made which is based on risk or critical functions identified for each scenario. It is important to recognize that this analysis establishes the priority for intervention by starting the investigation with the most critical functions. This underscores the possibility and necessity of different and additional priorities and analyses. The rationale behind this is that critical functions possess the greatest potential to propagate functional resonance within the system, consequently leading to emergent events. In particular, the OFCV of each function was prioritized and ranked for each scenario using the scree test following the approach by Grabbe et al. (2022b). The risk functions represent critical paths, which will be explained below. When comparing the OFCVs for the risk functions in each scenario, different spots of high destabilization on the functional level can be identified.

The previous analysis provides an overview of the criticality of functional variabilities within the system, albeit in a condensed and simplified manner. However, this criticality comprises two distinct dimensions: functional variability and system resonance. Consequently, the subsequent analysis delves into these dimensions separately for risk functions to achieve a more comprehensive understanding. This is proposed by a matrix that represents the criticality of functions and their potential for functional resonance along the two dimensions of functional variability and system resonance, which make up the Functional Variability-System Resonance Matrix (FVSRM) (Grabbe et al., 2022b). The FVSRM assigns functions into nine different areas, considering five colours (see Figure 13.1):

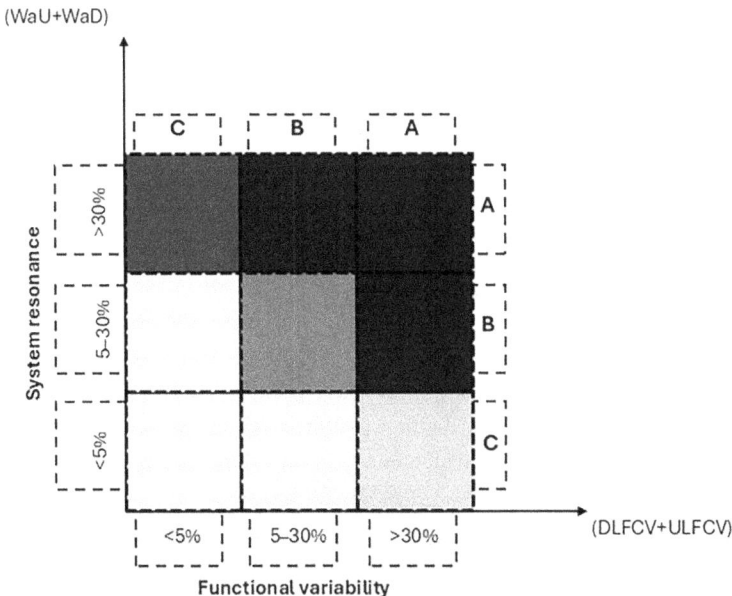

FIGURE 13.1   The functional variability-system resonance matrix (FVSRM).

- light grey (C-C, C-B, B-C) for non-critical functions,
- grey (A-C) for high variable functions with low system resonance,
- dark grey (B-B) for medium variable functions with medium system resonance that are between uncritical and critical functions,
- darker grey (C-A) for low-variable functions with high system resonance, and
- black (B-A, A-A, A-B) for critical functions.

The quantifications serve to provide an overview of the influence and impact of system functions, along with their variabilities and interactions, across different scenarios. This comparison identifies areas with a high potential for functional resonance, serving as leverage points for system design. According to Meadows (1999), leverage points are places within a complex system where small shifts or adjustments in one or a few element(s) (i.e., functions) can produce significant changes in the system and thus represent points most effectively to intervene in the system. In this work, patterns are established using critical paths of risk functions, which are defined as the direct couplings between a risk function and its upstream and downstream functions, excluding all indirect connections to avoid complicating the analysis. Following the Pareto principle, the direct couplings of a risk function represent the interactions with the highest leverage to improve the resilience of the system in comparison to the indirect couplings. While indirect couplings can also contribute positively to improved system resilience, their impact is considerably less, such as fine-tuning. Ultimately, the patterns represent critical paths occurring repeatedly over several scenarios, which is in accordance with the "pattern-centered inquiry" from Alexander et al. (1977), which states that a pattern is rather general but expressed in many different situations and settings. In essence, the identified patterns serve as pivotal points in system design aimed at enhancing resilience, owing to their significant system impact or effect. In principle, upstream functions can adapt to facilitate the performance of the risk function, while downstream functions can adapt to compensate for potential adverse effects of the risk function. Conversely, the risk function can support the performance of their downstream counterparts, potentially enhancing overall system stability. These patterns represent two sides of the same coin as providing a positive or negative contribution to road safety. Effective coordination of adaptations ensures system safety, avoiding potential accidents; however, inadequate coordination may lead to accident development and probably its occurrence. Thus, patterns exhibit a strong stabilizing or destabilizing nature, contingent upon the fitting of performance variability. This highlights why an activity conducted with identical variability in two distinct contexts may result in acceptable system performance in one scenario but potentially lead to an accident in another. Thus, context and the interactions in the whole system matter.

RESULTS

In the following, the main findings by Grabbe et al. (2022b) and Grabbe (2024) are shown to capture key insights.

OVERALL FRAM MODEL OF OVERTAKING

The generic FRAM model for overtaking on rural roads (cf. Grabbe et al. 2022b) comprises 285 functions (210 foreground functions and 75 background functions) with 799 couplings.

Domain Expert

Table 13.4 presents a synthetic view of the functions: in addition to the agent and stage assignment, the driving functions are categorized into three levels of information processing – perception, cognition, and action – utilizing the framework of types and levels of automation based on the four-stage model of human information processing presented by Parasuraman et al. (2000). This facilitates function allocation between humans and automation. More elaborate information can be found in Grabbe et al. (2022b).

TABLE 13.4
A Brief Description of the Main Functions of the Overall FRAM Model for Each Agent and Stage

Stage	EV	LV	RV	OV
Follow	to follow LV through (i) recognizing the following situation, (ii) keeping the lane, and (iii) maintaining headway separation; to decide to overtake or not, mainly based on (i) assessing the opportunity to overtake safely, (ii) judging whether overtaking is permitted, and (iii) evaluating the reasonableness of overtaking	to drive free by (i) keeping the lane and (ii) adjusting adequate speed; to react to being followed by EV through (i) observing EV's intention to overtake as well as its (ii) following distance	to follow EV through (i) recognizing the following situation, (ii) keeping the lane, and (iii) maintaining headway separation	to drive free by (i) keeping the lane and (ii) adjusting adequate speed
Swerve	to adopt the overtaking position by (i) lane keeping, (ii) reducing headway from the normal following, and (iii) adjusting the speed to that of LV; to swerve completely to the oncoming lane afterward, (i) check any hazards behind or in front, (ii) assessing the overtaking opportunity is still safe, and (iii) use the left indicator	to detect EV's swerving into the oncoming lane; to maintain speed; to react to being passed by (i) responding to potential passing problems of EV (optional)	to detect EV's swerving into the oncoming lane; to react to being passed by (i) responding to potential passing problems of EV (optional)	to detect EV's swerving into the oncoming lane; to maintain speed; to react to being passed by (i) responding to potential passing problems of EV (optional)

(Continued)

TABLE 13.4 (Continued)

A Brief Description of the Main Functions of the Overall FRAM Model for Each Agent and Stage

Stage	EV	LV	RV	OV
Pass	to perform the overtaking through (i) accelerating LV decisively or (ii) merging back into the starting lane if the manoeuvre is unsafe, and (iii) abandoning the manoeuvre	to detect the passing vehicle in peripheral vision; to react to being passed by (i) responding to potential passing problems of EV (optional)	to react to being passed by (i) responding to potential passing problems of EV (optional)	to react to being passed by (i) responding to potential passing problems of EV (optional)
Merge	to merge progressively into the starting lane by (i) adjusting EV's speed in relation to other traffic, (ii) assessing the situation to enter safely, and (iii) using the right indicator	to prepare to provide a larger opening for EV to merge back; to react to being passed by (i) responding to potential passing problems of EV (optional)	to prepare to provide a larger space to LV in case of EV's manoeuvre abandoning, or to catch up to LV; to react to being passed by (i) responding to potential passing problems of EV (optional)	to prepare for braking; to react to being passed by (i) responding to potential passing problems of EV (optional)
Get in lane	to complete the overtaking through (i) positioning into the starting lane, (ii) evaluating the driving situation, and (iii) resuming at the desired speed	to follow EV; to react to being followed by RV	to follow LV	to drive free

Source: Grabbe et al. (2022b). The wording "(optional)" means that this function or task is not necessarily fixed to the assigned stage and rather can be executed in the Swerve, Pass, or Merge stage or not at all if not required. The underlined words refer to the function's name.

GLOBAL LEVEL

Figure 13.2 depicts the GSV for the overall system and per stage between each scenario. It has to be noted that the GSV here is a relative rather than an absolute value, meaning that, for example, the GSV per stage shows the GSV in relation to

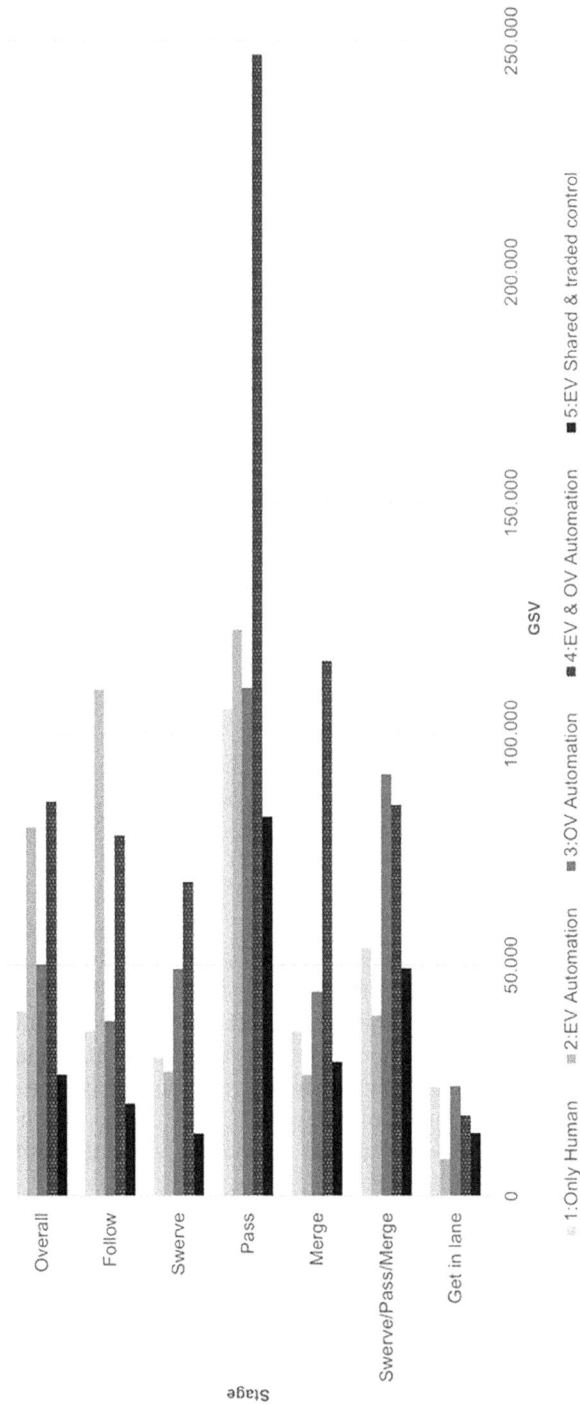

FIGURE 13.2 Comparison of the global system variability for the overall system and per stage between each scenario.

the number of functions within a respective stage. Also, the higher the GSV, the higher the potential instability in the system. Overall, scenario five exhibits the highest level of stability, whereas scenario four demonstrates the highest degree of instability. Interestingly, the proportions among the stages remain largely consistent across all scenarios. For instance, the Pass stage demonstrates the highest GSV, while the Get-in-lane stage shows the lowest GSV within each scenario. Consequently, it is noteworthy that the Pass stage emerges as the most critical one across all scenarios.

Moreover, it becomes evident how the interactions and interdependencies lead to different consequences regarding the GSV. For instance, in scenario four, when both the OV and EV are automated, there is a notable increase in GSV during the Pass stage. Conversely, in scenarios two or three, solely automating either the EV or OV results in a marginal rise in GSV compared to scenario one. Instead, in the Follow stage, automating both OV and EV in scenario four leads to a considerable reduction in GSV compared to solely automating the EV in scenario two. These instances illustrate how the automation of OV alongside EV can yield both positive and negative effects on the functional resonance within the system.

Potential for LoDA 4 is discernible primarily for EV during the Merge and Get-in-Lane stages, as well as for EV functions associated with the performance of LV, RV, and OV during the Swerve, Pass, and Merge stages. In other cases, adopting a shared and traded control concept for EV is advisable.

FUNCTIONAL LEVEL

A more nuanced analysis of risk functions for both human drivers and automation was conducted, focusing on the two dimensions of functional variability and system resonance, as illustrated in Figure 13.3. In Figure 13.3a, the functional variability (represented by DLFCV and ULFCV as stacked columns on the left y-axis) and system resonance (indicated by WaU and WaD as stacked markers on the right y-axis) of risk functions are depicted for human drivers, while Figure 13.3b represents these dimensions for the automation of EV. Certain risk functions for human drivers are highlighted and elucidated below. Functions highlighted in light grey are deemed most critical due to their combination of high functional variability and high system resonance. It should be noted that in the following qualitative description of the results, the notions of "high", "low", "little", and "much" are related to the percentual assignments in the nine areas of the FVSRM.

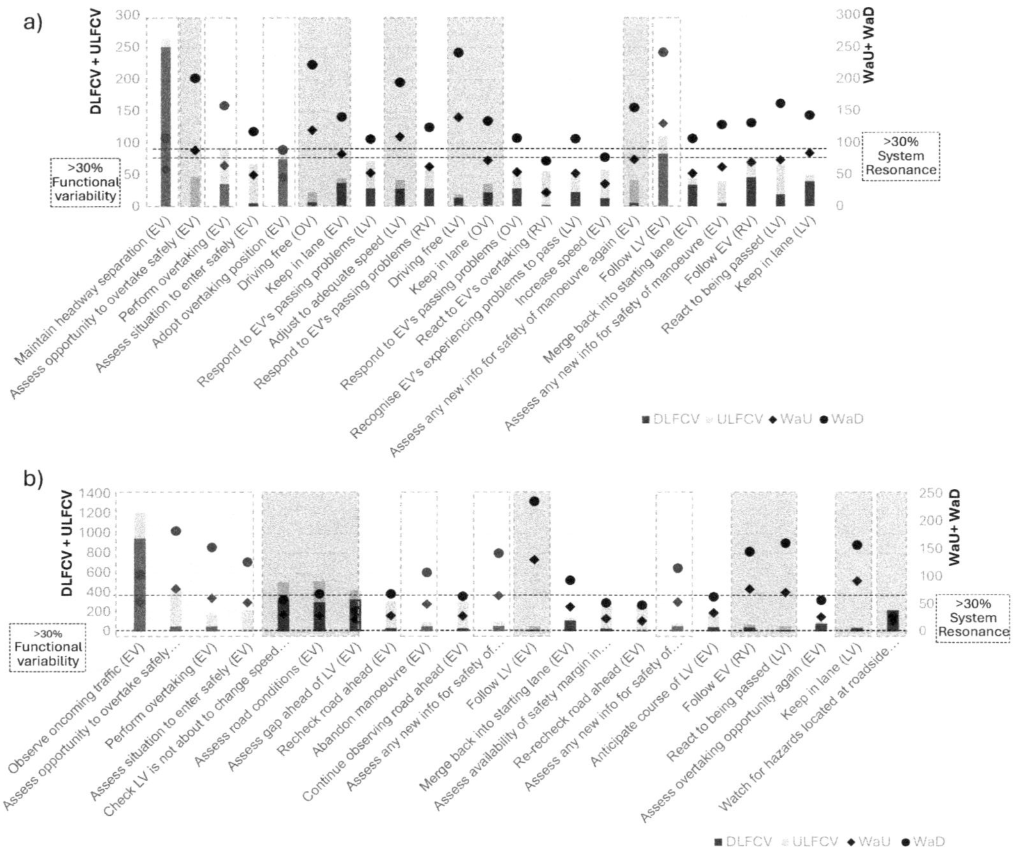

FIGURE 13.3 Risk functions for human drivers (scenario 1) (a) and automation of EV (scenario 2) and (b) composed of functional variability and system resonance.

Domain Expert

Here, <maintain headway separation (EV)> and <follow LV (EV)>, in particular, stand out, with high variability and system resonance values, whereby they transfer variability for the most part and receive very little. The darkest grey highlighted functions are risk functions that have relatively low variability but are combined with a strong system resonance. It can be argued that these functions are success factors demonstrating resilience because, despite their strong system effect and affectedness, they have little variability and are therefore stable. In particular, <driving free (OV)> and <driving free (LV)> with very high system resonances are noteworthy here. These functions must nevertheless be viewed with caution, especially under different scenario conditions, as a sudden increase in variability in these functions may have a large systemic effect. The function <assess the opportunity to overtake safely (EV)> is also notable because it is strongly influenced by the system and receives a relatively large amount of variability, but transfers very little variability into the system. Further, the functions <assess opportunity to overtake safely (EV)> and <merge back into starting lane (EV)> exhibit fairly high system resonances but with relatively low variability. So, variability rarely occurs here, but if it does, then it often results in accidents.

Compared to the automation of EV in Figure 13.3b, it can be seen that humans have significantly lower variability values and that overall, significantly more risk functions in automation have high functional variability.

Domain Expert

In the automation of EVs, various risk functions are colour-coded as well. This results in seven critical functions (light grey), with <observe oncoming traffic (EV)> standing out. Furthermore, four risk functions can be identified as success factors (darkest grey), for example, <follow LV (EV)> and <keep in lane (LV)>, each with high systemic resonance and low variability. In addition, there are risk functions in automation that have a relatively low systemic impact but are highly variable (medium grey), especially <watch for hazards located at the roadside environment (EV)> or <assess road conditions (EV)>. It can be argued that these high functional variabilities are somewhat irrelevant because of their low system resonance and therefore low potential for functional resonance, and they rarely lead to adverse events. However, it is important not to underestimate this variability, particularly in the event of changes in scenario conditions, which could potentially alter the system's resonance.

Patterns Analysis

An illustration of two exemplified patterns, A and B, representing the critical paths of the risk functions < follow LV (EV) > in scenarios one and three and < perform overtaking (EV) > in all five scenarios, respectively, is provided (see Figure 13.4). Abstractly, these patterns visually capture a set of relationships between functions

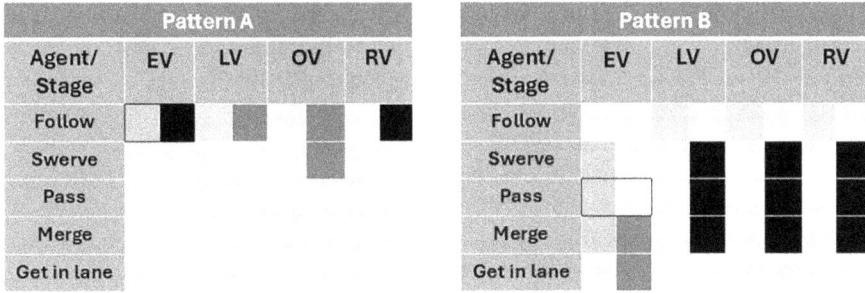

FIGURE 13.4   Visual depiction of two identified patterns on a stage and agent level.

within the time-space/agency framework, where the pattern emerges from these interactions, representing emergent properties and a strong potential for functional resonance due to adaptive behaviour. In detail, the patterns show the affectedness of the respective risk function by upstream functions (light grey cells) and the impact of the respective risk function by downstream functions (black cells) on a stage and agent level holistically. The intensity of colour represents the extent of affectedness or impact: the more intense, the higher the affect or effect. The location of the respective risk function is visualized through black-framed cells. In general, the two patterns differentiate in their spot of functional resonance represented by the temporal stage and agent. For example, the functional resonance in Pattern A is mainly concentrated on the Follow stage and the agents EV and RV. Instead, the functional resonance in Pattern B extends over several temporal stages and all agents.

Pattern A exhibits high interrelatedness (i.e., the interaction between agents) but low interrelatedness (i.e., the interaction between stages). The system faces a significant destabilization risk in the Follow stage, based on the following behaviour of EV. This behaviour is predominantly influenced by the EV itself and, to a lesser extent, by the LV within the Follow stage. It affects both the overtaking decision-making process of the EV and the driving performance of the other three agents, particularly the RV, within the Follow stage. It can be argued that Pattern A establishes the initial conditions for a successful overtaking manoeuvre by influencing the information-gathering process related to the overtaking decision. Specifically, the assessment of the opportunity to overtake safely can directly impact the subsequent process, resulting in a tight coupling. While the EV primarily influences the initial situation, the other three agents can potentially mitigate any adverse effects to maintain stability and resilience within the Follow stage.

Pattern B demonstrates both a high interrelatedness and interrelatedness. The system can be strongly destabilized in the Swerve, Pass, and Merge stages. This process is primarily driven by the EV's overtaking performance in passing the LV. In these stages, the EV's behaviour is primarily influenced by operational tasks such as overtaking position preparations, swerving into the oncoming lane, passing the LV, abandoning the manoeuvre, and merging into the starting lane, as well as tactical tasks like iterative safety checks. Furthermore, the EV's behaviour is moderately influenced by the operational driving performance of the other three agents in the Follow stage. Additionally, it affects the reactive driving behaviour of the other three agents in the Swerve, Pass, and Merge stages. This results in a strong interdependence among the three other agents, with simultaneous impacts in the Follow stage and effects in the Swerve, Pass, and Merge stages, leading to a high degree of "cascading process" that can either amplify or dampen depending on variability behaviour. Moreover, it slightly affects the evaluation of manoeuvre safety in the Merge stage and the completion of overtaking in the Get-in-lane stage of the EV. Due to the strong interdependent nature of Pattern B, a resonant situation can quickly propagate throughout the system during the overtaking process, with system-wide implications if any single agent destabilizes the system due to mutual resonance between all agents. However, this pattern also possesses high resilience potential, as each agent can compensate to prevent adverse events. Additional patterns can be retrieved in Grabbe (2024).

SYSTEM DESIGN AND FUNCTION ALLOCATION

The shared and traded control concept of EV shows the best performance in terms of the GSV analysis, followed by the manual driver of EV. Interestingly, automating the OV only marginally increases the GSV compared to scenarios one and five. However, full automation of either the EV alone or both the EV and OV would significantly destabilize the system. Hence, automating the entire scenario is not advisable, but selective automation of individual stages could involve transferring authority. LoDA 4 presents considerable potential specifically for the EV during the Merge and Get-in-Lane stages, as well as for EV functions related to the actions of the LV, RV, and OV during the Swerve, Pass, and Merge stages. Otherwise, a shared and traded control approach for the EV is preferable (cf. Grabbe et al., 2022b). The proposed design suggests that humans predominantly handle the Follow and Pass stages, while automation handles the Swerve, Merge, and Get-in-lane stages, culminating in full automation in the final stage. Shared control is only involved in 12% of functions across all three levels of information processing or driving.

Moreover, through pattern identification, critical leverage points within the system emerge as the most efficient means to enhance the system's resilience in overtaking scenarios through automation or technological support. Consequently, specific functions, such as <follow LV (EV)>, could benefit from automation, ideally supported by an Adaptive Cruise Control (ACC) system, which also aids in adopting an appropriate overtaking position. This approach helps mitigate the critical path within Pattern A. Simultaneously, the close connection between <follow LV (EV)> and the cognitive assessment of <evaluate a safe overtaking opportunity (EV)> can be uncoupled, potentially supporting system resilience. Regarding Pattern B, employing automation to support the EV driver during overtaking, aside from ensuring a proper overtaking position, seems impractical. However, it becomes apparent that automation opportunities exist for other agents to enhance the EV's overtaking performance or mitigate the potential negative impacts of its manoeuvres. For the former objective, maintaining a stable drive for the LV and OV is crucial to prevent sudden braking, acceleration, or overspeeding. For the latter purpose, automation should ensure that the LV maintains or decreases its speed while being overtaken, the OV brakes or maneuvers to the roadside if the EV is still on the opposing lane within a critical distance, and the RV maintains a sufficient gap with the LV in case the EV needs to abort the maneuver and merge back.

DISCUSSION

This discussion revolves around two main elements: initially, it explores the practical benefits of the approach for practitioners, and later, it delves into methodological advancements, discussing new metrics and analytical frameworks introduced in the study.

PRACTICAL BENEFITS OF THE FRAM

This research offers novel insights into overtaking safety in road traffic, contrasting with previous approaches which are less system-oriented (e.g., Hegeman et al., 2005; Näätänen & Summala, 1976; Reichart, 2000):

- Recognizing the positive roles of both drivers and automation, highlighting areas of system resilience rather than solely focusing on drivers' negative contributions.
- Identifying potential risks associated with automation's negative impacts alongside its benefits.
- Highlighting strategic intervention points for improving overtaking safety through automation interventions.
- Adopting a systemic perspective that considers all agents and their interactions, rather than solely focusing on the EV in isolation.
- Emphasizing a functional perspective over a structural one, aiming to guide decision-makers by allocating functions between drivers and automation without prescribing a specific design or structure for the physical system.

The design space acknowledges traffic complexity through a joint cognitive system (JCS) approach (Hollnagel & Woods, 2005), fostering driver-automation collaboration for safe overtaking. This demonstrates how the FRAM enables a more conscious function allocation, and the need for co-agency between humans and automation, considering complex dynamics and interactions within the system. FRAM serves as a critical tool in early-stage concept development, supporting system designs and empirical tests. It also integrates well with other analysis methods, such as failure tree analysis and system-theoretic process analysis, creating a multi-level analytical approach.

As expected, the FRAM is particularly suited for analyzing complex socio-technical interactions, focusing on interactions within the traffic and vehicle domains. Traffic interactions include merging and overtaking, while in-vehicle interactions cover driver-automation collaboration. Concepts like the H-mode (Flemisch et al., 2014) and manoeuvre-based driver-vehicle cooperation (e.g., Walch et al., 2016) are also relevant for any FRAM analysis. Additionally, the FRAM can be employed to analyze scenarios involving remote assistance or remote driving (SAE J3016, 2021), aiding in identifying potential conflicts within these intricate systems (cf. Arenius, 2017; Parasuraman, 2011).

Methodological Advancements

Quantification in the FRAM can aid with specific issues, such as promoting more reliable and valid "answers" derived from the analysis once a FRAM model is built, particularly in interpreting FRAM models related to large-scale complex systems such as the road system by preventing overwhelming qualitative representations. However, it is important to note that quantitative results are relative rather than absolute metrics. They serve as indicators of where to look and must be reviewed qualitatively by the model, requiring careful interrogation to comprehend and anticipate potentially useful interventions.

Then, the introduced Space-Time/Agency framework, along with various function types based on levels of information processing, enables the systematic structuring of systems across different dimensions and resolutions of analysis. This is aligned with previous uses of the FRAM (see Steen et al., 2021), the Abstraction/Agency framework (Patriarca et al., 2017a) and the JCS framework (Adriaensen et al., 2022), which offer comparable approaches to tackling complexity in STS analysis.

Additionally, the quantitative extension of the FRAM renders the concepts of couplings and complexity to characterize STS, as originally proposed by Perrow (1984), more explicit and measurable. These metrics are intertwined with metrics for variability rather than set in isolation. This results in comprehensive global system stability metrics encompassing complex and emergent behaviours, representing safety as a system property that emerges from how elements interact and fit together. In more detail, the new metrics facilitate comparisons of stability, resilience, and adaptive capacity across various scenarios or system designs. These metrics offer quantitative indicators for safety where safety is viewed in terms of positive outcomes, as for modern system safety management and RE. Moreover, by separating functional variability and system response, it becomes feasible to assign a weighting factor to a function's variability. Accordingly, it is possible to identify the tipping or leverage points within the system. Also, the GSV has the potential to provide a system-wide performance metric, that is, adaptive capacity, making it ultimately possible to compare the driver and automation by one metric, as claimed by Winner (2016).

Overall, the application of FRAM in this work represents an advancement in explicitly identifying critical paths within a system that may result in functional resonance – the adverse combination of multiple functions' "regular" performance variability over time and space, aligning with FRAM's primary goal. Notably, these recent enhancements, particularly incorporating quantitative approaches, render FRAM especially pertinent for modelling complex engineering processes involving human and social agents, such as automated vehicles. This relevance is underscored by the industry's heavy reliance on quantitative data in such contexts. In the future, the proposed semi-quantitative approach can be directly implemented in FMV using the metadata functionality (cf. Hill & Slater, 2024).

Lastly, for data collection, combining literature reviews, simulator experiments, and interviews was essential to capture functional variability systematically. Advanced techniques like eye tracking and neuroergonomics may further enhance the accuracy of perceptual and cognitive assessments, bridging qualitative and quantitative insights and improving system adaptability and road safety.

CONCLUSION

This chapter highlights the significance of the FRAM as an invaluable approach for evaluating the safety of AD in road traffic, particularly by examining interactions to uncover emergent patterns – a dimension often overlooked in existing methodologies but crucially important. The FRAM's white-box modelling offers a comprehensive understanding of how systems typically function, shedding light on why failures rarely occur. This understanding enables targeted interventions to amplify what drivers currently do well in coping with complexity. Advocating for analytical research over indiscriminate empirical testing, the FRAM warns against the increasing reliance on AI-driven black-box models in safety-critical contexts, which may lead to solutionism (Morozov, 2013) – a dangerous blind flight by trial-and-error. Overall, a FRAM model guides where to direct attention within a system by clarifying the actual problem and prompting more precise questioning.

The FRAM presents a valuable complement to existing approaches in addressing the multifaceted issue of road safety, offering diverse opportunities and unique advantages. There is no one-size-fits-all solution; instead, a comprehensive toolbox integrating various perspectives and methodologies is needed. This way, it is here advocated, as in several chapters of this book, that the FRAM has the potential to fill an essential gap in managing the present and future challenges associated with assessing the safety of AD within the complex and dynamic road system.

The central goal must be to enhance the safety of the entire system by fostering effective interaction among drivers, machines, and road users. Rather than solely prioritizing the safety of vehicle automation and demonstrating its safety, the pivotal inquiry shifts towards leveraging automation to craft a traffic system that optimizes multiple competing objectives like safety, efficiency, and comfort. This study explored potential solutions utilizing FRAM to address this fundamental question.

REFERENCES

Adriaensen, A., Berx, N., Pintelon, L., Costantino, F., Di Gravio, G., & Patriarca, R. (2022). Interdependence analysis in collaborative robot applications from a joint cognitive functional perspective. *International Journal of Industrial Ergonomics*, 90, 103320.

Alexander, C., Ishikawa, S., Silverstein, M., Jacobson, M., Fiksdahl-King, I., & Shlomo, A. (1977). *A Pattern Language: Towns, Buildings, Construction* (Vol. 2). Oxford University Press.

Anfara Jr, V. A., Brown, K. M., & Mangione, T. L. (2002). Qualitative analysis on stage: Making the research process more public. *Educational Researcher*, 31(7), 28–38.

Arenius, M. (2017). *Identification of Change Patterns for the Generation of Models of Work-as-Done using Eye-tracking* (Vol. 22). Kassel University Press GmbH.

Bainbridge, L. (1983). Ironies of automation. In *Analysis, Design and Evaluation of Man–Machine Systems* (pp. 129–135). Pergamon.

Bellini, E., Ceravolo, P., & Nesi, P. (2017). Quantify resilience enhancement of UTS through exploiting connected community and internet of everything emerging technologies. *ACM Transactions on Internet Technology (TOIT)*, 18(1), 1–34.

Bengler, K., Winner, H. Wachenfeld, W. (2017). No human – no cry? Automatisierungstechnik. Methoden und Anwendungen der Steuerungs-, Regelungs- und Informationstechnik. 65(7), 471–476.

Creswell, J. W., & Miller, D. L. (2000). Determining validity in qualitative inquiry. *Theory Into Practice*, 39(3), 124–130.

Endsley, M. R. (2023). Ironies of artificial intelligence. *Ergonomics*, 66(11), 1656–1668.

Endsley, M. R., & Kiris, E. O. (1995). The out-of-the-loop performance problem and level of control in automation. *Human Factors*, 37(2), 381–394.

Eurocontrol, A. (2021a). *The Systemic Potentials Management: Building a Basis for Resilient Performance*. A white paper.

Eurocontrol, A. (2021b). *Patterns in How People Think and Work Importance of Patterns Discovery for Understanding Complex Adaptive Systems*. A white paper.

Falegnami, A., Costantino, F., Di Gravio, G., & Patriarca, R. (2020). Unveil key functions in socio-technical systems: Mapping FRAM into a multilayer network. *Cognition, Technology & Work*, 22(4), 877–899.

Fastenmeier, W. (2015). Fahrerassistenzsysteme (FAS) und Automatisierung im Fahrzeug– wird daraus eine Erfolgsgeschichte? *Zeitschrift für Verkehrssicherheit*, 61(1), 15–25.

Flemisch, F. O., Bengler, K., Bubb, H., Winner, H., & Bruder, R. (2014). Towards cooperative guidance and control of highly automated vehicles: H-Mode and Conduct-by-Wire. *Ergonomics*, 57(3), 343–360.

Fuller, R. (2005). Towards a general theory of driver behaviour. *Accident Analysis & Prevention*, 37(3), 461–472.

Geiser, G. (1985). Man-machine communication in motor vehicles. *Automobiltech. Z. (Federal Republic of Germany)*, 87(2).

Grabbe, N. (2024). The Contribution of Automated Driving to Road Traffic Safety – A Resilience Engineering Approach (Doctoral dissertation, Technische Universität München). https://mediatum.ub.tum.de/1720274

Grabbe, N., Arifagic, A., & Bengler, K. (2022a). Assessing the reliability and validity of an FRAM model: the case of driving in an overtaking scenario. *Cognition, Technology & Work*, 24(3), 483–508. https://doi.org/10.1007/s10111-022-00701-7

Grabbe, N., Gales, A., Höcher, M., & Bengler, K. (2022b). Functional resonance analysis in an overtaking situation in road traffic: Comparing the performance variability mechanisms between human and automation. *Safety*, 8(1), 3. DOI: 10.3390/safety8010003

Grabbe, N., Kellnberger, A., Aydin, B., & Bengler, K. (2020). Safety of automated driving: the need for a systems approach and application of the functional resonance analysis method. *Safety Science*, 126, 104665.

Hegeman, G., Brookhuis, K., & Hoogendoorn, S. (2005). Opportunities of advanced driver assistance systems towards overtaking. *European Journal of Transport and Infrastructure Research*, 5(4).

Hill, R., & Slater, D. (2024). Using a metadata approach to extend the functional resonance analysis method to model quantitatively, emergent behaviours in complex systems. *Systems*, 12(3), 90.

Hollnagel, E. (2016). The nitty-gritty of human factors. *Human factors and ergonomics in practice: Improving system performance and human well-being in the real world* (pp. 45–64). CRC Press.

Hollnagel, E., & Woods, D. D. (1983). Cognitive systems engineering: New wine in new bottles. *International Journal of Man-Machine Studies*, 18(6), 583–600.

Hollnagel, E., & Woods, D. D. (2005). *Joint cognitive systems: Foundations of cognitive systems engineering*. CRC Press.

Hollnagel, E., Woods, D. D., & Leveson, N. (Eds.). (2012). *Resilience engineering: Concepts and precepts*. Ashgate Publishing.

Larsson, P., Dekker, S. W., & Tingvall, C. (2010). The need for a systems theory approach to road safety. *Safety Science*, 48(9), 1167–1174.

Ma, Z., & Zhang, Y. (2022). Driver-automated vehicle interaction in mixed traffic: Types of interaction and drivers' driving styles. *Human Factor*. https://doi.org//10.1177/00187208221088358

Meadows, D. H. (1999). *Leverage points: Places to intervene in a system*. Sustainability Institute.

Morozov, E. (2013). *To save everything, click here: The folly of technological solutionism*. PublicAffairs.

Näätänen, R. & Summala, H. (1976). A model for the role of motivational factors in drivers' decision making. *Accident Analysis & Prevention*, 6(3–4), 243–261.

Noy, I. Y., Shinar, D., & Horrey, W. J. (2018). Automated driving: Safety blind spots. *Safety Science*, 102, 68–78.

Parasuraman, R. (2011). Neuroergonomics: Brain, cognition, and performance at work. *Current Directions in Psychological Science*, 20(3), 181–186.

Parasuraman, R., Sheridan, T. B., & Wickens, C. D. (2000). A model for types and levels of human interaction with automation. *IEEE Transactions on Systems, Man, and Cybernetics*, 30(3), 286–297. https://doi.org//10.1109/3468.844354

Patriarca, R., Bergström, J., & Di Gravio, G. (2017a). Defining the functional resonance analysis space: Combining abstraction hierarchy and FRAM. *Reliability Engineering & System Safety*, 165, 34–46.

Patriarca, R., Di Gravio, G., & Costantino, F. (2017b). myFRAM: An open tool support for the functional resonance analysis method. In *2017 2nd International Conference on System Reliability and Safety (ICSRS)* (pp. 439–443). IEEE.

Perrow, C. (1984). *Normal Accidents: Living With High-Risk Technologies*. Basic Books.

Preuk, K., Stemmler, E., Schießl, C., & Jipp, M. (2016). Does assisted driving behavior lead to safety-critical encounters with unequipped vehicles' drivers? *Accident Analysis & Prevention*, 95, 149–156.

Rasmussen, J. (1997). Risk management in a dynamic society: A modelling problem. *Safety Science*, 27(2–3), 183–213.

Read, G. J., Beanland, V., Lenné, M. G., Stanton, N. A., & Salmon, P. M. (2017). Systems thinking in transport analysis and design. In *Integrating Human Factors Methods and Systems Thinking for Transport Analysis and Design* (pp. 3–17). CRC Press.

Reichart, G. (2000). Menschliche Zuverlässigkeit beim Führen von Kraftfahrzeugen–Möglichkeiten der Analyse und Bewertung (Doctoral dissertation, Dissertation am Lehrstuhl für Ergonomie der TU München).

Risser, R., & Brandstätter, C. (1985). Die Wiener Fahrprobe. Freie Beobachtung. KL FACHBUCHR KFV, 21.

SAE International (2021). Taxonomy and Definitions for Terms Related to Driving Automation Systems for On-Road Motor Vehicles (J3016).

Steen, R., Patriarca, R., & Di Gravio, G. (2021). The chimera of time: Exploring the functional properties of an emergency response room in action. *Journal of Contingencies and Crisis Management*, 29(4), 399–415.

Wachenfeld, W., & Winner, H. (2016). The release of autonomous vehicles. In Maurer, M., Gerdes, J., Lenz, B., Winner, H. (eds) *Autonomous driving* (pp. 425–449). Springer, Berlin, Heidelberg. https://doi.org/10.1007/978-3-662-48847-8_21

Walch, M., Sieber, T., Hock, P., Baumann, M., & Weber, M. (2016). Towards cooperative driving: Involving the driver in an autonomous vehicle's decision making. In *Proceedings of the 8th international conference on automotive user interfaces and interactive vehicular applications* (pp. 261–268).

Walker, G. H., & Stanton, N. A. (2017). *Human Factors in Automotive Engineering and Technology*. CRC Press.

Wiener, E. L., & Curry, R. E. (1980). Flight-deck automation: Promises and problems. *Ergonomics*, 23(10), 995–1011.

Wilde, G. J. (1982). The theory of risk homeostasis: Implications for safety and health. *Risk Analysis*, 2(4), 209–225.

Winner, H. (2016). Quo vadis, FAS? In *Handbook of Driver Assistance Systems* (pp. 1557–1584). Springer Vieweg.

Woods, D. D., & Branlat, M. (2017). Basic patterns in how adaptive systems fail. In *Resilience engineering in practice* (pp. 127–143). CRC Press.

14 Machine Learning to Support Human Learning from Variable Performance in FRAM

Hideki Nomoto

CONTEXT: ALLOCATING BEDS IN A HOSPITAL

In England, the National Health Service (NHS) has been facing a long-term trend of decreasing hospital beds. The total number of NHS hospital beds in England has more than halved over the past 30 years, from around 299,000 in 1987/88 to 141,000 in 2019/20, making it increasingly difficult for each hospital to assign new patients to each ward (Ewbank et al., 2021). To optimize bed allocation, bed meetings are held several times a day. In addition to these regular meetings, ad hoc coordination takes place as needed at various locations, which makes it difficult to track overall bed availability. Occasionally, a shortage of beds occurs, necessitating the urgent conversion of ward beds into critical care beds, reassignment of staff and, in some cases, the transfer of patients to other hospitals. This situation is occurring more frequently.

Cook and Rasmussen (2005) say

> with bed demands intense, beds are available for new patients only as old patients are discharged… The ICU's (Intensive Care Units) were continuously working at capacity - the units were full… A patient undergoing surgery needed to go to an intensive care unit when the surgical procedure was complete. The surgical procedure was finished before the bed was available. But after a few hours in a recovery room the patient was eventually transferred to the ICU.

To address this issue, the bed assignment process needs to be analyzed, visualized, and understood in a detailed way: this is the area where the FRAM shines [Nomoto].

Bed allocation is determined by the two supply–demand balances: Critical Care (CC) beds demand/supply and ward beds demand/supply.

The total demand and supply of CC beds and ward beds can be reliably calculated by tracking the following variables:

1. The demand for Critical Care (CC) beds is calculated by the sum of:
   a. The number of new CC beds approved by the Emergency Department (ED) for newly approved surgical patients
   b. The number of new CC beds for internal patients that occurred as an Internal Collapse (EI).
   c. The number of new CC beds for patients transferred from other hospitals as External Referrals (ER).
   d. The number of new CC beds for Elective Surgery (EL) patients
2. The supply of CC beds is calculated by the sum of:
   a. The number of patients discharged from CC beds and assigned ward beds.
3. The demand for ward beds is calculated by the sum of:
   a. The number of new ward beds approved by the ED for newly approved surgical patients.
   b. The number of new ward beds for patients transferred from other hospitals to the ER.
   c. The number of new ward beds needed as EL.
   d. The number of patients discharged from CC beds and newly assigned ward beds.
4. The supply of ward beds is calculated by the sum of:
   a. The number of patients being discharged.

These values are determined by the regular bed allocation meetings held several times a day. However, between the regular meetings, each variable is coordinated ad hoc in various places (meeting rooms, corridors, or by phone), making it difficult to accurately grasp all the movements of the numbers, that is, the actual status of the system.One possible resolution for this ad hoc coordination problem is to place the latest data on a server and to allow the relevant stakeholders to update it when any changes are necessary. When the currently available bed numbers are updated, the prediction value could be automatically updated. If the current and predicted data are always available from anywhere in the hospital, bed allocation coordination will be able to use the correct values. As a result, it becomes possible to eliminate some official bed meetings, keeping them only for exceptional circumstances.

This is what this chapter aims to achieve, building on a systemic modelling of the intertwined process of bed allocation, relying on a version of the FRAM that is strengthened by machine learning to support traditional learning processes. This way, ultimately, stakeholders can save time and resources they have been spending on bed allocation coordination.

More specifically, in order to perform an accurate prediction, the reliability of the data is the most important requirement. The reliably available data are i) the bed margin after the movement of patients decided at the regular meetings (Let's call it "data x"); and ii) the bed margin just before the regular meeting (that is, the final result of all ad hoc coordination before the regular meeting. Let's call it "data y").

In an attempt to trivialize the problem at hand, if the data x is given and if the data y just before the next regular meeting can be accurately predicted, everyone can know the forecast value of the expected surplus or deficit of beds up to the next regular meeting. This allows for more planned personnel allocation, greatly contributing to the improvement of the quality of life for both patients and staff.

However, understanding the tight relationships between these variables is nothing but a challenging task. Thus, what specifically should be done to enhance resilience by introducing learning and prediction? The answer is to strengthen the four resilience functions (Monitor, Anticipate, Learn, and Respond), whose modelling is supported by the FRAM in actual operations. This is what Resilience Engineering and this chapter advocate (Hollnagel, 2011).

THE "FOUR FUNCTIONS" FOR RESILIENT PERFORMANCE

In Resilience Engineering, the word "resilience" is defined as the ability to "adjust its functioning prior to, during, or following events (changes, disturbances, and opportunities), and thereby sustain required operations under both expected and unexpected conditions" (Hollnagel, 2017a). This is frequently considered possible to obtain a proper combination of four abilities, that is, resilience cornerstones: monitor, anticipate, respond, and learn.

To adjust its functioning, a system needs to incorporate feedback from its elements and acquire new ones from the environment. For the feedback process, it is essential to repeat a cycle of learning from past experiences, predicting future changes based on those experiences, monitoring the difference between predicted and observed values, and improving one's actions before the changes in circumstances become unmanageable. This learning cycle relies on the idea of "feedback loop": a system that can effectively use these feedback loops usually has a higher potential to demonstrate resilient performance, whenever needed, flexibly responding to changes.

In the context of bed allocation, it is important to continually refine this feedback so that it can adapt to various sudden events, such as neighbourhood accidents, abnormal weather, and disasters, in addition to normal seasonal fluctuations.

Next, the most important fact in this chapter is that the four functions mentioned above are not only useful to enhance resilience but also essential when performing machine learning consciously. Machine learning always requires a cycle of learning from past data in order to predict the future based on it, continuously monitoring the difference between predicted (i.e., anticipated) and observed values, and fine-tuning the control parameters used for prediction, responding as a function of the envisioned difference. Conversely, without these four functions, the feedback loop necessary for machine learning cannot be created in a meaningful way.

In other words, making the system resilient and making a model capable of executing machine learning share a common foundation.

To make the system resilient, it is necessary to strengthen these four functions, and in order to properly strengthen them (optimize the objective variable), it is necessary to know what value to give to which explanatory variable. The optimization result of the value of this explanatory variable is the quantitative success factor for this system to become resilient.

Our aim in this chapter is not to create a universal machine learning system for any problem but to visualize the "way of working" necessary to make tasks resilient and to execute work successfully. That is, the identification of factors required for function execution, such as control parameters and resources. In other words, if the function you want to strengthen in the process of the analysis target is identified, then the Input, the Control parameters, Time constraints, Preconditions, or Resources become explanatory variables, and the Output of that function becomes an objective variable. By performing machine learning for optimization, our goal is to enhance the resilience of the process, supporting learning as traditionally achieved by humans only. Eventually, a well-functioning FRAM machine learning model able to produce optimal Outputs is a good example of an improved form of an actual process.

When coming to the specific context of this chapter, our objective is to improve the bed allocation process, proposing a new process model that can forecast today's demand and supply from yesterday's results, and determine the control parameters necessary for forecasting through machine learning.

DATA COLLECTION

The data table shown in Table 14.1 was defined by interviewing a healthcare professional, Prof. Ralph MacKinnon, at Manchester Academic Health Science Centre Academic Chair, who also performed research work related to the topic discussed here (cf. MacKinnon et al., 2023).

Based on the identified data items and sample real values, the content of the data was generated by software to mimic the real sample data. This is due to the fact that a great amount of internal hospital data cannot be disclosed, and we were interested here in evaluating the potential for our methodology to function, starting from realistic data. While this choice represents a limitation for the operational value of the chapter, it only minimally interferes with the logical foundation behind the proposed approach. Accordingly, the total number of data points generated is 100 million. This large number of data cases was used to statistically determine the optimal parameters for machine learning.

The way the data content was created was not completely random but was generated with periodic components by adding or subtracting a random number within a certain range to the previous day's data (the choice of addition or subtraction is also random). By doing this, we were able to generate data with mixed macro trends, similar to statistical data of economic indicators such as currency exchange rate trends, where the previous day's value affects today's value, and micro noise-like fluctuations. As a result, we were able to create data that, while being a natural progression in terms of the number of available beds, sometimes may fluctuate significantly. This allowed us to create a dataset with sufficient diversity for machine learning, which was validated before its usage.

TABLE 14.1
Variables Used for Machine Learning

ID	Variable Name	Description	Type
1	CC_ED	New CC (Critical Care) beds approved by the ED (Emergency Department) for new surgical patients	INPUT
2	CC_EI	New CC beds for internal patients that occurred as EI (Internal Collapse)	INPUT
3	CC_ER	New CC beds for patients transferred from other hospitals as ER (External Referrals)	INPUT
4	CC_EL	New CC beds needed as EL (Elective Surgery)	INPUT
5	Discharge_CC	The number of patients discharged from the CC bed and assigned to a ward bed	INPUT
6	CC_demand_morning	The total demand for CC beds at the regular meeting (1+2+3+4)	INPUT
7	CC_supply_morning	The total supply of CC beds at the regular meeting (5)	INPUT
8	CC_Margin_Morning	The total CC supply – CC demand at the regular meeting (7-6)	INPUT
9	More_CC_ED	Additional demand for CC for ED	INPUT
10	More_CC_EI	Additional demand for CC for EI	INPUT
11	More_CC_ER	Additional demand for CC for ER	INPUT
12	More_CC_EL	Additional demand for CC for EL	INPUT
13	More_CC_by cancelling EL	Additional number of patients discharged from the CC	INPUT
14	CC_Margin	The total CC supply – CC demand at the end of the day (13 – (8+9+10+11+12))	INPUT
15	CC_mod	The prediction parameter for CC	OUTPUT
16	CC_prediction	CC prediction result	OUTPUT
17	CC_gap	CC prediction error (14-16)	OUTPUT
18	WARD_ED	New ward beds approved by the ED	INPUT
19	WARD_EI	New ward beds for internal patients that occurred as EI	INPUT
20	WARD_ER	New ward beds for patients transferred from other hospitals as ER	INPUT
21	WARD_EL	New ward beds needed as EL	INPUT
22	Discharge_WARD	The number of patients discharged from ward bed	INPUT
23	WARD_demand_morning	The total demand of ward bed at the regular meeting (1+2+3+4)	INPUT
24	WARD_supply_morning	The total supply of ward bed at the regular meeting (5)	INPUT
25	WARD_margin_morning	The total ward supply – ward demand at the regular meeting (7 - 6)	INPUT

(Continued)

TABLE 14.1 (Continued)
Variables Used for Machine Learning

ID	Variable Name	Description	Type
26	More_ward_ED	Additional demand for a ward for the ED	INPUT
27	More_ward_El	Additional demand for the ward for El	INPUT
28	More_ward_ER	Additional demand for a ward for ER	INPUT
29	More_ward_EL	Additional demand for ward for EL	INPUT
30	Fail_to_discharge_ WARD	Number of patients who could not be discharged from the ward	INPUT
31	WARD_margin	The total ward supply – ward demand at the end of the day (24 - 25+26+27+28+29+30+5))	INPUT
32	WARD_mod	The prediction parameter for ward	OUTPUT
33	WARD_prediction	The ward prediction result	OUTPUT
34	WARD_gap	The ward prediction error (31-33)	OUTPUT

Our goal is to predict the final bed availability (data y) before the next regular meeting as accurately as possible from the bed availability determined at the latest regular meeting (data x). The required number of beds fluctuates between the meetings due to various unexpected events (implicitly assumed in the data-driven variations modelled in the data stream). What kind of learning should we do to make predictions even under such circumstances?

As a result of the FRAM analysis, it was found that in order to predict data y from data x, it is necessary to analyze the following trends:

- Seasonal fluctuation elements (i.e., which season of the year tends to increase the number of patients)
- Sudden fluctuation elements (accidents or disasters occurred in the neighbourhood)

To obtain these trends or fluctuation elements, it might be useful to record environmental data such as daily weather information and accidents in the neighbourhood. However, even if we take one parameter like "temperature" data, it is not easy to formalize how it contributes to the increase or decrease in the number of patients. And it is very difficult to definitively say which parameter in the weather information should be used. Does lower temperature contribute to an increase in patients when the humidity is high? Or does the strength of the wind matter? We must answer such difficult questions, which *ex-ante* seem to be not answerable. Moreover, knowing all accidental elements and modelling their relationships is far from being a trivial task.

When it is not clear what the dominant factor is, all the collected data can be of interest, especially if ingested into a black-box machine learning model. But in this way, the significance of each input parameter is unknown in advance to us. Determining the significance of each input data is left to the software, making our machine learning model largely opaque. As long as sufficient computing resources are provided, it is expected to build a model which produces accurate predictions. This type of machine learning is called "brute force machine learning".

*** Domain Expert ***

This strategy is unrealistic for tackling our problems for at least four dimensions:

- For seasonal fluctuation elements, it is necessary to extract countless parameters from our social environment, such as temperature and humidity, events such as Christmas, or an increase in virus spread. There is no established theory about what data is likely to contribute to the increase or decrease in the required number of beds.
- For accidental elements, on top of the above, there is a diversity that is difficult to determine, from car accidents to the victory or defeat of the local soccer team, to socio-cultural dimensions.
- Naturally, there is no guarantee that all important information has been successfully identified. Thus, after all these efforts of collecting hundreds of data points for thousands of days, the most critical data could still be ignored.
- Finally, even if the number of parameters was relatively small (e.g., 30), there is no prospect that humans can grasp the causal relationships between all those parameters and the result of the number of beds. This is because there are infinite variations of the combination of 30 parameter values. Even if they are the simplest discrete parameters only (for example, the temperature is higher or lower than the threshold: TRUE or FALSE), there are more than 1 billion variations of combinations because it is 2 to the power of 30. Preparing all possible combinations of input data is, to date, nearly impossible or probably harder than what an analyst would like to deal with.

In such a situation, if the target is a prediction of very short-term data with some kind of trend, we humans temporarily set aside the detailed contributing factors and perform a simple estimation that combines prior information and observed data. This prediction method assumes that today's bed availability has some relationship with yesterday's availability. As a result of this assumption, any prediction parameter is updated by adding a compensation value to yesterday's parameter using the prediction gap and the learning rate. That is because today's availability depends, to some extent, on yesterday's value. If the increase and decrease of beds are completely random, or if randomness completely overshadows periodic components, the above method will not be able to predict well.

However, if the data has some kind of periodic trend due to seasonal fluctuations, regardless of the frequency and the amplitude, even if it has noise components due to the occurrence of sudden accidents or the outbreak of epidemics, it should repeat up and down as a macro trend, and by repeating parameter adjustments every time at regular meetings to compensate for the prediction gap, it is possible to gradually improve the prediction accuracy. Essentially, our systems are not completely stochastic, but they are based on their functional properties, which are aimed at being captured by dedicated and representative machine learning models.

Domain Expert

Finally, it should be mentioned that the bed margin exists only because of various efforts made by hospital staff to somehow keep the final bed margin of the day positive. The number of margins that were somehow squeezed out by this effort can be obtained by, for example, the cancellation of lower-prioritized elective surgery. In other words, in the daily bed allocation, a situation has occurred where more beds than predicted are needed, and as a response to this, relatively low-priority scheduled surgeries are cancelled, and the beds left over by this choice are allocated to high-priority patients. Therefore, in the system we aim for, we should build a prediction system that can operate the hospital without such surgery cancellations. If this elective surgery cancellation number is subtracted from the "CC margin" and "ward margin" values at the end of the day, we can know yesterday's true "CC margin" and "ward margin". This true margin is the number of bed surpluses in the case of not cancelling the patient's surgery, and originally, beds should be secured so that this margin is always positive. Our machine learning data takes this point into consideration, and from the balanced data (data y) at the end of the day, the cancellation number of elective surgery has been subtracted.

METHODOLOGY

To operationalize the logic described above, a set of specific phases has been proposed to extend the traditional building steps of a FRAM model, as in the following list:

- Phase 1: Model the current process via FRAM, envisioning functions for the four resilience potentials, that is, check the possibility to include explicitly in the FRAM model developed the four cornerstones of resilience (extending the traditional Steps 0-1 by means of functions that mimic <To learn>, <To anticipate>, <To monitor>, and <To respond>)
- Phase 2: Prepare data for machine learning simulations (extending the traditional Step 2 of the FRAM)
- Step 3: Develop and run the machine learning simulation (extending the traditional Step 3)
- Step 4: Assess the results and iterate; that is, iterate until sufficient prediction accuracy of the results can be obtained, and it becomes useful to manage the inherent variability of the process to be performed in the daily work (empowering the traditional Step 4)

PHASE 1: MODEL THE CURRENT PROCESS USING THE FRAM

Figure 14.1 presents the current process as an FRAM model, obtained from the data collection process described above.

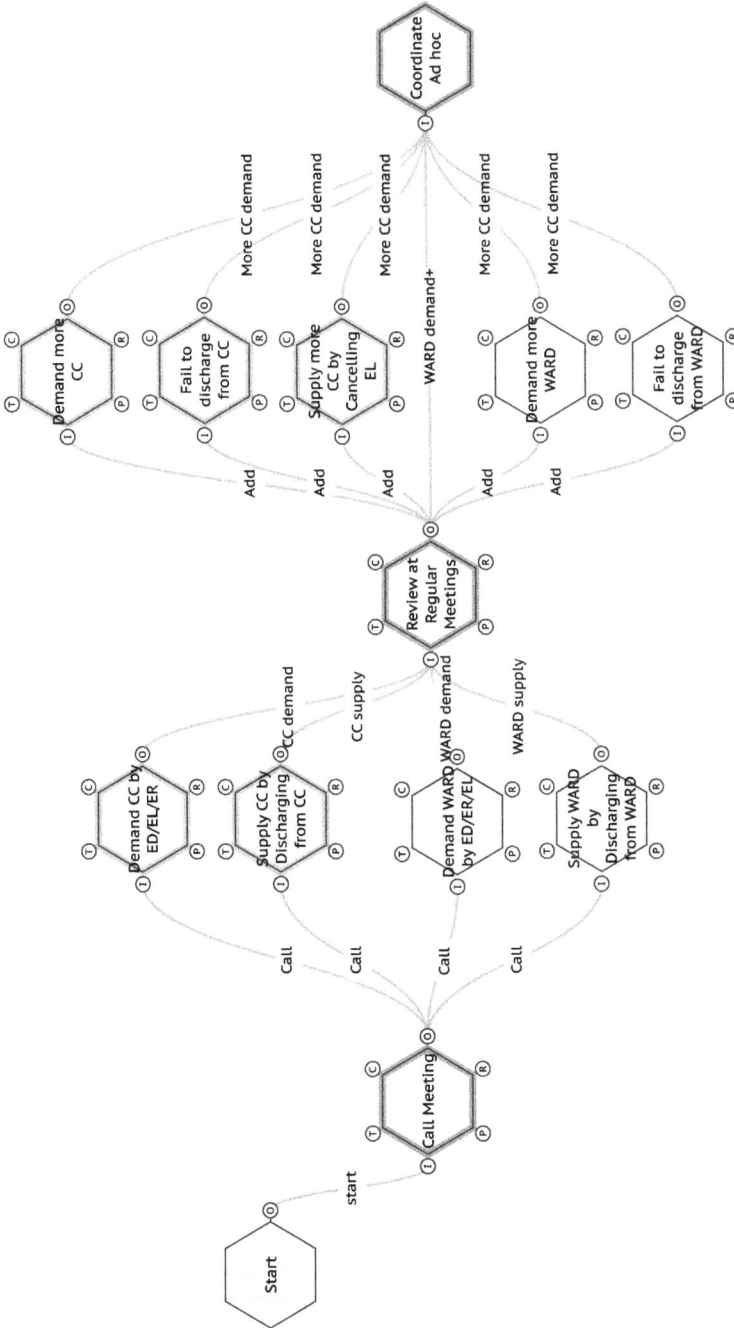

FIGURE 14.1 Current process of bed assignment.

Some functions can be highlighted as central to the process of bed assignment, as they initiate the diverse loops of actions described below, as for the following sections.

To Call Meeting

The <Call Meeting> function, located at the far left of the diagram, summons the relevant parties to the daily regular meeting. The second column from the left presents the following data: (i) the demand for Critical Care (CC) beds, (ii) the supply of CC beds, (iii) the demand for ward beds, and (iv) the supply of ward beds.

The demand for Critical Care (CC) beds is calculated by the sum of the followings and is output from the "Demand CC by ED/EI/ER/EL" function: (i) new CC beds approved by the ED (Emergency Department) for new surgical patients, (ii) new CC beds for internal patients that occurred as EI (Internal Collapse), (iii) new CC beds for patients transferred from other hospitals as ER (External Referrals), and (iv) new CC beds needed as EL (Elective Surgery).

The supply of CC beds is calculated by the following: (i) the number of patients discharged from the CC bed and assigned to a ward bed.

The demand for ward beds, like CC, is the total number of ward beds needed by ED/ER/EL and is output from the <Demand WARD by ED/ER/EL> function.

The supply of ward beds, like CC, is the number of discharges from the ward bed and is output from the <Supply WARD by Discharging from WARD> function.

In the third column from the left, there is a function called <Review at Regular Meetings> that reviews these supply and demand figures.

To Coordinate Ad Hoc

From the <Review at Regular Meetings> function, the CC demand (Morning) and ward demand (Morning) data are output to the <Coordinate ad hoc> function. These are the supply and demand data announced at the regular meeting in the morning. However, in the ad hoc coordination conducted when new demand is identified after the regular meeting, this supply and demand data is referred to as an indicator. In other words, if there are available beds secured today, new demand is approved at the ad hoc coordination. However, since ad hoc coordination is an irregular adjustment, the overall aggregation results remain unknown until the next regular meeting. Depending on the situation, all available beds might be filled before the next regular meeting, and furthermore, demand that cannot fit in the available beds may have been approved. In such a situation, as mentioned earlier, it becomes necessary to remake or rearrange beds, cancel surgeries, etc. Such situations are represented by the group of functions in the fourth column from the left.

The <Demand more CC> function is, literally, a request for additional CC beds. The <Fail to discharge from CC> function represents a situation where a discharge from a CC bed, which was scheduled by the <Supply CC by Discharging from CC> function in the second column, could not be performed for some reason (for example, deterioration of the patient's condition). The <Supply more CC by Cancelling EL> function is one that cancels relatively low-priority elective surgery to meet the additional demand for CC beds.

Furthermore, as you go down the fourth column, you will find the <Demand more WARD> function and the <Fail to discharge from WARD> function. These are the same as the identical functions for CC beds.

In the fifth column, the aforementioned <Coordinate ad hoc> function performs adjustments by inputting these new supply and demand numbers. Because it is difficult to look at the overall supply and demand balance and make ad hoc adjustments, there is a possibility that it will be found at the next regular meeting that there are not enough beds.

This model, however, does not capture the four functions for resilient potential, which are discussed in the following sections and are added to the complete model that is depicted in Figure 14.2. These functions have already been discussed as being central to understanding actual processes from a socio-technical perspective.

To Anticipate

The CC demand (Morning) and ward demand (Morning) revealed in the regular meeting are input into the <Anticipate> function. This function predicts the final supply-demand balance of the day (the bed margin remaining at the end of the day) from today's latest supply-demand balance data. The bed margin calculated from this morning's latest supply-demand plan is called "data x", and the bed margin remaining at the end of the day is called "data y".

The <Anticipate> function is a feature that predicts data y for tomorrow morning from data x for this morning. In the past, there were attempts to predict the number of beds by learning data from multiple factors (such as weather and timing) that could affect the results, but these attempts failed due to the difficulty of data collection and the resulting prediction accuracy. Therefore, we will not try to collect all relevant environmental data and conduct brute-force machine learning.

Instead of predicting by accumulating evidence and considering all the influencing factors, we perform a "rough adaptation" as humans do. That is, we only provide the morning data and the end-of-day data and predict based on their numerical relationship alone. This is close to the way humans usually do things for short-term prediction, following the intuition that comes from the empirical rule that there is continuity between yesterday's trend and today's trend, and that if we continue to predict by slightly modifying yesterday's prediction parameters, things will usually go well. This is similar to the empirical rule that it is more reliable to predict today's value of, for example, a stock exchange rate, simply from the short-term trend rather than predicting using all the factors, such as long-term profit trend, that could affect the company's long-term performance.

In Resilience Engineering, this "rough adaptation" is called ETTO (Efficiency-Thoroughness Trade Off), which can be defined as follows:

> In their daily activities, at work or at leisure, people (and organizations) routinely make a choice between being effective and being thorough, since it rarely is possible to be both at the same. If demands to productivity or performance are high, thoroughness is reduced until the productivity goals are met. If demands to safety are high, efficiency is reduced until the safety goals are met. The ETTO principle refers to the fact that people (and organizations) as part of their activities frequently - or always - have to make a trade-off between the resources (primarily time and effort) they spend on preparing to do something and the resources (primarily time and effort) they spend on doing it.

(Hollnagel, 2017b)

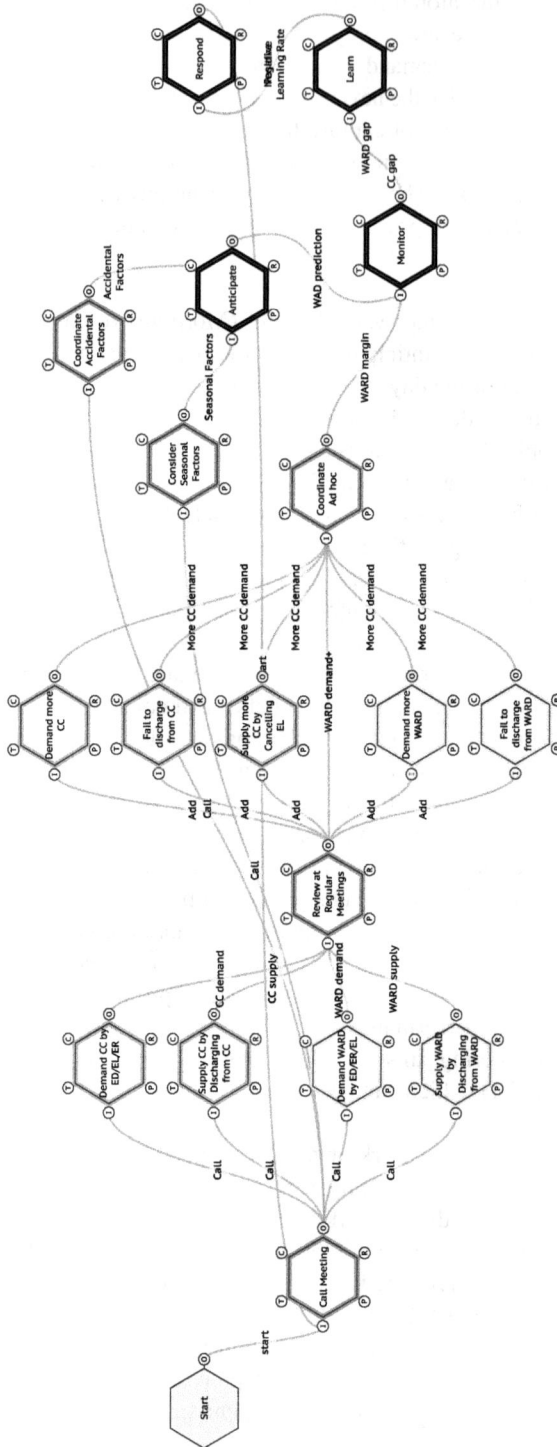

FIGURE 14.2 Improved process of bed assignment by adding four resilient functions: Anticipate/Monitor/Learn/Respond.

This time, what we aim for is high-performance prediction using only reliably available data. At the same time, we added a mechanism which realizes conservative prediction to minimize emergency rearrangement of bed assignment or cancelling surgery, which will be described later. We predict today's "data y" from yesterday's "data x" using the "parameter a" which was modified by yesterday's prediction gap.

To Monitor

The "CC prediction" and "ward prediction" generated by the "Anticipate" function are input into the <Monitor> function. There, the next morning, they are compared with the final remaining bed margin values ("CC margin" and "ward margin"). In other words, the <Monitor> function is about finding the difference between the predicted value and the actually observed value. This difference is named "CC gap" and "ward gap". These different data are used to improve the accuracy of the <Anticipate> function in the next cycle. If there is a difference, two prediction parameters ("CC mod", "ward mod") will be updated in the <Respond> function to fill the gap and to improve the next cycle's prediction accuracy using "learning rate" described in the next subsection.

To Learn

In this function, the most important parameter is called "learning rate". The "learning rate" parameter is used to compensate for the gap between the observed value and the predicted value. In machine learning, the learning rate is a numerical value ranging from zero to one. When we try to update a prediction parameter to minimize the gap, we use this learning rate to decide how much you fill the gap.

Method Geek

When you do not want to fill it, you will use "0" to be multiplied by the compensation value. When you want to fill 100% of the gap, you use "1". When you want to fill 50%, you will use "0.5" and so on. In many cases, this value is set to be a small percentage, such as 0.1. That is because typical machine learning uses a large amount of learning data. When you have a sufficient amount of data, it is wise to fill the gap slowly to avoid overshooting.

In the <Learn> function, we tried multiple values of the learning rate to find the just-right value to compensate for the gap that matches the environment. In our case, factors such as sudden changes in climate, the occurrence of traffic accidents, emergency transfer of patients from other hospitals, or short-term factors such as the arrival of the Christmas season have a considerable impact, so it is effective to set the learning rate slightly higher and quickly adapt to changes. This learning rate is called a hyperparameter in machine learning terms. Optimization of the hyperparameters is one of the most critical tasks to build a machine learning model. Furthermore, the most important thing for hospital management is patient safety. We need to make the learning rate not only accurate but also trustworthy to use, that is, safer for the decision maker.

To achieve this purpose, we apply dual learning rate mechanisms to always respect the following statement: the predicted value should always have an appropriate positive margin. In other words, when predicting next vacant bed numbers (data y) from the current vacant bed numbers (data x), the difference between the predicted value and the actual value (actual value minus predicted value) should always be positive. This constraint is necessary since if the delta becomes negative, more beds than predicted will be needed, and measures such as cancellation of elective surgery and transfer to other hospitals will have to be taken to secure the insufficient beds, which carries the risk of compromising patient safety. So, the accuracy of the model can be sacrificed for guaranteeing a safe positive margin.

The learning rate for predicting on the safe side should not only be set to a single value that enables the most accurate prediction but also needs to be adjusted to make a safer prediction when the environment changes. Specifically, if the prediction gap was positive (more beds are available than the ones predicted), the learning rate should be set small so as not to drastically change the coefficient of the prediction to keep the good positive margin, and conversely, if the prediction margin was negative (more beds are needed than the ones predicted), the learning rate should be set large to quickly catch up with the change.

This allows for rapid improvement in prediction accuracy if the situation continues to deteriorate (negative margin), and at the same time, even if the situation continues to improve (positive margin), it does not try to fill up all the gap, and it is always prepared for the next "deterioration period", implementing a behaviour similar to conscious human decision-making that prioritizes safety by maintaining sufficient positive margin.

To summarize, by having not one but multiple learning rates in the <Learn> function, it is possible to achieve safer learning and thus establish a prediction system that can solve the problem of hospital bed placement optimization, which prioritizes safety.

It should be pointed out that the same effect can be achieved by introducing the "Reinforcement Learning (RL)" method, which navigates the output toward a safer side with the technology to search for bigger rewards (more positive margin). However, RL is often considered to have a more "black box" nature compared to traditional supervised learning (the learning which has the "correct answer" in the learning data, in this case, "data y"). The reasons for this are its inherent modelling complexity and the lack of explainability (Kirsch et al., 2021; Selim et al., 2022). In reinforcement learning, an agent learns the optimal policy through interaction with the environment. This process usually involves very complex neural networks, making it difficult to understand the reasons for the agent's decisions and the basis of its actions. In contrast, in supervised learning, the correspondence between inputs and outputs is clearly defined, making it easier to analyze the behaviour of the model.

On the other hand, the machine learning of the FRAM model is a white box model that is understandable to humans and does not have a complex layered structure, and the meaning of each node (hexagon indicating function) and each parameter (data transferred between functions) is also clear. It is not universally applicable to any problem, but everything in the model has its own meaning and role.

To Respond

When the <Learn> function can set the learning rate to a just right value, the <Respond> function finally uses the determined learning rate, multiplies it with the difference between the predicted value and the actually observed value ("CC gap" and "ward gap" generated by the <Monitor> function), and updates the control parameters "CC mod" and "ward mod" to be used in the next cycle's <Anticipate> function. This cycle is repeated many times, and after going through multiple cycles and the occurrence of several unexpected events, the converged "CC mod" and "ward mod" can grow into prediction parameters that are optimal for the current seasonal trend and can withstand past experienced accidental events.

Phase 2: Prepare Data for Machine Learning Simulations

In this phase, it is necessary to prepare the data for the machine learning simulation. Given the context of the study described in Section 2 of this chapter, here it is only worth recalling that the data obtained from a mixture of real data and computer-generated data, for a total of 100 million entries with respect to the following key variables linked to supply and demand of beds (cf. Table 14.1 for the full list of variables).

Phase 3: Develop and Run the Machine Learning Simulation

To build a working machine learning model that is able to leverage the FRAM, first, it is necessary for humans to construct a tailor-made feedback loop that can be used as a machine learning model. Next, the way of learning and predicting shall be modelled to match the human task if there exists a learning task. If there are no such tasks, it is required to add them to improve the current process. The good news is that the FRAM is very good at mimicking expert behaviours because it explains how the expert's actions (function) can be executed successfully by using a variety of parameters such as control parameters, time constraints, preconditions, and resources.

By taking this effort, it becomes possible to construct a machine learning model that is largely understandable to humans, and it can be used in places where human lives are at stake, such as hospitals.

Method Geek

At this stage, we can thus look specifically at the key algorithm stages, shown in Table 14.2.

TABLE 14.2
Overview of the Machine Learning Algorithm

Stage	Description	Algorithm
1	Get the latest bed allocation plan data (data x) at the regular meeting.	
2	Based on data x, make a final balance prediction up to the next regular meeting (data y').	Prediction Value y' = Data x + Prediction Parameter

(Continued)

TABLE 14.2 (Continued)
Overview of the Machine Learning Algorithm

Stage	Description	Algorithm
3	Get the observed bed balance result (data y) at the next regular meeting.	
4	Calculate the difference between the observed value (data y) and the predicted value (data y'). We call it "the prediction gap".	Prediction Gap = Observed value y - Prediction Value y'
5	Update the next cycle's prediction parameter using the prediction gap and the learning rate.	Prediction Parameter = Prediction Parameter + Prediction Gap * Learning Rate
6	Repeat from Step 1	

PHASE 4: ASSESS THE RESULTS AND ITERATE

The machine learning-enriched FRAM model obtained has been then iterated several times, with the outcomes of the iteration and the final support for decision-making being presented in the next section, along with a critical discussion on its implications.

RESULTS

Figure 14.3 represents the prediction results. As shown in the legend, the dotted line represents the predicted value, the bold line represents the vacant beds planned at the latest regular meeting (data x), and the thin line represents the vacant beds observed at the next regular meeting (data y). In this section, we present the outputs of the

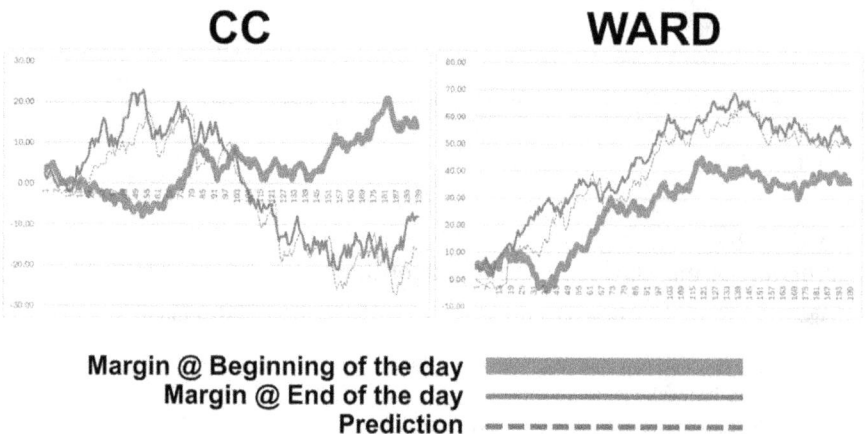

FIGURE 14.3 Machine learning results.

simulation, emphasizing two key topics, that is, analytical accuracy of the prediction and operational implications for safety management.

ANALYTICAL ACCURACY

As can be seen from the time series data sample referred to the ward beds, the periodic fluctuations of data x and data y of the ward beds are somewhat synchronized. In other words, there is a tendency that when data x rises, data y also rises. Conversely, for the data of the CC bed, this correlation is weak. That is, when data x rises, data y may fall, and vice versa. For the CC bed, data x and data y have a strong asynchronous tendency. In light of the actual situation, for the ward bed, if a certain number is added to the data x at the regular meeting, the data y at the next regular meeting can be easily predicted. In other words, data x and data y are relatively simple relationships. It is typical data where things have proceeded as planned without any sudden accidents.

On the other hand, the weak correlation for CC bed between data x and data y indicates that things did not go as planned. In other words, the data shows the first half, where the situation significantly deteriorated from the initial plan, and conversely, the second half, where the situation greatly improved from the initial state.

Despite these example inputs having significantly different trends, the prediction accuracy is equivalent for both CC and ward beds. That is, the difference level between the prediction line (dotted line) and the final balance (thin line) is almost identical. This trend remains unchanged regardless of the synchronization between the data x margin and the data y margin, indicating a high robustness of the prediction accuracy. The prediction formula is a simple one where the "parameter a" (CC mod and ward mod) is optimized by machine learning so that *(data x + parameter a = data y)*. From such a simple calculation, it was possible to achieve high prediction accuracy that adapted to irregular environments, like the results of the CC bed.

Such a good result is achievable due to two factors described below.

- Inherent adjustment of the learning rate: when there is a prediction error (CC gap, ward gap), this is compensated using the appropriate learning rates, which were determined by extensive data learning.
- Data used for learning: we used the only data that can be obtained with 100% certainty, with 100% reliability of the contents, which is the "remaining number of beds". As a result, learning is not influenced by missing data or hard-to-use data (for example, the weather data is hard to use because it changes many times in a day).

Method Geek

These factors make the estimate comparable to human assessment. However, let's look at an example of human predictive ability. What we want to know is a short-term prediction, namely, the number of beds for today. To achieve this kind of very short-term prediction, we usually estimate it based on the trend data up to now. This means that we assume today's data is related to yesterday's data. For example, if the currency exchange rate suddenly drops to one-fiftieth of its value tomorrow, it won't be enough to just make a fuss. In other words, everyone assumes that the exchange rate will only increase or decrease around today's value. This is a rule that we naturally assume for changes in values that have some kind of macro trend. And in most cases, events proceed as expected. That's why graphs of transitions are always used in explanations of currency exchange market conditions, and we can survive without losing everything. When predicting today's rate, considering environmental parameters such as a country's economic or political stability is for longer-term forecasts.

Our target task, which is today's bed count prediction, is conducted in the same way as short-term currency exchange rate forecasts. Instead of using many parameters for long-term trend analysis, we simply refer to the trend up to now to predict today's value. This method is not reliable for predicting one year or ten years ahead, but for tomorrow's forecast, this method is the most effective.

OPERATIONAL IMPLICATIONS

This example shows that we tend to judge situations not so much by the combination of values of many factor variables but by looking at the "change" in the final result value that appears as a result (bed allocation). Debunking this way of thinking is a central aspect for FRAM modelling, that is, "capturing variability" to explain the underlying mechanisms of working phenomena. If we can clarify what changes should be noted and what functions produce those changes, we can model the success factors of ETTOing that we humans are doing, not just the success or unsuccess themselves.

Accordingly, if the success factors can be modelled (and this is assumed possible via a dedicated FRAM model), it is of great significance to quantify the parameters needed to realize them through a simulative approach. In the case of the bed arrangement optimization problem, if the change in the parameters used in today's and yesterday's prediction results exists, it needs to "follow the changes" by adjusting the prediction parameters. Because our system is one where ensuring a proper safety margin is paramount, we need to change according to the type of changes, whether they are desirable or dangerous ones. To this end, we set two learning rates and implemented a system that responds slowly to changes when desirable changes occur while responding quickly when dangerous changes occur. The reason why humans are strong against changes in situations is largely due to their high capability of making such flexible responses. According to the principle of ETTO, it can be said that the property necessary to enhance resilience is that it is usually rough and therefore has high flexibility and that it can react quickly in dangerous situations where it should be alarmed.

LEARNING THE LEARNING RATE

Finally, let's discuss the method of determining the learning rates. When conducting machine learning, the learning rate is usually set to a fixed value imposed by hyperparameters optimization, and then learning is performed by varying the input values. In this example, we varied the input via a grid search while exhaustively testing combinations of learning rates in increments of 10% (with a total data count of 100 million) and statistically determined the optimal combination. This kind of comprehensive iteration for hyperparameters optimization is a commonly used technique (Bergstra and Bengio, 2012). As a result, it was found that setting the learning rate to 1.0 when the prediction gap is negative and 0.1 when the prediction gap is positive provides the best balance between safety and prediction accuracy. Furthermore, it was found that by subtracting 1 from the bed prediction number, the prediction gap will always be on the positive side. We call this a "safety margin". Because the prediction number always falls below the actual number of vacancies, we will rarely be in situations like emergency reassignment of beds or cancellation of surgeries. You can imagine this safety margin as a buffer for cases when an increase in accidental factors is expected, such as heavy snowfall.

DISCUSSION

In this final section, we discuss the relationship between machine learning and FRAM as a method from Resilience Engineering. Machine learning, in a nutshell, is an approach that inputs data, compares the results of the estimation made from it with the observed results, and feeds back the difference to improve the accuracy of the next cycle estimation. The reason why organisms are resilient to environmental changes is that they are constantly repeating this feedback, which allows them to quickly respond to new environmental changes. Constantly repeating feedback is synonymous with continuously conducting online machine learning. In other words, being resilient is the same as continuously repeating feedback-type learning throughout one's life. In order to measure whether a socio-technical system is resilient or not, it is important to see whether the system is constantly conducting learning (Hollnagel, 2011).

This is why Resilience Engineering recognizes the feedback process of learning, by a combinations of Learn/Anticipate/Monitor/Respond, to be paramount.

We argued in this chapter that visualizing the learning process is something the FRAM can handle successfully. If the system does not learn at all, it is impossible to quickly respond to environmental changes, and if the system is learning incorrectly, it may react abnormally to unexpected changes instead of becoming stronger against changes. Therefore, the following two conditions are essential for a socio-technical system to meet:

- Correctly learning
- Continue learning constantly

So, what does "correct learning" mean? The hospital bed allocation optimization problem we are dealing with is a problem that requires more safety than many of the optimization problems in the world. Therefore, "correct learning" here should embed safety as be prioritized.

CORRECTLY LEARNING

To ensure the safety of learning methods, it is essential that the learning process is always presented in a form visible to us humans and that the method is convincing in its validity. In this regard, black-box type learning, such as deep learning, is disadvantaged. While it is possible for humans to understand the mechanism of deep learning, its learning process cannot be presented in a visible form. What deep learning does is not updating variables that convey meanings understandable to humans, but adjusting the binding strength values of each variable so that the model's estimated results match the answer, as a result of updating a large number of variables that do not have a meaning to humans on their own.

On the other hand, in the FRAM-based machine learning example, the variables to be updated are "learning rate when a plus margin is achieved", "learning rate when a minus margin is achieved", and "how many beds to add to yesterday's bed count to estimate today's bed count", all of which can be understood by anyone and do not allow for subtle differences in interpretation.

As a result of the FRAM analysis, if we can clarify the logic that leads from explanatory variables (variables that provide the condition to the system) to objective variables (the goals the system aims for), we can stand at the starting point for performing "safe machine learning". For this, it is necessary to clarify what the function of estimating the number of beds (the "Anticipate" function) does. FRAM analysis revealed that "the transition of the number of beds follows a cyclical movement along a certain seasonal and time variation pattern, and even if there is a sudden event that exceeds expectations, such as a local accident, there is a relationship between data x (the planned value at the regular meeting) and data y (the actual value before the next regular meeting) that follows this pattern. Therefore, we were able to find a simple estimation formula:

$$y_{(t)} = x_{(t-1)} + a$$

Such a "mechanism" needs to be clearly articulated and explained as a set of functions and variables. Furthermore, for optimization methods based on machine learning to be understandable to us, two things are necessary: the method must be clearly explainable, and the method must be based on a concept that humans can use. Even if the logic used in the method is fully explained and there are no ambiguous parts, if the logic itself completely overturns human common sense, it is difficult to have confidence in its safety. Conversely, no matter how much we are persuaded that the logic is based on common sense, if we cannot actually see the logic, we still cannot have confidence. Especially when the act of prediction involves human life, it is a prerequisite that the prediction logic is understandable to humans.

CONTINUOUSLY LEARNING

This issue will pose a difficult question for us. We understand that for learning to be correct, there must be safe learning methods. However, when we finally establish correct learning, if we continue to "learn" further, isn't there a possibility that the learning data could gradually lead us in an unsafe direction?

Resilient systems enhance their ability to adapt to environmental changes through learning. On the other hand, the environment is always changing. This is a fact guaranteed by the second law of thermodynamics (the law of increasing entropy). Therefore, if learning does not continue constantly, we will not be able to respond to changes. This is the reason why we must always continue to learn.

Typically, a machine learning system is tested after learning. If it passes the test, the learning ends there. We tend to assume that once a machine learning system passes the safety test criteria, that system is safe forever or for a long time. However, reality does not work that way. A very clear example is a cyber virus detection system. If it learns all the known attack patterns and passes all the tests, the detection system is guaranteed its safety. But because the environment is constantly changing, within a few months, new attack patterns are invented, and the learned detection system becomes useless. To break through this, the detection system must always continue learning. Similarly, our bed placement optimization system must always continue learning from data and continue to adapt to the emergence of new patterns. In that sense, our system is excellent because it embeds continuous learning by design. As shown in the two learning result graphs in Figure 14.2, our learned model can adapt to the newly emerged pattern and frequencies with high accuracy and a sufficient safe margin because it never stops learning. If it stops learning for even one day, the prediction functionality of this system will not work properly because it always uses the "previous day's gap" value. If environmental changes are not completely random and always have continuity along the arrow of time, the form of learning we should aim for must naturally be continuous learning.

OVERFITTING ISSUE

Black box-type machine learning is known to cause a troublesome phenomenon called "overfitting". This is a phenomenon in which, due to biases in learning data, etc., it is optimized only for a certain pattern and loses its versatility. If you continue to learn data, the model that was learned just right may lose its versatility and become optimized only for environmental data that happened to occur frequently. If that happens, it will surpass the pattern learned in the past, and the problem that was properly solved in the past will not be solved correctly. The biggest reason for this phenomenon is the bias of learning data. For moderate-sized machine learning, such as 256 parameters, the size of the state space for 256 parameters is at least 2 to the power of 256, which is a 77-digit number. In contrast, the total number of molecules in the universe is estimated to be explainable by about a 22-digit number. In other words, you will never be able to prepare the data with all possible combinations of parameters. That means you cannot make unbiased data. No matter how hard you try, the data you make will only be a small part of the whole state space. Even if you use gigawatts of power per second to do machine learning, you cannot finish learning during the history of the universe. But interestingly, the energy our brain uses is said to be only about 20 watts (Kovác, 2010). What an efficient computer! The machine learning that the FRAM aims for should be like the learning that the human brain is doing. If it is a human brain, it should get smarter as it continues to learn.

CONCLUSION: HUMAN BRAIN VS AI

Why humans have such excellent learning ability is, of course, not yet well understood. One thing that can be said is that it is impossible for a small brain to learn a number of data that exceeds the total number of molecules in the universe. In other words, the estimation our brains are making is not from brute-force machine learning like modern black box-type artificial intelligence, but it can be assumed that it is saving a large number of parameters in some way. The method is unknown to date. However, to reduce the number of parameters, it may be wise to use this excellent "brain" instead of the slow "computer". In other words, the input data applied to machine learning should be a minimum that humans have carefully selected.

Method Geek

Let's look at a hint on how our brain can reduce the state space significantly (Hirose et al., 2024; FAA, 2022). A pilot of an aeroplane is trying to land on a runway. He uses a wide range of parameters, such as altitude, ground speed, air speed, roll attitude, pitch attitude, yaw attitude, wind speed, rainfall, runway constraints, and possible approach routes, to estimate the current state and control it to approach the desired state. The parameters listed here are all 10. If you get all of these, you should be able to land somehow. However, even if a situation occurs where the instrument cannot be used, the pilot is trained to be able to land. Here is how they do. The pilot can land without any numerical input of all the parameters listed above by looking at the flowing scenery, seeing that the runway continues to stay in a fixed place in the field of view, and that the shape of the runway is the expected shape of a "trapezoid". In this case, the parameter is only two. If you try to realize this landing by machine learning with the 10 parameters, most of them listed are continuous values, so the number of combinations of learning data is infinite. Of course, infinite learning cannot be completed within the history of the universe. But what about the two parameters listed later (whether the runway is moving or not and whether the trapezoid is the expected shape or not)? Since these two parameters have only two values (TRUE or FALSE), the number of combinations is only four. Don't you feel that learning with just these two parameters is possible even with 20 watts of power?

The brain's great ability to save parameters is advocated in a cognitive theory called the "Ecological Approach" of perception (Gibson, 1979). The brain, like landing of an aeroplane, dramatically saves the energy used for cognition in various everyday scenes and performs tasks. If you are a great captain, you can safely make a manual landing even if the instruments are broken.

The machine learning we conduct should be implemented with reduced parameters. Our form of saving parameters is what we introduced for bed assignment. The input parameter is only one, the latest bed usage plan (data x). If so, the state of overfitting cannot occur. As for the placement of the number of beds, the only learning item is what percentage of the learning rate will balance the prediction accuracy and safety, so the more you learn, the more you can refine the adjustment, especially if a new pattern is born as time goes by. And no matter what value the input data takes, if there is some continuity between yesterday's data and today's data (in other words, it is not completely random), the predicted value derived from there also applies some kind of continuity. Therefore, we have good prediction accuracy. The drop in accuracy occurs when the result of the previous day and today have changed more than expected. If the changes are larger than expected in a risky way, the learning will proceed to increase the learning rate, and if the changes are in a good way, the learning rate will be decreased. To make this learning rate safe and accurate, all you must do is follow the latest data. Therefore, the problem of bias in learning data that occurs due to a huge state space never occurs.

When you are asked what the best data for your prediction is, you can always answer the question with "The next one". Continuous learning is a necessary condition for our survival as living beings. There are still many things we don't know about the decisive difference between artificial intelligence and the human brain that we have been considering so far. However, it seems that it is not inappropriate to think that this surprising fuel efficiency of our brain is produced by learning that is fundamentally different from the learning of modern artificial intelligence.

The FRAM is certainly a valuable tool for imitating the "parameter saving" that humans are doing. If the function in the created FRAM model takes in too many inputs, we may find it worthwhile to stop and ask ourselves. Are humans truly processing all these complex inputs, or is there another input that substitutes for them to reduce the load?

By conscious and understandable machine learning, which was devised to build artificial intelligence, we may be able to awaken to the wonderfulness of the natural intelligence that we humans have.

REFERENCES

Bergstra, J., Bengio, Y. (2012). Random search for hyper-parameter optimization. *Journal of Machine Learning Research*, 13, 281–305.

Cook, R. I., & Rasmussen, J. (2005). Going solid: A model of system dynamics and consequences for patient safety. *BMJ Quality & Safety*, 14, 130–134.

Ewbank, L, Thompson, J., McKenna, H., Anandaciva, S. (2021). NHS hospital bed numbers: past, present, future. *The King's Fund*. https://www.kingsfund.org.uk/publications/nhs-hospital-bed-numbers, (Access: April 30, 2025) 2021

Federal Aviation Administration. (2022). *Airplane flying handbook (FAA-H-8083-3A)*. https://www.faa.gov/regulations_policies/handbooks_manuals/aviation/airplane_handbook (Access: May 29, 2024).

Gibson, J., 1979. The theory of affordances. *The ecological approach to visual perception* (pp. 119–137). Taylor & Francis.

Hirose, T., et al. (2024). *Success factor analysis of great ship captain onboard based on ecological approach. Proceedings of the FRAMily 2024*, Lund (Sweden).

Hollnagel, E. (2011). *RAG* – The resilience analysis grid. In: E. Hollnagel, J. Pariès, D. D. Woods & J. Wreathall (Eds). *Resilience Engineering in Practice*. Ashgate.

Hollnagel, E. (2017a). *Safety-II in practice: developing the resilience potentials*. Routledge.

Hollnagel, E. (2017b). *The ETTO principle – Efficiency-thoroughness trade-off: Why things that go right sometimes go wrong*. CRC Press.

Kirsch, L. et al. (2021). *Introducing symmetries to black box meta reinforcement learning. Proceedings of the Thirty-Sixth AAAI Conference on Artificial Intelligence (AAAI-22)* (arXiv:2109.10781).

Kovác L. (2010). The 20 W sleep-walkers. *EMBO Reports*, 11(1), 2. DOI: 10.1038/embor.2009.266.

MacKinnon, R., Barnaby, J., Nomoto, H., Hill, R., Slater, D. (2023). Adding resilience to hospital bed allocation, using FRAM. *Medical Research Archives*, 11(10), 2023. DOI: 10.18103/mra.v11i10.4265

Selim, M., et al., (2022). Safe reinforcement learning using Black-Box reachability analysis. *IEEE Robotics and Automation Letters*, 7(4), 10665–10672.

15 The FRAM-IA to Understand and Assess Distributed Cognition and Control

Arie Adriaensen and Paulina Zurawska

CONTEXT: THE INTERPLAY OF CAPACITY AND DEPENDENCY

There is a fundamental relation between capacity and dependency, whereby 'capacity is the total set of inherent things (e.g., knowledge, skills, abilities, and resources) that an entity requires to competently perform an activity individually' and 'dependence exists when an entity lacks a required capacity to perform an activity in a given context competently' (Johnson et al., 2014, p. 49).

Dependency originates from some lack of capacity in which one agent cannot fulfil a function without the support of an activity from another agent. Capacity can be present at the team level, whereas dependencies exist between the individual agents of the team to achieve the capacity. Interdependence can subsequently be defined as 'the set of relationships used to manage these dependencies' (Johnson et al., 2014, p. 49). Hence, interdependence is the successful management of these capacity deficiencies by the support of an activity of another agent (human or technical).

An essential breakdown of different types of interdependence (i.e., pairs of a capacity and a dependency) has previously been defined by Johnson et al. (2014) as being composed of the qualities of Observability, Predictability, or Directability, within the context of Interdependence Analysis (IA).

In FRAM, these labels can be used to make an abstraction of interdependence types by characterizing what coordination category (Observability, Predictability, or Directability) is locally propagated by any concrete coupling between any two consecutive FRAM functions performed by two different agents (Adriaensen, Berx et al., 2022). Distributed cognition posits that cognition and knowledge are not confined to an individual's mind but are distributed across objects, individuals, artefacts, and tools within the environment (Hutchins, 1995). The Observability, Predictability, and Directability can be considered as characteristics of system readability that apply to human agents but also to system affordances. This type of distributed

DOI: 10.1201/9781003518167-19

cognition is not restricted to human cognition but describes the cognition of a system as a whole, described as the theory of Joint Cognitive Systems (Hollnagel & Woods, 2005). Interdependence Analysis can be considered as an analysis of multi-agent affordances for human and non-human agents in such Joint Cognitive Systems.

DEFINING OBSERVABILITY, PREDICTABILITY, AND DIRECTABILITY IN IA

In this section, we briefly introduce Observability, Predictability, and Directability, with a listed overview in Table 15.1 at the end of this section.

(i) *Observability* is defined by transmitting and receiving individual status, knowledge of the team, the task, and the environment. It does not necessarily apply to 'what' the other agent is observing but describes 'how' the other agents reveal status and intention or the interpretation of pertinent signals (Johnson et al., 2014).

(ii) *Predictability* is defined as the degree to which others can rely on outcomes when considering their own actions. It involves the use of a priori agreements, the use of models, and the synchronization of actions (Johnson et al., 2014), as well as time-projected status predictions.

(iii) *Directability* is defined as the ability to direct behaviour and to have one's behaviour directed by others. It involves explicit commands, such as how tasks are allocated and roles assigned, but also implicit commands, such as providing instructions or directive warnings (Johnson et al., 2014). Henceforth, the use of Observability, Predictability, and Directability labels can be referred to as OPD labels in short.

Method Geek

A structured overview summary of OPD labels and their definitions is provided in Table 15.1 below, together with possible forms.

TABLE 15.1
Interpretation of OPD Principles

OPD (Principle and Symbol Used in This Chapter)	Description	Can Take the Following Forms
Observability _ O _	Transmission and receipt of status, knowledge of the system, team, task, and environment (real-time Observability of the system or its parameters)	reveal status reveal intention interpretation of signals observable transformation of information across media observable transmission of physical states and forces monitoring actions

(*Continued*)

TABLE 15.1 (Continued)
Interpretation of OPD Principles

OPD (Principle and Symbol Used in This Chapter)	Description	Can Take the Following Forms
Predictability _ P _	the degree to which others can rely on outcomes when considering their own actions, including the degree to which a system predicts future states or reveals future system status/or reveals the boundaries of failure or critical system performance before these boundaries are actually reached by the system or its parameters	a priori agreements the use of models synchronization of actions use of time-projected information (similar to Observability projected over an added time dimension) cues that support the prediction of boundary conditions for users (guidance, suggestions, and warnings)[a]
Directability _ D _	the ability to direct physical control, processes, agents, or system states	task allocation role assignment cues that support redirection of actions (guidance, suggestions, and warnings)[a] control transfer corrective actions (incl. outputs of monitoring actions)

Source: Adapted and extended from Johnson et al. (2014).

[a] 'Guidance, suggestions, and warnings' appear twice depending on whether the coupling displays boundary information or is used to redirect agents' behaviour. The two different sentences preceding it delineate the different contexts of use.

IA IN THE CONTEXT OF FRAM MODELLING

Recently, Interdependence Analysis (IA) was proposed as an additional layer of investigation to complement the traditional FRAM steps (Adriaensen, Berx et al., 2022) in support of variability management once undesirable performance variability and functional resonance are identified. Therefore, principles of multi-agent goal coordination are added by applying IA as an additional layer of analysis for the concrete functions and couplings already described in a traditional FRAM model. IA looks at the qualitative characteristics of the relationship produced by any two functions performed by two or more agents.

In traditional FRAM, an Output can represent matter, energy, or information. Typically, think of a decision, a command issued, or the changed state of a material. Traditional phenotypes (i.e. the type and quality of an Output) represent any relevant quality of 'what' the function produces, manufactures, or simply what it 'outputs'. Therefore, qualities of these traditional FRAM Outputs, such as matter, energy, or information, are usually designated as degrees of 'timing', 'precision', 'speed', 'force', etc. (Hollnagel, 2012). We call these qualities phenotypes. Whereas traditional Outputs of functions represent the potential variable performance of 'what' is embodied by the coupling, IA adds a different label type that characterizes 'how' the coupling acts as an interdependency between upstream and downstream agents. Interdependency is explained in more detail in the section below.

While this chapter primarily focuses on outlining the methodological approach, it also provides two simplified, illustrative use cases to guide readers through the proposed solution in a more accessible, step-by-step manner.

DATA COLLECTION

We show the feasibility of Interdependence Analysis as an extension of the traditional FRAM by applying it to (i) a well-known stable control structure used in aviation when a pilot takes manual control when it becomes unsafe to use the autopilot and (ii) the activation of an emergency stop in an industrial robot system. These well-known functional control structures represent normal work because, for many decades, the course of expected actions has been deeply embedded in procedures required by aviation procedures and machine safety regulations. For many other socio-technical systems in the world, we still need more research to understand the full functional control structure. The proposed method in this chapter can help achieve equal levels of functional understanding for socio-technical systems that are not yet fully comprehended. Interdependence Analysis is added as a layer supporting system readability and functional control, complementing traditional FRAM models in which outputs typically represent transformations of matter, energy, or information.

The FRAM model used in the results section of this chapter presents the functional requirements for human intervention in a generic automated control structure. The identification of functions and couplings is based on the previous professional experience of one of the authors. In accordance with procedural and regulative documentation (ISO, 2006, 2016; The Boeing Company, 2002), a FRAM model was built by self-ethnographic experience based on the first author's previous airline pilot and industrial safety manager expertise. The models were additionally validated by domain experts, that is, respectively, one active airline pilot and one active industrial safety manager, both safety science researchers with some degrees of FRAM experience. The specific FRAM-IA describes a system in which the choice between the automated and manual alternatives is supported by system readability and system control from human and machine agents.

METHODOLOGY

IA can be applied to any socio-technical domain which involves multiple agents. The IA method mirrors the traditional FRAM building steps but extends them, as summarized below.

STEP 0: RECOGNIZE THE PURPOSE OF THE ANALYSIS

This step includes the usual FRAM modelling, yet paying attention to the model's scope should be relevant for understanding dependencies and interdependencies (i.e., the qualities of coupling between any consecutive functions performed by two agents). For the IA scope, two additional important scope elements are defined:

(i) The system's mission success: A socio-technical system is a purposeful structure that consists of interdependent social and technical elements influencing one another, maintaining the system's existence while pursuing the system's or subsystem's functional goal. In the subsequent steps of the analysis, each FRAM function can be assessed individually for its positive functional contribution to the system's mission success.

(ii) The system integrity of the system: the functions are performed by human and technical agents, which can all show failure modes or conditions that impede safety or performance. In subsequent steps of the analysis, each FRAM function can be assessed individually for its functional negative impact on the system integrity.

STEP 1: IDENTIFY AND DESCRIBE THE SYSTEM'S FUNCTIONS

Although this step is identical to the first building step of the traditional FRAM, a suitable functional granularity should be chosen to reflect inter-agent exchanges on the level of functional identification. Whereas at first sight, a function such as <to monitor> appears to be a single-agent function, it becomes a series of human–system exchanges between human agents and system interfaces when zooming in on what <to monitor> actually means in a particular socio-technical system.

STEP 2: IDENTIFY VARIABILITY

At this stage, the introduction of IA plays the first relevant role. In addition to the usual variability, a dedicated approach should be chosen for OPD information gathering. As described in the data collection phase, the scenario in the results section of this chapter utilized auto-ethnographic data collection supported by domain expert validation. Each function is individually assessed for having an Observability, Predictability, or Directability role in the FRAM model, while simultaneously, the particular Outputs' *mission success* and *system integrity* are assigned for each function individually as traditional FRAM Outputs. The functions are constructed bottom-up, and their phenotype values are assigned individually based on their local contribution or hindrance to the system's functioning.

The first difference with traditional FRAM is that by the end of Step 2, Observability, Predictability, or Directability labels are added to couplings between functions performed by different agents.

An expert FRAM user will notice that (January 2025) in the FMV, they are added to the receiving function for practical restrictions of the FMV application.

STEP 3: AGGREGATE VARIABILITY

This step is the core of IA and accounts for assessing coupling values in a more sophisticated way, delving into the actual agent interactions between functions. A specific guide for the application of the gathered data in the demonstration example from this chapter is covered in the Results section. At this point, it suffices to remember that the previous phenotype values for OPD, as well as for mission success and technical integrity, are propagated downstream to one or more so-called 'pivotal functions' described in Step 4 below.

By the end of Step 3, an essential difference with traditional FRAM is the introduction of the pivotal functions and the propagation and aggregation of the upstream OPD labels displayed on the pivotal function. Alternative outputs at the pivotal function are selected based on the aggregation of upstream phenotypes and OPD labels.

STEP 4: ASSESS THE CONSEQUENCES OF THE ANALYSIS

In this step, the outcomes of the ODP assessment help to explore their propagation and aggregation in the domain-specific problem space and functional structure. This step consists of studying OPD patterns as expressions of linear or complex control action-feedback loops (OPD configurations), whereby selected functions that receive OPD propagation at pivotal moments are typically used to make choices between different functional propagation paths. This assessment can be used to debrief operations and define ways in which suboptimal (or even dangerous) ODP couplings could be managed more safely and efficiently. The method's principles laid out in this chapter can also be used to better understand functional control in complex socio-technical systems, from which the control structure is not yet fully understood.

Step 4 provides a structured analysis of functional propagation based on the rules from Steps 2 and 3, which are not prepared in traditional FRAM, whereby the scope is on distributed system readability and system control.

Given the scope of this chapter, a dedicated section on the actual application substeps of introducing OPD into a FRAM analysis is provided, reaching a technical yet more practitioner-oriented representation.

Method Geek

EXPERT GUIDE FOR FRAM-IA IMPLEMENTATION

The following guiding rules can be used to apply the *Interdependence Analysis* to the traditional FRAM developed, that is, Steps 2 and 3 of the FRAM application steps described above:

THE IDENTIFICATION OF INTERDEPENDENCY

(1) 'Can the upstream *<function description>* produce the supporting activity required by the downstream *< function description>* to efficiently manage the |*coupling description*|?'
  • Note: If the answer is yes, an (upstream) capacity exists (and a downstream dependency is implicitly assumed)
(2) 'Does the downstream *<function description>* need activity (support, assistance, resource, etc.) from the upstream function involved in the coupling to manage the |*coupling description*|?'
  • Note: If the answer is yes, a downstream dependency exists.
    If the answer to both (1) is no, and (2) is no:
    – An interdependency does not exist. Such couplings do not require further assessment in terms of *IA*.

(3) An interdependency does not exist but might have been originally expected for the mission purpose but has failed on both agent sides (failure of capacity and dependency). For an interdependency to be successfully met, there needs to be a yes to answer (1) and (2).

(4) If the answer to (1) is no and to (2) is yes, the interdependency exists but has failed.

(5) If the answer to (1) is yes and (2) is no, an interdependency exists in which the upstream agent still offers a capacity that is not strictly required but can produce system redundancy.

THE QUALIFICATION OF THE INTERDEPENDENCIES

(6) Whenever an interdependency exists, continue with answering the subsequent question: **'What type of interdependency (O, P, or D) is produced by the |coupling| at the receiving level of the** <*downstream function description*> **in relation to the state or performance received from the** <*upstream function description*>?**'** In simpler words, which of the OPD labels does the |coupling| produce for the receiving <*downstream function description*>?

* Note that in traditional FRAM the question 'what is produced' by the coupling provides the coupling label name as the result of something that represents matter, energy or information. The traditional phenotype assessments are therefore the qualities of how these couplings are produced (i.e., differing degrees of speed, accuracy, and timing).
* The guiding rule described in (7) ultimately represents a fundamentally different quality and produces the OPD label as an addition from the IA extension.

THE QUANTIFICATION OF THE INTERDEPENDENCIES

(7) To rate the interdependency quantitatively, rate the quality of the OPD quality from step (6) with a number from 0 to 1, representing a range with 1 being the optimal value and 0 the total absence of the required OPD property.

GRAPHICAL REPRESENTATION AND LOGICAL INTERPRETATION OF FRAM-IA MODELS IN FMV

The graphical representation and interpretation are supported by Figure 15.1.

(8) Top values: The 3 top values are reserved for OPD labels.

(9) Bottom values 'left-right': The two remaining bottom positions (left-right) are reserved to display traditional phenotype values. They are specifically allocated to *technical integrity* and *mission performance* values. Technical integrity displays the integrity of the system or subsystem represented by the FRAM model, whereas mission performance provides the corresponding value for the socio-technical purpose of the subsystem under

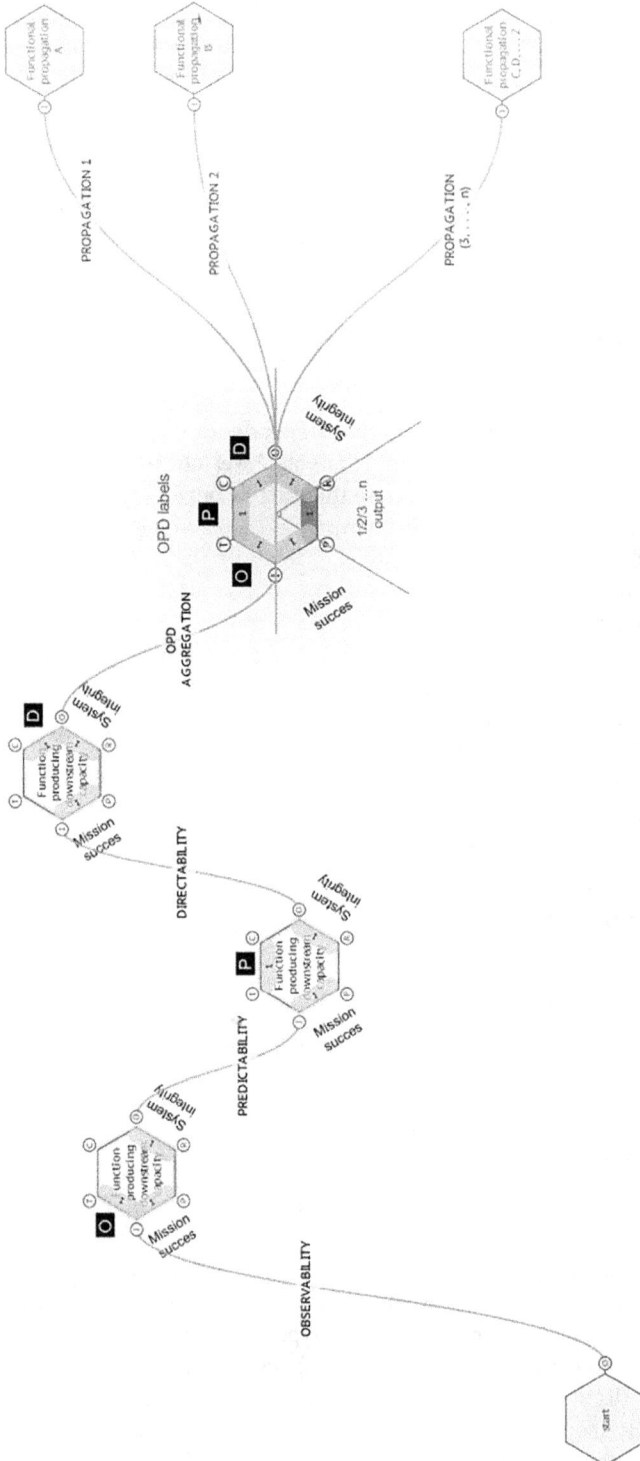

FIGURE 15.1 Dummy FRAM with positions of OPD values and pivotal function (shaded background). Pivotal functions show propagated aggregation values, an algorithmic selection outcome (bottom value) and multiple possible outputs/propagations.

consideration. These two performance parameters are paramount to assess the criticality of the system output, whereas the OPD labels simultaneously assess the inter-agent readability and controllability of these parameters.

(10) Bottom value 'center': In FRAM-IA, pivotal functions are functions in which a selection needs to be made on whether a functional propagation plan A, plan B, etc., will be executed. This can be recognized by functions that show two or more outputs leading to different functional propagation alternatives. A sixth bottom position is reserved for showing the decision algorithm outcome that shows the propagational alternative.

(11) Analogous to traditional phenotype values, quantification between 0 and 1 can also graphically be assigned to the OPD qualities, with 0 being the absence of the quality and 1 being the optimal fulfilment of the quality. Determining OPD labels is typically derived from information provided by Subject Matter Expert inputs, for example, via questionnaires, interviews, or the Critical Decision Method (see Klein et al., 1989).

(12) Phenotype values at one or more of the six hexagonal positions are only displayed for the OPD labels and traditional phenotype values that are relevant at the receiving function.

(13) When applying IA to FRAM, a sixth hexagonal position is required for displaying the decision or algorithmic outcome of pivotal functions in which a decision needs to be made whether a functional propagation plan A or plan B will be executed. This can be recognized by functions that show two or more outputs leading to different propagation alternatives. The sixth (centre-bottom) position [1st, 2nd, …, n output] is a decision algorithm based on an aggregation of the three OPD values (readability and control), as well as the mission success (positive system outcome) and the mission integrity values (positive system outcome hindered).

RESULTS

This section describes the usage of the application steps in a case study describing an operator-initiated override that can be applied to different domains, like aircraft handling in aviation or robot handling in industrial automation, with regulation-required emergency stops. In both cases, the common theme is transitioning control from automated systems to a higher-level human-supervised control mechanism to ensure safety and operational continuity.

RESULTS OF STEP 0: THE PURPOSE

The purpose of the analysis is to examine the functional control involved in a real-world example describing human interaction with automation. The mission success covers all the functions that contribute to maintaining the existence of the system itself while pursuing the system's or subsystem's functional goal. In the aviation example, this is keeping the aircraft within the safety envelope. In the industrial system, this is keeping the robot within operational boundaries and within a previously

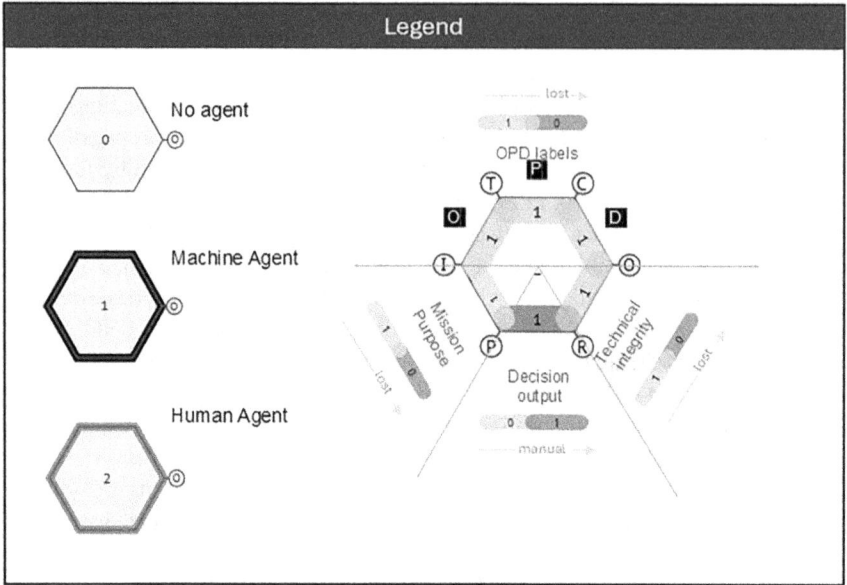

FIGURE 15.2 Legend with greyscales for different agents and the position of the values for the pivotal function.

defined cartesian space. The mission integrity covers all the functions that in this system could individually fail, including both technical and human agents. For simplification, we have assumed in the case presented in this chapter that mission success is equal to mission integrity. This is credible for the example under consideration, but might differ in other case studies. This means that as long as all the different subcomponents keep mission integrity, the safety envelope or operational boundaries are also assumed to be within limits.

RESULTS OF STEP 1: THE FUNCTIONS

The model describes functions and their couplings, configured in line with the conditional transitioning from the automated to human-control structure. The legend in Figure 15.2 shows the different human and technical agents, their corresponding greyscales, and a summary of the different aggregation values in the pivotal function. The pivotal function receives the propagated values from upstream functions as previously represented in Figure 15.1.

Domain Expert

This legend will subsequently be applied to Figure 15.3. The latter can be read as follows: An aircraft cockpit or industrial automated robot system will <provide automated control #2> as the result of receiving an input to <initiate automated operations #1>. The automation will detect environmental inputs <#3> such as barometric pressure, temperature, aircraft/robot speed, robot force, etc., through

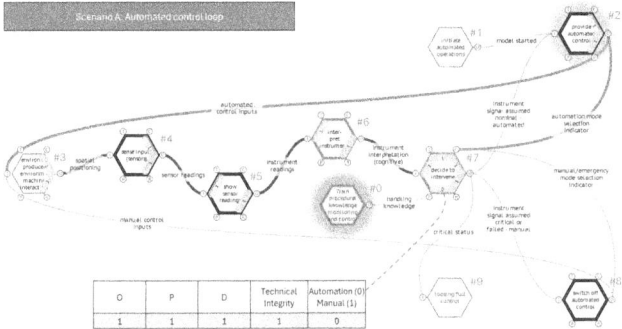

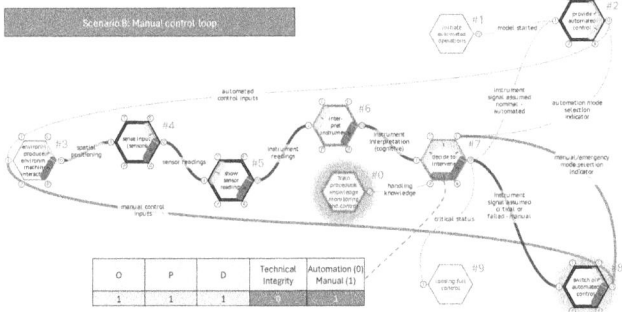

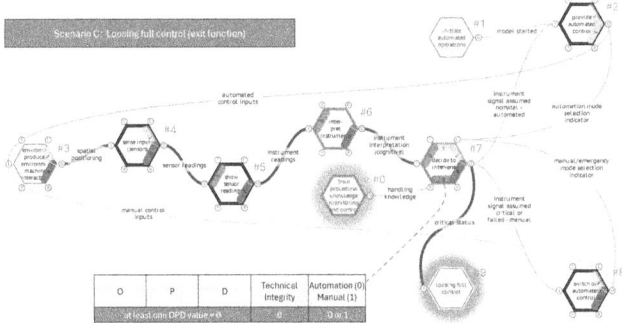

FIGURE 15.3 Application of Steps 2 and 3 – the identification and aggregation of performance variability in automated vs manual control in aviation as well as in industrial robot systems.

<sense inputs #4>. These inputs are subsequently translated into semantical information through instrument readings and display monitors <show sensor readings #5>. This semantically transformed information is interpreted by human operators <interpret instruments #6>. Typically, the system can maintain automated control if the instrument signal is assumed to be reliable or system readings are considered to be within safe limits through or select mineral control by <decide to intervene #7>. When these conditions are not met, the human operator is required by training

<train procedural knowledge for monitoring and control #0>. <switch off auto-mated control #8> reflects the system activation of an emergency stop in indus-trial installations, whereas, in the aviation example, manual control of the aircraft will be prioritized over autopilot whenever safe system boundaries are exceeded. <Losing full control #9> depicts an exit function representing a loss or partial loss of control. The fact that Figure 15.3 covers two fundamentally distinct domains illustrates that goal coordination follows recurrent patterns of system control.

Table 15.2 shows OPD labels applied to the couplings in this model to support the interpretation of IA to the instructive example of this section in correspondence with Figure 15.3 with OPD labels assigned as O (upper left), P (upper centre), and D (upper right). In this particular case, values 0 and 1 were used as a binary choice

TABLE 15.2
OPD Labels Applied to the Use Case at Hand

Function #	OPD Label	OPD Label Explanation
<0- 1>	n/a	Entry boundary functions do not receive aspects upstream and can consequently not be assigned an OPD label. A receiving agent is required to determine the OPD label (no inter-agent exchange). Training function #0 is always active, as training is supposed to happen before starting operations.
<2>	Directability	Function 2 is the technical agent receiving the physical outcome from the decision made by the operator to not intervene and thereby characterizers Directability.
<3>	Directability	Function 3 depicts the environment receiving the control outputs from the technical agents (flight controls influence aerodynamics, our robot navigation changes its position relative to a cartesian space)
<4>	Directability	The novice Interdependence Analyst might be misled into thinking that a sensor is by definition, producing Observability, but what is actually received by the sensor is a change or redirection of its physical inputs and therefore represents Directability.
<5 – 6>	Observability	What the downstream agent receives is, in both cases <#5-6>, what the upstream agent revealed as available information concerning a local system output.
<7>	Observability & Predictability & Directability aggregation	Predictability is received from the training function #0. It corresponds with the 'a priori agreements' from Table 15.1, being how one will act in the future by previous training and is the only Predictability function in the model. The Observability and Directability values result from the propagation of the most critical value found upstream from this function.
<8>	Directability	This function displays the imminent need for the transfer to manual control. The role of this coupling is, in its functional context, not the observation of the signal but the action triggered by its presence. This is a consequence of the instructions and training previously introduced in function <#0>.
<9>	n/a	No label is assigned, as this is an exit function

instead of using a range between 0 and 1. The value 1 represents optimal performance, and 0 represents suboptimal performance, whereby a 0 value could also represent the complete absence of the required OPD property.

RESULTS OF STEP 3: THE POTENTIAL AGGREGATION OF THE VARIABILITY

In traditional FRAM, endogenous variability (the type of variability triggered within a function) can be added by assigning the expected variability under conditions of normal work. For example, sensor inputs in industrial systems can be degraded when covered with dust. In aviation, speed or altimeter sensors can become frozen due to extreme weather phenomena or become obstructed after a bird strike.

Domain Expert

The multi-agent goal coordination on the systems level can best be understood by looking at three instantiations, that is, particular functional scenarios, from Figure 15.3. The inset tables for each individual scenario, pointing to the pivotal function depicted in the inset, show the values that are used for the computation of the bottom selection value. The executable model of Figure 15.3 with randomization of the different instantiations according to the functional resonance can be accessed through the following hyperlink.[1]

(i)   In scenario A (where the executable model follows the upper loop in Figure 15.3), all values work optimally. The OPD values show that all agents have optimal readability and control, while technical integrity is additionally maintained. The bottom selection value remains at zero, which means that the operation, therefore, runs in a continuous automated loop.

(ii) In scenario B (where the executable model follows the lower loop in Figure 15.3), all OPD values work optimally, but there is an issue with technical integrity. The bottom value changes to 1, which means that the operation is expected to go through manual control to maintain mission success. This means the functional propagation enters a loop of manual flight in aviation. In the robot system, an emergency stop would be activated when going to manual control, whereby the functional propagation would reach a stopping condition. Note that under these conditions, technical integrity as part of mission integrity is degraded, but system control is still maintained because these scenarios have been prepared for.

(iii) Scenario C (where the executable model follows the lower loop in Figure 15.3, but ends with the stop function <losing full control>, describes a stop function where full system control is lost because the mission integrity parameter equals 0 while not having full system readability and control, that is, any OPD equals 0.

This demonstrates that in analogy to traditional FRAM, FRAM-IA models do not represent functions merely as a sequence of tasks but as a network of potential functions encountered in a range of normal work operations. Therefore, the different propagation alternatives in Figure 15.3 represent different instantiations of a single FRAM model, showing the functional resonance potential for either automated or manual system control.

Note that the described FRAM model thereby encompasses human–machine goal coordination for both reliable and unreliable sensor outputs to maintain control over a range of situations. In this particular functional control model, mission success can thereby even absorb technical failures as long as the OPD properties are providing optimal performance. This is in line with the FRAM philosophy of representing how things work and how control can be maintained instead of representing accident scenarios, as in many traditional safety assessment methods. The example of unreliable sensor inputs just leads to a different functional propagation, essentially because this possibility is covered within the goal coordination ensemble of observing changes (system readability) and adapting control (system control) accordingly.

RESULTS OF STEP 4: MANAGEMENT OF PERFORMANCE VARIABILITY

Particular configurations and patterns of OPD labels usually require a combination of system readability in terms of Observability or Predictability on the one hand and a control capability, that is, Directability on the other hand. The pivotal functions are used to decide what normal propagation alternatives look like and which combination of OPD values, in combination with mission success and mission integrity. In systems where historical data is readily available or in centralized systems, patterns of goal coordination are deliberately engineered into existing systems. On the other hand, patterns of goal coordination can also act as an expression of self-organization and emerge rather than being the product of deliberate engineering. Even engineered systems can drift into failure over time. One notorious example is the Boeing 737 MAX, which crashed twice in the years 2018 and 2019 (Herkert et al., 2020). These crashes represent a deviation from a previously well-engineered control model for previous B737 models. The B737 Max accident instantiations are, therefore, not part of the scenarios from Figure 15.3. This is in line with the fact that FRAM does not usually represent accident instantiation but normal work or systems under control. However, the B737 MAX instantiation can be explained as an accident instantiation in which manual and automated output loops were both acting simultaneously (see function <#2> and <#8> in Figure 15.3) while producing opposing control outputs. In addition, pilots were not aware that this engineering was functionally allowed by the aircraft systems. The <provide training #0> function, therefore, did not align with the assumed system outputs. This means that the Predictability label received in the downstream function <decide to intervene #7> additionally impeded the pilots from being able to know what was going on.

Method Geek

ADDITIONAL METHODOLOGICAL REMARKS

Note that the quality of the capacity produced by the upstream agent is ignored in the examples. Rather, the capacity is simply assumed to be present. This means that only the dependencies are shown. This chapter also assumes the rule that the capacity-dependency direction of the interdependency is unidirectional, that is, always upstream–downstream. This is, in reality, a simplification, but it is safe to assume this simplification when assessing an existing operational reality. This is because, from a Work-as-Done perspective, we are producing a descriptive assessment in which re-engineering is not put into question. This can change when we are managing performance variability and when we start to define system requirements, in which upstream capacities are part of the investigation during the engineering phase, in which we change to a Work-as-Intended perspective. Such bi-directionality of the interdependencies has already been introduced elsewhere (Adriaensen, 2022).

We remind the reader that IA is only applied to inter-agent functional exchanges. Two consequent functions performed by the same agent do not produce an interdependency, as there is no lack of capacity in need for 'another' agent. Contrarily, the latter is, therefore, called an intra-agent functional propagation. Accordingly, no OPD labels are assigned in this case. In human agency, intra-agent functional propagation, for example, constitutes cognitive processes. Analysts or researchers only have access to the action produced by the agents, and we cannot directly assess the cognitive mechanisms in functional terms. It is, therefore, consistent with the focus of analysis in this chapter on merely labelling inter-agent exchanges.

In reality, the OPD labels belong to the coupling (as received by the downstream function) and are graphically positioned in the downstream function. This is simply a graphical FMV restriction, as the FMI feature of the FMV does not currently permit adding the metadata labels graphically to couplings, but only to the functions.

DISCUSSION

A systematic analysis of socio-technical systems is made possible by the transformation of task-relevant information into its subsequent state from a functional perspective. This approach avoids the issue that safety analysis and researchers lack full access to the internal representation that underpins human cognition or sense-making (Adriaensen et al., 2019). Internal cognitive representations are subsequently ill-suited for being correctly described by a functional analysis. Instead, this demonstration example to which we applied FRAM-IA is well-suited for studying the ecological perspective specifically described by Hutchins (1995) as the role of the material media in which representations are embodied and in the physical processes that propagate representations across media.

By creating a functional model, one may see how systems that are bigger than a single person or agent can have unique characteristics that are not reducible to the cognitive traits of people or functional traits of individual machine agents. This aligns with previous work from Hutchins (1995) and the theory of Joint Cognitive Systems (Hollnagel & Woods, 2005). Rather than trying to model and assess the cognitive approach (e.g., assessing the cognitive constructs of individual decision-making, situational awareness, and complacency), the IA approach deliberately focuses on the conditions that shape, facilitate, or hinder such cognitive processes in terms of system readability and system control, in combination with functional outputs that contribute to mission performance and mission integrity.

One essential challenge remains that the relationship between the understanding of the functional control structure and the systemic aggregation of interdependencies is inherently dynamic. Attempts to gain system understanding from a pure reductionist perspective in which all functions and couplings are a priori known, and aggregation is the simple sum of all individual couplings, will not be useful for complex system control. Instead, one can derive more abstract patterns in complex technical systems and perform a deconstruction assessment by the principles proposed in this chapter.

POSSIBLE AREAS OF INTEREST FOR THE FRAM-IA

The current proof of concept of the different application steps of FRAM-IA, described in this chapter, is applicable to many different domains. Categorizing these patterns for different system goals can be considered ongoing research, partially supported by the FRAM literature covering different application domains, systems, and case studies (Patriarca et al., 2020; Salehi et al., 2021). A full categorization of complex system controls has, to the best of our knowledge, not been undertaken, but many examples can be imagined. Think of the introduction of autonomous agents such as self-driving cars and the emerging challenges of self-organization and adaptive behaviour. As another example, consider the increased polycentric control from decentralized systems for energy generation, where households exchange their energy with the power grid depending on dynamic demands. Or think of the changed nature of control from hyper-connected systems through the Internet-of-Things in today's manufacturing industries and supply chains. All of these systems require more complex functional control structures than the one demonstrated in this chapter, although they ultimately conform to similar FRAM-IA underpinnings, which can be applied to the OPD labelling of individual functions and the aggregation of upstream outputs in the pivotal functions that demarcate the selection points for functional propagation alternatives.

While the example in this chapter explains how FRAM-IA can be applied to a well-known system, it can be suggested that the IA perspective could, in the future, also be suitable for supporting or even engineering serious games about complex socio-technical challenges. One can think of those serious games whose design philosophy relies on exploring complex behaviours, making them visible, and using them as a discussion standpoint for improving system properties.

In this context, IA could be used to better understand goal coordination and system control by engineering certain game elements into different game scenarios. It would be possible to compare the perceived system readability and cognitive limits for collaboration on team collaboration while also measuring the actual system load by adding or removing particular game parameters.

The game outcome can be calibrated to represent mission success, depending on multi-agent collaboration. In serious games, actuators could be logged and could be linked to couplings and their traditional performance phenotypes in the underlying FRAM model. Mission integrity can be manipulated by deliberately introducing technical failures or information bottlenecks. The perceived aggregated OPD effects can be compared by the actual level of comprehension of specific instructions, artefacts, actuators, and tasks, which is measurable by the player's game experience, or in the system engineering language, their perception of the system readability and control by assessing their actual task understanding and game navigation choices. This can serve the purpose of consequent player game experience steering, depending on the purpose of the specific gameplay, as well as the scientific understanding of distributed decision-making under semi-laboratory conditions that simulate real-life scenarios. The combination of player debriefings and well-structured questionnaire results conducted after each gameplay, further translated into a structured IA assessment, is the way to support a better understanding of how past success results from static design features and dynamic self-organization strategies between players. The serious game case is only an example of a joint approach of FRAM-IA and engineering tools that enrich the research with an increased understanding of distributed system readability and control.

CONCLUSION

The FRAM-IA method provides support for assessing and managing the potential for goal coordination in socio-technical systems. The method has previously been applied to research concerning collaborative and surgical robots (Adriaensen, Berx et al., 2022; Adriaensen, Pintelon et al., 2022). It is a domain-agnostic approach to study human–human, human–machine, or human–system interactions. Goal coordination can be seen as a way of distributed system cognition and control, which can describe and inform the redesign of more complex control structures. Interdependence Analysis adds an extra layer of analysis to FRAM in addition to defining functions, assigning couplings, and phenotypes. It could be argued that this remains an important dimension to explore in FRAM analysis, as interdependency could be seen as providing essential atoms of a functional resonance analysis.

Neither the description of traditional FRAM Outputs (in terms of traditional phenotypes like timing, precision, speed, and force) nor the addition of goal coordination principles as essential characteristics of dependencies and collaboration will, in isolation, permit a full understanding of how system control is maintained in complex socio-technical systems. Only when understanding how these different dimensions behave in relation to each other can system control, and for that matter, the benefits of IA, be fully appreciated.

NOTE

1   https://zenodo.org/records/16724900

REFERENCES

Adriaensen, A. (2022). *Guide for the combined application of the FRAM and IA'*. https://1drv. ms/b/s!AiFbvpjSA_UXg6E1sf7OOs2quyfOYg?e=Efbrin

Adriaensen, A., Berx, N., Pintelon, L., Costantino, F., Di Gravio, G., & Patriarca, R. (2022). Interdependence Analysis in collaborative robot applications from a joint cognitive functional perspective. *International Journal of Industrial Ergonomics*, *90*. https://doi. org/10.1016/j.ergon.2022.103320

Adriaensen, A., Pintelon, L., Costantino, F., Di Gravio, G., & Patriarca, R. (2022). *Systems-theoretic interdependence analysis in robot-assisted warehouse management*.

Herkert, J., Borenstein, J., & Miller, K. (2020). The Boeing 737 MAX: Lessons for engineering ethics. *Science and Engineering Ethics*, *26*(6), 2957–2974. https://doi.org/10.1007/ s11948-020-00252-y

Hollnagel, E. (2012). *FRAM, the functional resonance analysis method: Modelling complex socio-technical systems*. Ashgate. https://doi.org/10.1017/CBO9781107415324.004

Hollnagel, E., & Woods, D. D. (2005). Joint cognitive systems: Foundations of cognitive systems engineering. In *Joint cognitive systems: Foundations of cognitive systems engineering*. CRC Press.

Hutchins, E. (1995). How a cockpit remembers its speed. *Cognitive Science*, *19*, 265–288.

ISO. (2006). *ISO 13850: Safety of machinery - Emergency stop function - Principles for design*.

ISO. (2016). *ISO/TS 15066 Robots and robotic devices — Collaborative robots*.

Johnson, M., Bradshaw, J. M., Feltovich, P. J., Jonker, C. M., Van Riemsdijk, M. B., & Sierhuis, M. (2014). Coactive design: Designing support for interdependence in joint activity. *Journal of Human-Robot Interaction*, *3*(1), 43–69. https://doi.org/10.5898/ jhri.3.1.johnson

Klein, G. A., Calderwood, R., & Macgregor, D. (1989). Critical decision method for eliciting knowledge. *IEEE Transactions On Systems Man And Cybernetics*, *19*(3), 462–472. http://ieeexplore.ieee.org/xpls/abs_all.jsp?arnumber=31053

Patriarca, R., Di Gravio, G., Woltjer, R., Costantino, F., Praetorius, G., Ferreira, P., & Hollnagel, E. (2020). Framing the FRAM: A literature review on the functional resonance analysis method. *Safety Science*, *129*, 1–23. https://doi.org/10.1016/j.ssci.2020.104827

Salehi, V., Veitch, B., & Smith, D. (2021). Modeling complex socio-technical systems using the FRAM: A literature review. *Human Factors and Ergonomics in Manufacturing*, *31*(1), 118–142. https://doi.org/10.1002/hfm.20874

The Boeing Company. (2002). *737 operations manual*.

Section IV

Step 4
Assess the Consequences of the Analysis

16 Dynamic Variability Modelling
Capturing Complexities in Variability

Doug Smith

CONTEXT: DYNAMIC VARIABILITY MODELLING

Are we safe? How can I improve operational performance? What is "causing" success and failure in my operation? What management strategies will help improve performance? How much will it cost?

These are a few of the fundamental questions that managers and researchers try to address as they manage their operations and develop management tools, respectively. Of course, these questions are not easy to answer, especially in the context of complex operations. One of the fundamental difficulties is the lack of tractability of operations. The FRAM provides an opportunity to address this issue by focusing on functionality rather than causality. Functional relationships are generally easier to track than causal relationships. However, in functional models, there are typically many more possibilities, which increase the amount of variability and complexity, and FRAM users often struggle to precisely characterize the variability of their models and ensure that it reflects the work-as-done in their applied system. The idea of Dynamic Variability Modelling (DVM) has been developed to help advance the study of variability using the FRAM. This chapter will outline some of the motivations for this approach, its connection to the standard building steps of a FRAM model, the current state of the developments, and select applications of the approach.

The motivation for DVM to apply the Work-As-Done concept to variability assessment. The FRAM offers some benefits for understanding socio-technical systems, including using functionality to promote tractability across system elements and non-linear visualization of the system structure (i.e., a FRAM model). These are a few of the advantages that help FRAM users gain value when it comes to understanding functional systems and their applications. FRAM users are encouraged to investigate Work-As-Done in their FRAM analysis. When describing a system, users should endeavour to become informed by system operators who actually perform the work. This is a safeguard against users producing an ill-informed model from imagined system performance (Work-As-Imagined).

DOI: 10.1201/9781003518167-21

A good FRAM model should describe all of (or as many as possible) the potential pathways that can be used for a system to execute. The FRAM model does not describe (at least directly) how a system outcome is achieved; rather, it describes all of the potential ways a system could achieve many of its outcomes. In effect, the FRAM user is visualizing all of the potential outcomes at once when looking at a model, instead of an instantiation.

Outcomes are realized in FRAM modelling through instantiations (or Variability/ Functional signatures). An instantiation of the model can be used to explain an outcome of the system. Essentially, the FRAM model describes the possibilities, but the outcomes are realized through variations in functional execution of the model (instantiations).

In practice, many of the systems which are modelled using the FRAM have an enormous amount of variability and diverse system outcomes. So, when the FRAM user finally produces a FRAM model, showing them all of the variability at once, it can be overwhelming and a challenge to decipher it all in manageable and/or useful ways. Because of these challenges, instantiations were largely used to provide a narrative to the model, usually through a single (or a small number of) case(s). This exercise helps to ground the model, but when applying it to a single case (or a small number of cases), it ends up ignoring a holistic view of the system, missing at least some possible occurrences of system performance. Your understanding of the system becomes thus biased by the cases you choose to consider. Furthermore, it appeared that collecting a large number of 'case studies' to develop more instantiations is not economical in terms of data storage and often lacks consistency of data, making it difficult to compare instantiations (i.e., you are often left with the problem of comparing "apples to oranges"). DVM was developed to help provide a systematic approach to collecting variability data, which may be stored in a more economical way and provide some consistency to the data quantities that are recorded, which can also be visualized with FRAM models.

Method Geek

DVM offers an approach to visualizing variability through instantiations (or functional signatures). When DVM was developed, instantiations were not thought of as a structured way to record quantitative and descriptive elements of case studies, so the term functional signature was developed to help describe the activity of recording variability contained in instantiations. As this approach has developed, the terms functional signatures and instantiations are becoming more synonymous. However, functional signatures may mean something specific to DVM, and the concept will be described in more detail in the next section of this chapter.

The DVM approach has been developed and published in scientific literature over the last half-decade. Some publications have dealt more generally with the method as an advancement to FRAM modelling, and others have been applied to applications in shipping and healthcare. Table 16.1 offers a summary of the sources available to date.

TABLE 16.1
Summary of Previously Published DVM Use Cases

Publication Title	Year	Domain
Understanding the complexity of a stroke care system through functional modelling and analysis	2024	Healthcare
Mapping the way: functional modelling for community-based integrated care for older people	2024	Healthcare
Visualizing Complex Industrial Operations Through the Lens of Functional Signatures	2023	Shipping
A reinforcement learning development of the FRAM for functional reward-based assessments of complex systems performance	2022	Healthcare
A Dynamic Version of the FRAM for Capturing Variability in Complex Operations	2021	Healthcare
Modelling and analyzing hospital-to-home transition processes of frail older adults using the functional resonance analysis method (FRAM)	2021	Healthcare
Integration of resilience and FRAM for safety management	2019	Road transportation
Visualizing and understanding the operational dynamics of a shipping operation	2019	Shipping
Using FRAM to better understand (Arctic) shipping operations	2018	Shipping

The earliest example of DVM was illustrated in Smith, Veitch, Khan, and Taylor (2018). At this stage, it would hardly be considered a method, but the concept of a functional signature was already used to describe the instantiation of the Exxon Valdez grounding in the context of a FRAM model that was created to describe ship navigation.

The next development in the DVM timeline was presented in Smith et al. (2020). This work looked to merge resilience measurement with functional modelling. It merged a technique proposed by Ayyub (2015) for measuring organizational resilience as a measure of system performance with the process mapping that could be achieved by coupling the FRAM and functional signatures with each system measurement. Each functional signature could serve as an explanation for why a system performance measurement was achieved. These early attempts to develop the DVM approach were then superseded by Salehi, Smith, et al. (2021). This work provides the most formalized description of the approach to date. Note that the publication dates associated with these papers may not exactly match the chronology of their conception.

This chapter will outline the steps of DVM, discuss some interesting use cases, discuss a software tool, DynaFRAM, which was developed to help visualize this technique, and describe some work that has been done in both shipping and healthcare with respect to applying this approach.

VARIABILITY AS A FUNCTIONAL SIGNATURE

Method Geek

DVM has been labelled as dynamic FRAM modelling in the past. The term DynaFRAM has also been used to describe DVM, as it is connected to the software tool DynaFRAM. These three terms can be used synonymously, but it feels appropriate to update the title of this approach to DVM at this time since this approach focuses on dynamically assessing the variability component of a FRAM analysis once a model is built.

The key concept you need to apply in DVM is the concept of functional signatures. As described previously, the goal is to understand many instantiations of a model, but those instantiations have to be recorded one at a time. The events and conditions of an instantiation that produced an outcome of the system can be thought of as a functional signature. This latter is the signature left behind by that system outcome. By understanding its signature, we gain insights into why and how this system outcome was produced.

Remembering the concept that a FRAM model does not describe single outcomes, but rather it outlines all the possibilities that could potentially be realized by the model itself, then it follows that we need to consider instantiations to fully understand why specific outcomes are realized. In other words, outcomes are produced in FRAM modelling by functional variations of the FRAM instantiations (i.e., ways in which the variability actualizes).

Method Geek

Functional signatures describe the variability that occurs in each instantiation of a model by three characteristic elements:

- The quality and quantity of the function's Output. The Output of individual functions has the potential to vary from instantiation to instantiation. Depending on the nature of the Output of the function, this could be quantitative, qualitative, or both. For instance, if a function is <To increase the temperature of a heating element>, the Output could be recorded as a quantity or as a temperature. If the function is <To assess a patient>, the most tangible Output may be the descriptive assessment provided by a physician in the patient's medical file. This Output could be purely qualitative, but there may also be significant quantities embedded in the description (e.g., cholesterol and systolic blood pressure).
- The time the function is executed. This refers to both the time at which the function becomes active and the time it takes to execute the function. For instance, if the function is <To assess a patient>, the time it takes to conclude the assessment will be variable. If the assessment takes a long time, it

could indicate inefficiencies in the system. However, a longer appointment time may also reflect the thoroughness of the assessment. Nevertheless, the time of the assessment has an influence on the overall system performance. Similarly, the time at which the assessment begins can also influence the overall system performance. If the appointment begins too early, it could begin before other downstream functional aspects are available. If the appointment begins too late, it could have the effect of producing the Output of this function too late or causing the assessment to be less thorough because the physician is trying to complete it more quickly to stay on schedule.

- The pathway variability is variability that does not belong to individual functions, but to the model and is related to the functions that are active during an instantiation.

The functions that are active and/or inactive during the instantiation need to be known and recorded. Functional activity should be recorded directly. On the other hand, functional inactivity could be recorded directly, too, yet it is more common to record it indirectly as those functions not present in the functional activity list for that instantiation.

The collection of the active functions forms a pathway through the model, or partway through the model for a case where the model cannot be fully executed. This pathway can also serve as a partial explanation for why the system outcome was realised.

In cases where a poor system outcome is realized, you may ask why some functions remained inactive. If most functions exist to promote functionality of the system, why were they not activated in this particular circumstance? This could lead to recommendations to improve system performance in the future. In cases where good system performance is realized, the inactive functions could indicate inefficiencies in the designed system. However, care must be taken when making this type of decision since it could be that the particular set of circumstances of that instantiation did not require those functions to be activated, but they may be needed if other conditions are present. As usual in the case of complex socio-technical systems: handle with care!

The guidance here would be to observe the system over many different sets of conditions (i.e., many instantiations) before trying to make such conclusions. This information can help inform such discussions about system management.

The three forms of variability (quality and quantity, execution time, and pathway) constitute the information that is required to create a functional signature for a model. The functional signature contains the information that helps explain why this system outcome was realized.

Figure 16.1 displays the FRAM model, which is used to provide context to help understand the concept of DVM. The FRAM model is a model of the 'hospital to home' transition process for older adults in a Canadian province.

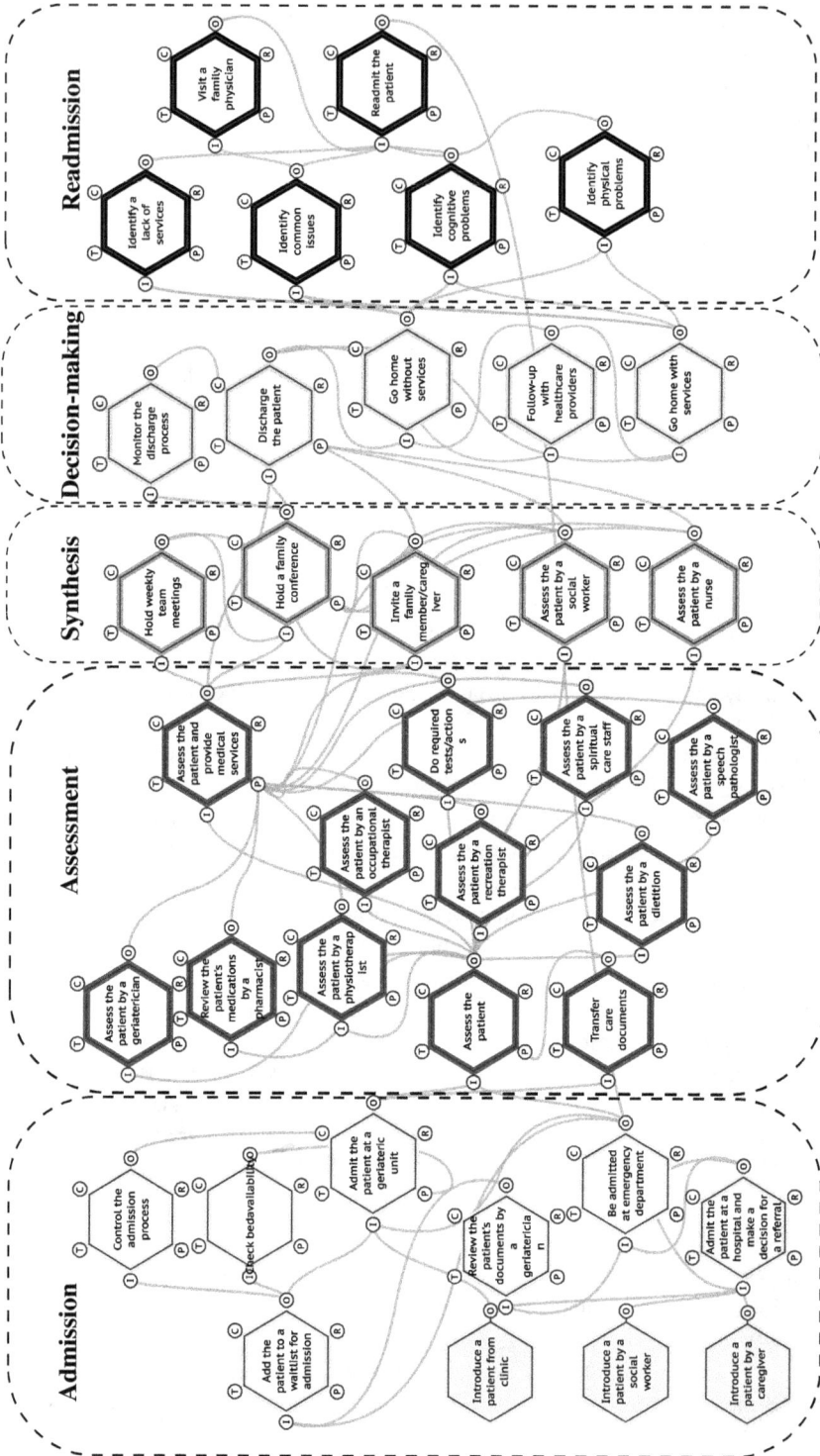

FIGURE 16.1   FRAM model of the 'hospital to home' transition process for older adults.

This model is organized in a multidisciplinary perspective, which is organized by 'macro-level functions' of the system. The colours and boxes surrounding groups of functions describe parts of the process to help orient audiences to the model, who may be more familiar with 'traditional process models' of the system. The model generally represents a system which consists of admission, assessment, synthesis, decision-making, and readmission processes. While this process generally moves forward by completing these steps in order, if you focus on the connectivity of the functions, you will see the level of connectivity is higher between these high-level processes and less linear than the 'traditional process models' may lead you to believe. This is an effective way to help communicate FRAM models to domain experts who have process model knowledge in their domain but are new to FRAM modelling. It helps them identify the parts of the process that they are familiar with and allows them to begin to comprehend the 'new' level of complexity and non-linearity that is presented by the FRAM model of their system.

It is also worth noting that the process refers to a 'hospital to home transition' model, but it extends to a part of the process labelled as readmission. This is because it was of particular interest to the healthcare professionals who have a stake in this model. There is a chronic problem that older adults keep showing up in hospital situations for similar or recurring reasons. Healthcare professionals wished to understand more about why older adults may get re-admitted to the hospital, so it was included in the scope of this study. The intention to transition the patient home (or to some other longer-term care situation) is made in the decision-making part of this model, but the process was extended to try to understand more about the functionality related to patient readmission. More details of the model and study can be found in Salehi, Hanson, et al. (2021).

Given a FRAM model, DynaFRAM allows displaying functional signatures of whatever size (e.g., a patient interacting with this system):

- Shaded functions represent the pathway variability up until the point where this snapshot was taken. A shaded function represents a function that was active in this patient's care history.
- Unshaded functions represent functions which were inactive during this patient's interaction with the system (or have not yet been active at the time the snapshot was taken).

Thus, the collection of active functions represents a pathway through the model which describes, in part, that patient's journey. Each active function has an Output label presented, which should be the specific state of the functional Output for that patient after the function was executed, that is, the function Output variability.

Nonetheless, the element of time variability is difficult to capture in a snapshot or as a picture. Video is the most effective way to display the time element of variability in DVM, which allows observing the execution rate of the function itself.

WHEN TO USE DVM: MODEL CHECKING

All models require some amount of checking after they are initially built. The FRAM Model Interpreter (FMI) offers the ability to check a FRAM model, as it checks that the modeller has followed the rules of the FRAM and that critical aspects are not omitted when the model is executed. This is an important part of checking an FRAM model, but there is a complementary way to check for models for an additional purpose using DVM, and that is checking that the FRAM model can describe the instantiations of the application realistically.

FRAM models are often built to represent complex sociotechnical systems. An FRAM model may never be complete, but the 'best models' are typically produced when FRAM users prioritize the Work-As-Done concept during their analysis. The rigorous consideration of Work-As-Done as it pertains to understanding functionality usually improves the model. Thus, models that can appropriately describe the specifics of multiple instantiations will be informed by more work as it is done than if the model is not subjected to this technique. In FRAM terms, it can amount to making the model more 'complete'.

DVM looks to be able to explain the real-world application as an instantiation of the model. It requires the user to create a scenario file that demonstrates consistency in both the realism of the scenario and its application to the FRAM model. One example of this is illustrated by describing the rationale for one of the features of the DynaFRAM software. The feature where the DynaFRAM software can create couplings that did not exist in the original model helps FRAM users identify when there may be a coupling missing from their model. It requires the FRAM modeller to put aside the expectations that are created by their own model and view the scenario in the real world, as it is.

Does the Output of a function really represent a Precondition as it was originally modelled, or might it be more appropriately labelled as an Input? Or did you identify the Resource as you are viewing the instantiation that you did not identify in your prior modelling?

WHEN TO USE DVM: UNDERSTANDING VARIABILITY INFLUENCE

In the FRAM, variability is said to propagate through a model and has the potential to produce unexpected system outcomes, especially when couplings combine non-linearly. This is analogous to stochastic resonance, where the FRAM derives part of its name. However, usually in stochastic resonance, all signals are quantifiable and can be somewhat easily assessed for resonance conditions. When applying the FRAM to sociotechnical systems, the Output of functions can have both qualitative and quantitative elements, which makes things more complicated than the analogy might suggest. Many of the tools which allow people to deal with stochastic resonance problems become inadequate for dealing with functional models of sociotechnical systems, but it is believed that the FRAM can still offer explanations for variable system performance which can be traced back to fundamental elements of variability as suggested by the three forms of variability in DVM.

In DVM, as in the FRAM, trying to prove that a particular system mechanism is responsible for a system outcome is difficult. Each function can be thought of as a variable that would exist in a traditional scientific experiment. Its Output can vary qualitatively and quantitatively, making it difficult to measure change in a consistent way, and usually, sociotechnical systems do not allow you to vary the variables in a controlled manner as you could in a laboratory setting. This makes it difficult to prove the statistical significance of cause-and-effect relationships within these systems and often requires an immense amount of data to do so. Often, more data than can be practically collected.

Nevertheless, DVM offers a framework to record variability data for FRAM models that could be collected over time to build larger databases incrementally. By collecting functional signatures that have been executed by the model, the elements of variability are recorded in a common way, allowing functional signatures to be later compared to each other. In theory, functional signatures that are different from each other could produce the same system outcome, thus indicating that the variability is somewhat insignificant in producing change in system outcomes. Likewise, functional signatures that are very similar have the potential to produce significantly different system outcomes. And there are many cases that exist between these two examples, which could be examined, provided enough variability data can be recorded.

WHEN TO USE DVM: MONITORING SYSTEMS

Many of the sociotechnical systems that have been investigated using the FRAM to date already have metrics that they record to monitor system performance. However, many of the metrics that are used are confounded by other system measures or lack explanations on how they actually influence the system outcomes. Many metrics do demonstrate probable cause in the system outcomes, but the strength of this signal can be quite low, leading to uncertainty. The measures lack deterministic relationships to the system outcomes. Thus, leading to management decisions that may be informed by uncertain system measures. Additionally, sometimes the metrics that are easiest to measure are the ones that are recorded, not the metrics that would be the most meaningful to the decision-makers.

Domain Expert

This can be illustrated by an example from healthcare. A common metric that has been used to measure the performance of hospital systems is the 'number of available hospital beds'. It is unclear how this metric is a good indicator for the quality of care of patients, the cost to provide health care, or the efficiency of the system. It is an easy metric to measure, as you can just count available hospital beds at any time. It may be better to look for an alternative metric or set of metrics by looking to the FRAM model and focusing on the functional outputs for that model to identify which outputs have quantifiable and measurable properties.

As explained already, a functional signature provides a matter-of-fact explanation for why a system outcome was realized. The system outcome was realized because these functional outputs were produced, it was executed at this time, and these functions were active during the execution. This provides a basis for FRAM users and decision-makers alike to discuss appropriate metrics for their system. In order to create a functional signature, you must determine an appropriate way to record the quality and quantity of the functional Outputs in your model, and that provides a map for connecting functional Outputs to the system outcome, directly or indirectly, as it is passed to the downstream functions.

METHODOLOGY

The application steps for DVM can be somewhat fluid and involve a level of pragmatism to apply them successfully. It depends on the nature of the system you are modelling, which may have varying levels of complexity, and that will influence the specific techniques you might have to use to apply this approach. Nevertheless, this section outlines the general process from start to finish, including the FRAM modelling steps. Tips and experiences related to practically applying this method can be found in the Results section.

STEP 0: RECOGNIZE THE PURPOSE OF THE ANALYSIS

It is important to be as precise as possible when defining the objectives(s), as under-specified objectives may lead to added uncertainty in the model. Also, during this step, it is helpful to select metrics and/or indicators that can help assess the variable performance of the system. You will need to have a performance measurement associated with every functional signature.

STEP 1: IDENTIFY AND DESCRIBE THE SYSTEM'S FUNCTIONS

You should do this step as described in the standard FRAM approach. DVM does not add much to this step, but some value can be gained in the model checking application. If you identify missing or incomplete functions, you may revisit Step 1 and update them.

STEPS 2 AND 3: IDENTIFY VARIABILITY AND AGGREGATE VARIABILITY

Since DVM focuses on the characterization of variability, it is relevant to Steps 2 and 3. Two of the three elements of variability contained in a functional signature are most relevant to Step 2, namely, i) the specific outputs of the active functions and ii) the time at which the function became active and the time it took to execute. These elements of variability can be determined by focusing on individual functions and recording the specifics for the instantiation you are considering. The third element of variability, the active function pathway, is most relevant to Step 3. The active function pathway is a result of aggregating the individual functional variability recorded for the instantiation.

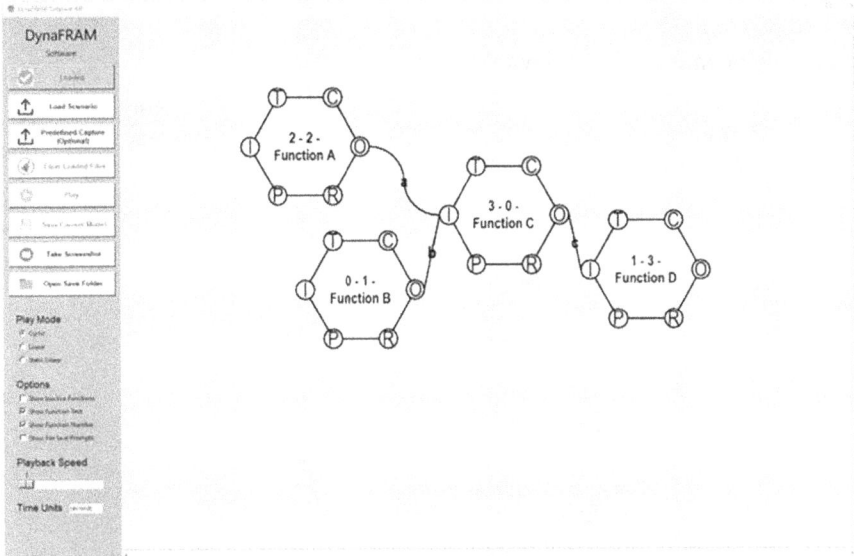

FIGURE 16.2   DynaFRAM software interface.

The system outcome should be able to be measured (and ranked) if a metric/indicator was defined in Step 0. After many functional signatures are produced, insights of how variability influences system outcomes should start to emerge.

STEP 4: ASSESS THE CONSEQUENCES OF THE ANALYSIS

You do this step as recommended in the standard FRAM approach. Make management recommendations as new system insights emerge. However, it is intended that by using DVM in Step 3, more (or at least different) insights may emerge as you investigate the variability at a deeper level.

DYNAFRAM USER INTERFACE

Figure 16.2 shows the user interface of the DynaFRAM software, which is used to visualize the functional signatures. Full details of the software interface can be gained by viewing the user manual (provided online with the software).

Method Geek

Just the 'model' and 'scenario' buttons will be discussed here. The model button allows any FRAM model that has been created by the FMV (.xfmv file) to be loaded. The scenario button allows an .xml or .csv file of the variability which is contained in the instantiation to be uploaded. This can be thought of as a set of commands for the DynaFRAM software to create a visualization of the functional signature as a video playback. More details on exemplary content for a scenario file in the shipping domain can be found in Table 16.2.

TABLE 16.2

Sample Scenario File for DynaFRAM

Time	Active Function	Active Function Output	Downstream Coupled Function	Coupled Function Aspect	Time Tolerance
3	1	No ice load	4	I	
3	5	Heading = 107.5 Speed = 0 kn	4	I	
10	4	Complete or partial assessment made	0	I	
30	0	Ice condition update	1	I	
33	2	Med FY ice < 1m	6	I	
33	7	Class 1C	6	P	
37	1	Medium ice load	4	I	
38	6	RI = -2	4	C	
43	4	Complete or partial assessment made	0	I	
48	0	Ice condition update	1	I	
48	0	Speed change	5	I	
48	0	Check ice condition	2	C	
51	2	Med FY ice < 1m	6	P	
52	5	Speed = 2.1108 kn	4	I	
54	1	Low ice load	4	I	

The columns 'Active Function' and 'Downstream Coupled Function' require numeric inputs. These numbers are the call signs of the functions from the original model file.

As you create functions in the FMV, a number gets assigned to them, and it is this number that you should use to call the function in DynaFRAM, not the function label.

The 'Time' column requires a numeric input, which is the time a function becomes active. This value is usually given in seconds by default, but units can be updated, and playback can be scaled to suit the video presentation preferences.

The 'Active Function Output' column represents the specific output of the active function after it has been executed. This column can have qualitative and quantitative elements depending on the nature of the function you are describing. It is important to be as specific as possible when recording this information, and small changes in the functional Output could have significant effects on the overall system performance, as per the notion of functional resonance.

The 'Coupled Function Aspect' column requires an alphabetic input, which indicates the aspect of the downstream coupled function where the connection should be made in the video. It is important to note that the connection does not necessarily need to be made in the original FRAM model for the DynaFRAM software to illustrate the connection in the scenario. This allows for additional coupling to be realized when describing scenarios that were not originally realized in the FRAM modelling stage.

The 'Time Tolerance' column is meant to accommodate time data that is recorded with a high sampling rate for a scenario that does not change very quickly. It can usually be left empty, though. However, if the time data is recorded to the tenth of a second, you may not want to create a new time step every tenth of a second in your video, so the time tolerance column could be used to say anything happening within a few seconds of each other can be considered the same time step. Many sociotechnical systems operate on a time scale such that functions happening within minutes of each other can effectively be modelled as happening at the same time. In those cases, this column may be useful to DynaFRAM users.

The 'Active Function Output' column indicates the quality and quantity of the function's Output, thus one element of the functional signature. It is up to the user to decide how precisely they should record this information, but the more precise the information, the greater the likelihood of observing variability when compared to another functional signature. The variability with respect to time has two elements: the first element is the time at which a function becomes active, and the software reads this directly from the time column of the row where the active function is listed. The second element of time variability is related to the time it takes to execute the function. This is calculated by the software by reading the time at which the function becomes active initially, considering its downstream coupled function, and then searching the scenario file for the next row where that downstream coupled function becomes the active function. The software then finds the time difference between the original function becoming active and the downstream couple function becoming active.

The resulting video then illustrates the functional execution time at an average rate determined by the two timestamps. The pathway variability will then be a result of the functional activity that is described in the scenario file. By indicating that a function is active, that indicates that it was used during this instantiation, and by leaving a function out, it indicates that it was not used.

RESULTS

This section describes two diverse domains taken from the various available use cases in the literature, where DVM has been tested and iterated.

Example 1: Ice-Clearing Maritime Operations

While the earliest example of functional signatures was presented in a shipping application, the Exxon Valdez grounding, it was largely a concept used to visualize an instantiation of a FRAM model at that stage. The functional signature concept was initially presented in Smith et al. (2020) and was then applied to a shipping application in Smith, Veitch, Veitch, et al. (2018), which focused on using a support ship to clear ice away from an offshore hydrocarbon installation during an emergency evacuation scenario. The ship navigation FRAM model used was a modified version of the model presented in Smith, Veitch, Khan, and Taylor (2018), which was built from ship navigator interviews.

The experiment took place in a ship bridge simulator modelled after an offshore support vessel operating near a hydrocarbon installation. The simulator featured ship-like controls and a 360° screen for full visibility. Participants were tasked with clearing ice near a lifeboat launch zone to enable emergency evacuation from the offshore installation. Two groups participated: experienced captains and beginner seafarers. While the primary aim was to compare their performance in ice-clearing operations, the recorded data was ethically repurposed to demonstrate the DVM approach.

Figure 16.3 shows a snapshot of an overhead playback file of a participant from the experiment. The smaller vessel represents the ship the participant is driving to affect the ice conditions. The larger ship represents the offshore installation, which is nominally fixed with respect to this scenario. The shaded square area represents the general area in which a lifeboat is required to be launched during the scenario. The polygons represent pieces of sea ice in the virtual environment. The pieces of sea have a slow drift speed, which is imposed by 'wind and current' in the scenario. But the ice pieces can be deflected directly by contact with the ship and indirectly by contact with other ice pieces. The participant should manoeuvre the ship in such a way that it will clear the lifeboat launch area of sea ice so a lifeboat could be safely launched.

The metric was chosen to measure system performance as the time it took for a circular area twice the diameter of the lifeboat's length to be cleared from under the davit launch area. This was determined to be a large enough ice-free area to launch a lifeboat in the presence of an ice-clearing vessel. There was then a functional signature that would belong to each participant that could be linked to that performance measurement. The metric would enable the overall performance of each participant to be quantified since the goal (Step 0) of the emergency evacuation process is to safely launch the lifeboat as quickly as possible. The performance could be measured by the time to launch the lifeboat, and the safety would emerge from the outcome of achieving the goal and the process that was followed to achieve that goal, which could be determined from the functional signature.

The time histories of the participant's interaction with the ship controls were analyzed to infer when ship navigational decisions and actions were made. The information helped form the functional signature for each participant. Five different functional signatures/participants demonstrated markedly better performance than the rest of the 26 participants, including one participant who violated the procedure by continuously exceeding a speed limit that was set for the ship in the modelled ice conditions.

A speed limit of 3 knots was imposed on the ship by the Polar Operational Limitation Assessment Risk Indexing System (POLARIS) guidelines for operating ships in ice-covered waters. The speed limit in ice-covered waters is calculated by considering the present ice conditions and the (structural) classification of the vessel.

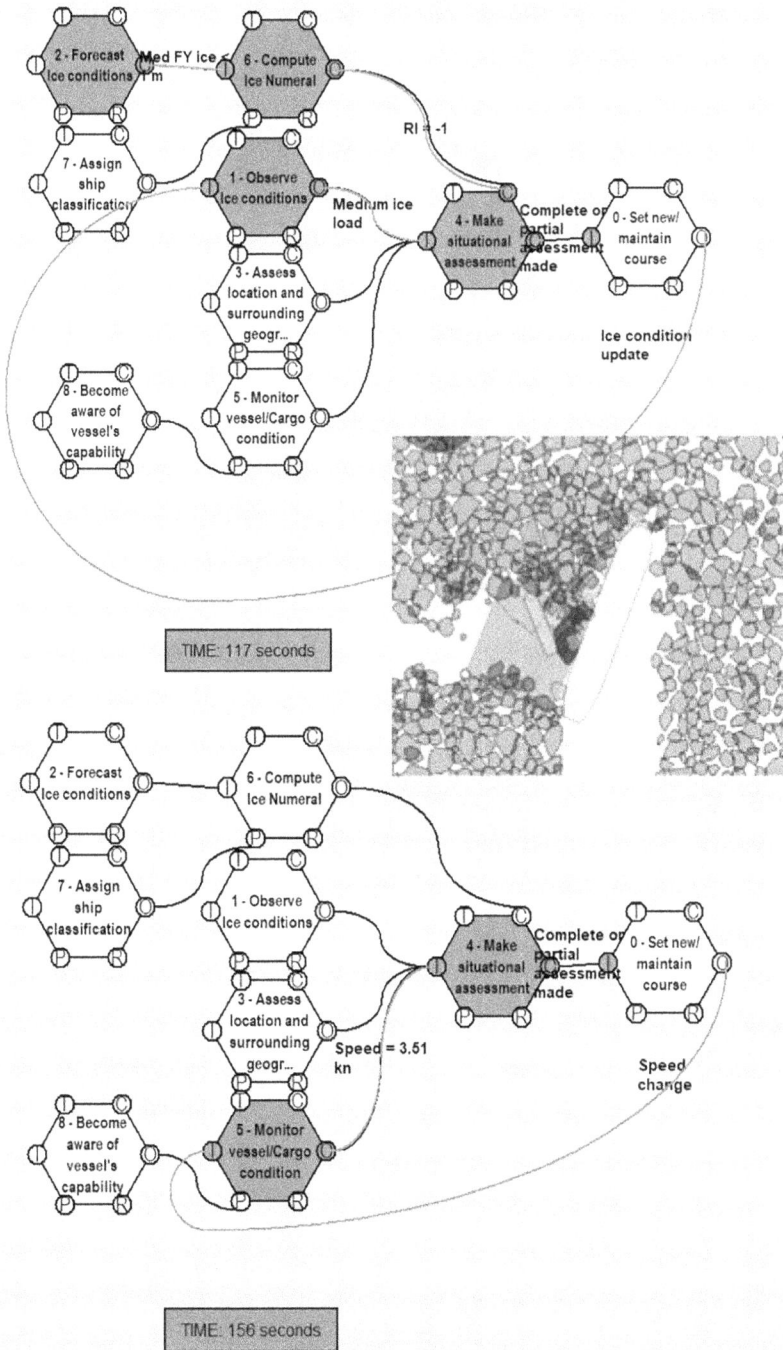

FIGURE 16.3 Functional signature of one participant for the shipping example.

The speed limit is intended to reduce the energy of the ship-ice collisions, thus lowering the risk of the ship suffering structural damage due to impacts with the ice. The participant who exceeded the speed limit indicated that they did it intentionally. They said that this was an emergency situation, and the lifeboat had to be launched from the offshore installation as quickly as possible. They indicated that any damage to the hull that may have resulted from excessive speeds would have been minor in nature. Their consideration of risk in this situation was that it would be worse to delay the time for the lifeboat to launch in this emergency situation than the risk of structural damage to their vessel, which they perceived to be minor in nature (possible denting of the hull). This participant did achieve one of the best performance scores of all participants in the experiment and was a great example of the difference between work-as-imagined vs. work-as-done.

Example 2: Hospital-to-Home Transition

DVM has also been applied to healthcare applications. Salehi, Hanson, et al. (2021) modelled the 'hospital to home' transition process for older adults. These patients often have complex care needs and will often end up back in hospital situations if care is not taken to properly transition them home. The FRAM was considered suitable to model this complex application.

Additionally, DVM was included as a way to reflect the 'patient journey' in this context. This model was highly interdisciplinary and had many different healthcare professionals performing work. The patient provided a point of consistency for all of the functions in the model. Although many of the functions were performed by different healthcare professionals, all of the functions exist to help provide care to the patient. By following the 'patient's journey', a functional signature could be created to describe the instantiation of care to that patient. In this work, after creating the FRAM model, six patients were examined to understand their journeys through the 'transition to home' system. Functional signatures were created of these patients' journeys.

McGill et al. (2024) built a FRAM model of the community-based geriatric medicine services system in a province in Atlantic Canada. This model consisted of over 200 functions. In this context, the model is so large that it is nearly indecipherable to non-FRAM users and anyone who was not deeply involved in the study and subsequent creation of the FRAM model. Some techniques, including DVM, were used to help explain this to readers and system stakeholders.

Figure 16.4 is a very large model which reflects a fairly large scope system. The system is the community-based geriatric medicine services system. The model focuses on multiple programmes, which are available to provide health care services to older adults. Health care for older adults can be quite complex, and the demand for health care services can exceed the available capacity for geriatric care specialists (GMS clinic in Figure 16.4). Additionally, the transportation involved in getting patients to appointments at the GMS clinic can be challenging, thus creating a need for some health care services to be delivered at home (community-based).

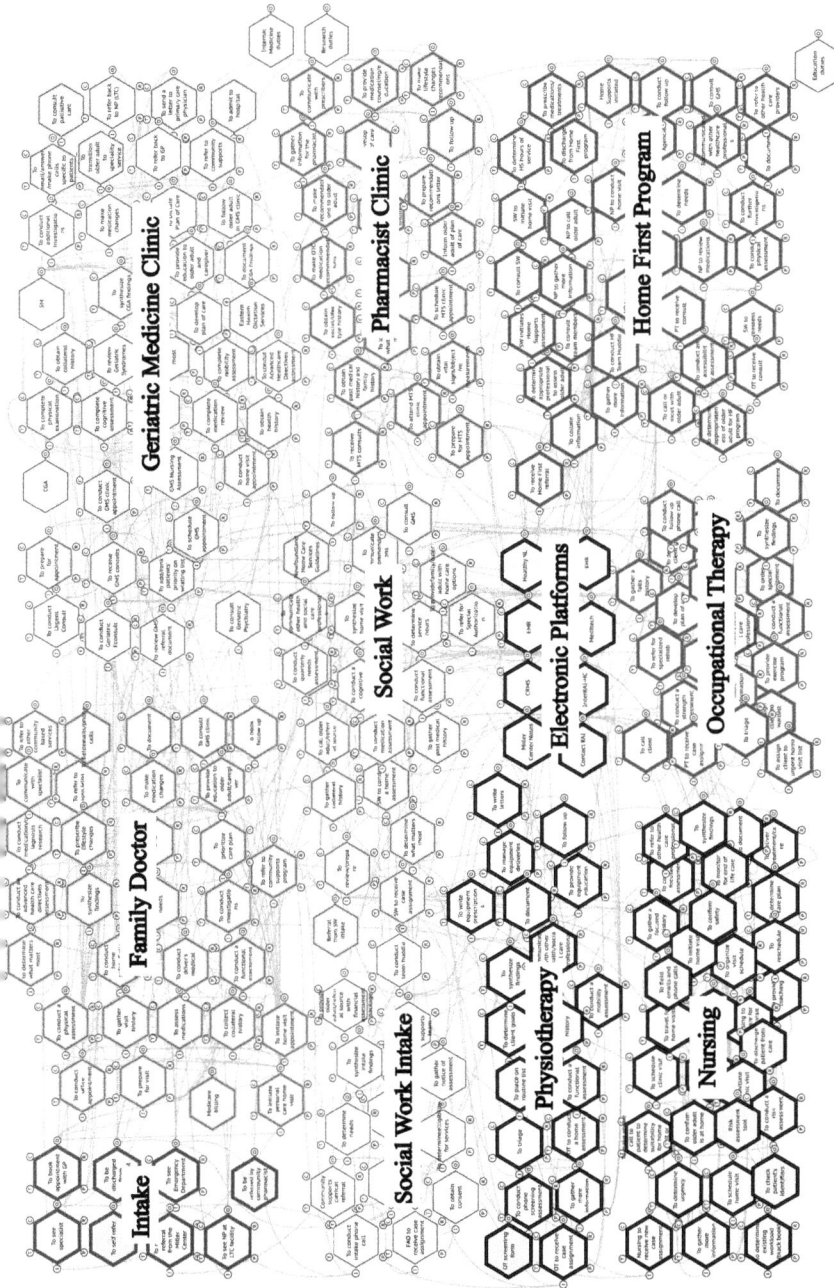

FIGURE 16.4 FRAM model of a community-based geriatric medicine system.

A video description of this model and a functional signature of a hypothetical patient, Fred, can be found at the link below. Readers are invited to view the video to see how the functional signature of Fred can help improve their understanding of the model presented in Figure 16.4. https://www.engr.mun.ca/~d.smith/cbgms.html

Figure 16.4 is organized by healthcare programmes/sub-systems that are present in this model. They consist of the geriatric medicine services (GMS) clinic, family doctor, home first programme, social work intake, social work services, physiotherapy, pharmacist clinic, nursing, occupational therapy, and the electronic platforms used for record keeping and information sharing by health care professionals. Labelling and colour coding indicate the functions that belong to each sub-system.

It can be a challenge (if not an impossibility) to describe a model of this size and complexity in a way that can be understood practically. One way to begin developing an understanding of this model is to understand how instantiations occur in the context of this model. DVM can be used to capture the functional variations which lead to a system outcome. This provides a more tangible understanding, which helps connect the FRAM model to practical considerations.

In this case, a single hypothetical functional signature was constructed from instances of variability collected during model-building interviews, and it was used to exercise the model. This provided an opportunity to describe an instantiation of a patient receiving care in this system and demonstrate that process using DVM to assess the FRAM model's ability to describe the situation. The narrative of the patient receiving care can be assessed by (system) users to determine if the scenario being considered is realistic. If the scenario is realistic (or realistic enough), the exercise of using DVM to describe the instantiation will either confirm the model's ability to describe such a scenario or illuminate discrepancies between the model and the 'real world'. In effect, this is the model-checking application of DVM. Time constraints of this project prevented a more detailed data collection campaign of functional signatures from being applied to this model.

CONCLUSION

DVM has been shown as an approach aimed at capturing Work-As-Done in the variability analysis part of the FRAM. By characterizing variability as specifically as possible, and for as many instantiations as possible (or practically possible), DVM can help understand the nuances of real-world complexities. To this extent, FRAM users gain the ability to check their model, enrich their understanding of how processes influence system outcomes, and potentially identify (specific) metrics that might improve system monitoring activities.

Some domain applications, that is, shipping and healthcare, have been briefly described in the chapter to illustrate both the evolution of this approach and previous uses. While it is acknowledged that more work is needed to fully explore the utility of DVM, the examples provided are intended to contribute context to the concepts discussed. It is also intended that the presented examples and respective discussions will shed additional light on the motivations behind the approach and the direction of advancement for DVM.

Complementary to the theoretical aspects of DVM, the chapter introduced the software DynaFRAM (available for free[1]), which is meant to support DVM. FRAM users may use this software as a companion to the FMV. You should be aware that it is currently unsupported, but user manuals and more information can be found at the link provided. Also, the recent 2024 additions to the FMV have allowed the FMV to produce similar results as the DynaFRAM software, so the DVM approach could be applied using the FMV in the near future.

NOTE

1 DynaFRAM download: https://www.engr.mun.ca/~d.smith/dynafram.html.

REFERENCES

Ayyub, B. M. (2015). Practical Resilience Metrics for Planning, Design, and Decision Making. *ASCE-ASME Journal of Risk and Uncertainty in Engineering Systems, Part A: Civil Engineering*, 1(3), 04015008. https://doi.org/10.1061/AJRUA6.0000826

McGill, A, Salehi, V, Smith, D, McCloskey R, Veitch, B. (2024). Mapping the Way: Functional Modelling for Community Based Integrated Care for Older People. *Health Research Policy and Systems*. In-review

Salehi, V., Hanson, N., Smith, D., McCloskey, R., Jarrett, P., & Veitch, B. (2021). Modeling and Analyzing Hospital to Home Transition Processes of Frail Older Adults Using the Functional Resonance Analysis Method (FRAM). *Applied Ergonomics*, 93, 103392. https://doi.org/10.1016/j.apergo.2021.103392

Salehi, V., Smith, D., Veitch, B., & Hanson, N. (2021). A dynamic version of the FRAM for capturing variability in complex operations. *MethodsX*, 8, 101333. https://doi.org/10.1016/j.mex.2021.101333

Smith, D., Veitch, B., Khan, F., & Taylor, R. (2018). Using the FRAM to Understand Arctic Ship Navigation: Assessing Work Processes During the Exxon Valdez Grounding. TransNav. *International Journal on Marine Navigation and Safety Od Sea Transportation*, 12(3). http://www.transnav.eu/Article_Using_the_FRAM_to_Understand_Arctic_Smith,47, 828.html

Smith, D., Veitch, B., Khan, F., & Taylor, R. (2020). Integration of Resilience and FRAM for Safety Management. *ASCE-ASME Journal of Risk and Uncertainty in Engineering Systems, Part A: Civil Engineering*, 6(2), 04020008. https://doi.org/10.1061/AJRUA6.0001044

Smith, D., Veitch, E., Veitch, B., Khan, F., & Taylor, R. (2018, October 25). Visualizing and Understanding the Operational Dynamics of a Shipping Operation. *SNAME Maritime Convention*. https://www.onepetro.org/conference-paper/SNAME-SMC-2018-032

17 Integrating the FRAM and Lean for Waste Analysis in Production Systems

Tarcisio Abreu Saurin, Flávio Sanson Fogliatto, Michel Anzanello, Natália Maciel Tocchetto, and Leonardo Bertolin Furstenau

CONTEXT: WASTES IN PRODUCTION SYSTEMS

A key step in FRAM application is analyzing the output variability of functions, considering both timing (e.g., too early, on-time, or too late) and precision (e.g., imprecise and precise) (Hollnagel, 2012). Such variability can impact several performance dimensions such as safety, quality, productivity, environment, and cost. The focus on a particular performance dimension depends on the nature of the modelled system and the purpose of the analysis. For example, in health services, the FRAM is often used to improve patient safety, highlighting the corresponding variabilities (Sujan et al., 2024). Safety is also a primary FRAM concern in sectors such as aviation, air traffic management, and construction (Patriarca et al., 2020). Conversely, in manufacturing plants, variabilities related to quality and productivity may be prioritized.

This chapter explores the use of the FRAM for analyzing variabilities that create waste in production systems, a purpose that has been relatively underexplored. Waste corresponds to the use of more resources than necessary or an unwanted output from a function (Bølviken et al., 2014). For instance, an accident can be considered a form of waste, as it represents an unwanted output with undesired implications such as injuries and direct and indirect costs (Calabresi, 2008). Waste elimination is a traditional goal of process improvement initiatives and is associated with benefits such as increased productivity, improved quality, enhanced safety, and shorter processing lead times (Jasti and Kodali, 2015). The lean production taxonomy of wastes, widely applied across several sectors, includes eight types: transportation, inventory, motion, waiting, overproduction (both in advance and in quantity), overprocessing, defects, and unutilized talent (Liker, 2004; Ohno, 1988).

Wastes are emergent outcomes of production systems, arising from partly uncontrolled and undesired interactions between multiple social, technical, and organizational factors and functions. This perspective suggests that wastes can be manifestations

DOI: 10.1201/9781003518167-22

of functional resonance, similar to accidents. It also implies that there may not be a root cause of waste that can be singled out and fixed, challenging oversimplifications in lean practices such as the five whys (Card, 2017). Indeed, wastes in production systems might be more accurately modelled as networks rather than linear chains (Fireman and Saurin, 2020). Moreover, the distinction between waste and necessary slack resources (and thus value, the flip side of waste) can be blurred, depending on factors such as risk tolerance, power relations, and stakeholder requirements (Browning and Heath, 2009). Consequently, there is a risk of short-sighted lean production initiatives that confuse value with waste, making the system vulnerable to both expected and unexpected variabilities (Potthoff and Gunnemann, 2023; Habibi Rad et al., 2021).

Lean and resilience management have several synergies, and there is a certain consensus that the resilience benefits of lean outweigh its drawbacks (Gayer et al., 2022; Uhrin et al., 2020; Williamsson et al., 2019; Soliman et al., 2018). Thus, when properly implemented, waste elimination tends to benefit both lean and resilience – for example, waste elimination improves efficiency and releases resources that can be used to enhance adaptive capacity.

The remainder of this chapter presents a case study that uses the FRAM to analyse wastes in the sterile unit of a hospital. This setting is particularly suited to the FRAM complexity-informed perspective due to several characteristics: (i) interactions with multiple hospital areas, particularly with the surgical unit, which requires sterilized instruments and other materials to be delivered on time, with the correct quality and right quantity, as variabilities in these aspects have life-saving implications; (ii) time pressures to manage urgent and unscheduled orders, resulting in tightly coupled functions; (iii) individual preferences of surgeons who demand customized instruments; and (iv) feedback loops, with instruments continuously circulating between the sterile and surgical units.

These and other complexities often lead to low performance in sterile units, which has prompted research into process improvement initiatives (Fogliatto et al., 2020). Such studies usually rely on lean production principles (Tortorella et al., 2017), which do not explicitly acknowledge resilience implications and therefore can offer incomplete solutions. The case study of the sterile unit emphasizes how the waste analysis perspective can impact each of the traditional FRAM steps.

DATA COLLECTION

The case study took place within a process improvement initiative at a medium-sized private hospital with approximately 200 beds, located in Southern Brazil. The project aimed to address perceived low performance in the sterile unit, which included delays in the delivery of instruments, and the provision of trays with missing, incorrect, or defective instruments. The catalyst for starting the project was an incident where surgery had to be cancelled due to a lack of instruments, even though the patient was already anaesthetized in the operating room (OR).

A process improvement team was formed, consisting of two managers from the sterile unit and five industrial engineers who acted as external consultants (i.e., the authors of this chapter). The team's tasks were to diagnose the current performance of the sterile unit, devise an action plan, and, to the possible extent within the four-month consulting contract, pilot some improvements. The consultants made 10 visits, each lasting an average of 6 hours, over the course of four months. During these visits, they collected data and discussed potential improvements with the hospital staff. The top management supported this initiative and was committed to implementing the action plan.

Method Geek

Data collection for the diagnosis, action plan development, and test-piloting of improvements included the following activities:

- A total of 30 hours of non-participant observations of the activities in the sterile unit. In this type of observation, the researcher watches the subjects with their knowledge but does not actively participate in the situation being studied (Scott and Marshall, 2009).
- A total of 20 hours of informal conversational interviews, with note-taking, were conducted by 10 workers and 2 supervisors from the sterile unit. Following Patton (2014), these interviews were held in the work environment and were guided by the flow of production activities occurring, lacking a predefined set of questions. Both the observations and interviews contributed to identifying the main functions, their variabilities, and the flows of materials and people. These flows were schematically represented in a spaghetti chart (Chiarini, 2013) that highlighted interferences and risks that could be addressed through layout planning.
- Analysis of records of missing instruments in trays delivered to the ORs, which helped identify priorities for new acquisitions.
- Analysis of databases of surgical procedures over a 12-month period revealed the most demanded instruments. This analysis was important for identifying mismatches between capacity and demand, as well as for determining the need for new instrument acquisitions.
- Analysis of databases of maintenance records to identify the surgical instruments in need of replacement.
- Analysis of everyday maintenance activities of critical equipment (e.g., sterilizers) that could be carried out by the machine operators. This would free up time for the maintenance department staff, who could concentrate on more complex maintenance activities. Such an initiative corresponded to autonomous maintenance, a lean practice (Guariente et al., 2017).
- Three focus groups with doctors and medical chiefs from major medical specialities to review the composition of surgical trays. This was the main improvement test piloted during the consulting period. The consultants led these focus groups to demonstrate how the review process should be conducted, enabling hospital participants to replicate the review procedures for all surgical trays. During each meeting, existing instruments were laid out,

and participants were asked to separate necessary from unnecessary items. This expert-based judgement approach proved effective for reviewing surgical trays, although mathematical modelling methods might also be useful (dos Santos et al., 2021). Each meeting lasted an average of 20 minutes, and participants reached a consensus in all instances. Doctors supported this initiative due to their dissatisfaction with instrument supply problems. Additionally, the presentation of positive results from similar initiatives led by the consultants in other hospitals (Fogliatto et al., 2020) helped mitigate resistance.

- An assessment of the level of multifunctionality of all workers in the sterile unit. This assessment was initially conducted by the unit's supervisor, and then individually reviewed with each employee until an agreement was obtained. The need for such assessment stemmed from anecdotal evidence indicating that the outputs of several functions used to be either imprecise or late due to workers not being properly skilled to carry out their tasks.

Data collection stopped once findings started converging and a comprehensive understanding of the system's functioning and wastes was achieved. This reflects the saturation criterion commonly used in qualitative research (Nelson, 2017). The saturation criterion is also applicable to FRAM modelling, which involves successive rounds of refinement, adding details as new information and insights emerge. In this study, although a draft model was developed immediately after the initial observations and conversations with the sterile unit supervisors, it was gradually refined throughout the data collection period. For instance, as our understanding of the system improved, we identified the need to include functions that had been overlooked, such as the function <store trays in shelves>. These functions were initially deemed trivial, but we later realized their outputs were imprecise and could cascade downstream to other functions, making their inclusion in the model important.

Domain Expert

MAIN CHARACTERISTICS OF THE STERILE UNIT

The sterile unit employed 34 workers and operated 24/7, supplying surgical instruments and gowns to several hospital units (e.g., emergency department, intensive care unit, and surgical unit) as well as to a smaller hospital (20 beds) owned by the same organization and located 20 km away. The surgical unit was the primary client and had eight ORs, including one designated for unscheduled urgent surgeries.

The sterile unit was located adjacent to the surgical unit, facilitating service flows. The production process at the sterile unit was divided into three main stages, each physically separated by walls to prevent cross-contamination. The process began with receiving the instruments returned from client units, sorting them according to tray composition (e.g., cardiology and orthopaedics), and performing a manual wash to remove excess contaminants (e.g., blood). Next, batches of instruments were prepared and loaded into a washing machine for initial disinfecting. All processes carried out in the receiving and washing area involved contaminated materials, which is the reason why this area was informally referred to by employees as the "dirty zone."

The exit door of the washing machines was located in the assembly area, where batches of instruments were unloaded and assembled into trays for each medical speciality. This assembly took place on benches, where workers visually inspected the quality of each instrument, verified the required number of instruments, and wrapped them before loading the batch in the sterilizer machine. The exit door of the sterilizers was in the storage area, where another worker unloaded the trays and allowed them to cool down. After cooling, each tray was transported to storage shelves designated by the medical speciality.

Scheduling surgical procedures was crucial for organizing work at the sterile unit. Every day, representatives from both the surgical and sterile units held a meeting to anticipate the surgeries scheduled for two days ahead. Additionally, there was a follow-up meeting later in the day to anticipate the surgeries for the next day. A key aspect of these meetings was comparing the schedule with the availability of supplies at the sterile unit. However, this check was entirely manual and relied on inaccurate information regarding inventory levels and the specific needs of instruments for each surgical procedure.

Furthermore, unscheduled surgeries frequently occurred throughout the day, disrupting activities at the sterile unit and stressing the workers. Staff were aware that many of these unscheduled surgeries were not true urgencies but rather regular procedures that had been omitted from the regular schedule. One reason for this was that the surgical unit had excess capacity, with idle ORs available on any given day. As a result, some surgeons chose not to schedule their procedures in advance, trusting that there would be empty ORs available at their convenience.

METHODOLOGY

Table 17.1 presents the waste analysis perspective for each of the traditional FRAM steps, prescribing additional activities or reflections to be undertaken. Table 17.2 serves a complementary role by defining the meaning of lean wastes within a FRAM model. This interpretation is based on logical associations between each waste type and the concepts of function output and output variability. For example, transport waste involves the movement of a physical entity from one location to another. In a FRAM model, this can be represented by the transportation of a function Output (e.g., an assembled surgical tray) to be used as an Input for another downstream function (e.g., surgery). Alternatively, it can be represented by the inclusion of a transportation function in the model.

Note that only three wastes occur in adding-value functions: overproduction in quantity, overproduction in advance, and overprocessing. Overproduction refers to the excess production of finished products or completed service orders, such as performing unnecessary exams on patients. In this case, <carry out exams> is an adding-value function, and waste occurs when the output exceeds what is needed, which, in

TABLE 17.1
FRAM Steps and the Waste Analysis Perspective

FRAM Steps	Waste Analysis Perspective
(0) Recognize the purpose of the analysis	• The purpose is the identification and control of wastes
(1) Identify and describe the functions	• Include functions related to both adding-value and non-adding-value activities (i.e., activities that are themselves wastes) • In a manufacturing system, adding-value functions and outputs are those consistent with one or more of the following characteristics (Liker, 2004; Ohno, 1988): (*i*) physical transformation of materials (e.g., assembly, cutting, and welding); (*ii*) fulfilment of the requirements of the end client in the modelled system (e.g., provision of sterilized instruments for patient care in a hospital); and (*iii*) client willingness to pay more for the activity and/or output.
(2) Identify variability	• Use the lean taxonomy of wastes to support the analysis of output variability (see Table 17.2)
(3) Aggregate variability, actual and potential	• Acknowledge relationships between types of wastes, as they may imply in couplings of functions. For example, overproduction can trigger other wastes such as transportation, motion, and inventory.
(4) Assess the consequences of the analysis	• Use lean principles for proposing waste control measures while remaining aware of unintended consequences that might hinder resilience; • Lean principles are mostly useful to address variabilities that can be eliminated or reduced. They can also be useful to eliminate non-adding value functions; • Resilience management principles are mostly useful for coping with remaining and inevitable variabilities as well as for enhancing desirable variability.

FRAM terminology, counts as imprecision. Overprocessing involves inefficient methods during the physical transformation of a product or service, such as administering intravenous medications when oral administration would suffice. This example also relates to an adding-value function (administering medication) with imprecise output. Overproduction does not apply to transportation or storage, as these are non-adding value activities. In these cases, these instances of waste are categorized simply as transportation and inventory wastes.

Defects and unutilized talent can occur in both adding and non-adding value functions. For example, defects may arise during the physical transformation of materials, such as the incorrect assembly of parts (an adding-value function), or they may occur in the storage of materials on the wrong shelf (a non-adding value function). Similarly, unutilized talent can manifest when an overqualified worker is assigned to perform a trivial assembly task or when they spend their time on material storage in a warehouse.

The other waste types – transportation, inventory, waiting, and motion – are typically associated with non-adding value functions, which are considered wastes regardless of their outputs. However, these "typical" associations between waste types and functions are true for manufacturing systems (Ohno, 1988) but may not apply to service systems. In fact, what constitutes waste and value ultimately depends

TABLE 17.2
Lean Wastes and FRAM Implications

Waste Type	Meaning in a FRAM Model
Overproduction in advance	Output of an adding-value function is produced too early.
Overproduction in quantity	Output of an adding-value function exceeds the number of products or services required by downstream functions. It can also mean an output that is not consumed by any function in the model.
Over-processing	Output of an adding-value function that used either more resources than necessary or inefficient production methods
Transportation	Output needs to be physically transported to the place where the downstream function occurs. It can also mean a transportation function.
Inventory	Output is composed of *materials* not immediately consumed by downstream functions, and therefore, the materials wait and pile up before being used. It can also mean a storage function.
Waiting	Output *other than materials* are not immediately consumed by downstream functions, and therefore they wait and form queues. It can also mean a waiting function.
Motion	Output that involves awkward postures, inefficient movement, and excessive walking by human agents. It can also mean a function related to the movement of human agents.
Defects	Output that does not meet technical requirements of downstream functions (e.g., surgical instruments still contaminated after cleaning)
Unutilized talent	The human agent who performs a function is over-qualified for it. It affects the Resource aspect in an FRAM function.

on customer requirements. For example, moving caretakers and medical supplies to a patient's home is considered as value to patients who are unable to visit the hospital; these patients might even be willing to pay more for such a service.

In addition to Tables 17.1 and 17.2, a note is important regarding how to account for interruptions in FRAM modelling from the lean perspective. In FRAM models of work-as-done in healthcare, the aspect of Time often includes interruptions as events that steal time from the function (Sujan et al., 2023). As interruptions disrupt the workflow, they tend to be seen as waste by lean production. The type of waste related to interruptions is context-dependent, varying with the nature and reasons for the interruptions. For instance, a machine breakdown leads to interruptions that result in waiting waste, as both the equipment and the operator may be idle for some time. Conversely, if an employee interrupts a coworker to seek information about resolving issues related to defective products, the interruption itself is not necessarily a waste. Rather, it is a consequence of defective production waste and could lead to additional waste if it hinders the productivity of the worker being interrupted.

RESULTS

This section describes the results and findings of the project in relation to the usage of the FRAM. From the identification of functions all the way up to the management of variability.

RESULTS OF STEP 1: THE FUNCTIONS

Regarding FRAM step 1 (identification of functions), 18 functions (see Table 17.3) were identified and described considering a scenario of everyday work at the sterile unit. Five of these are background functions that represent the model boundaries.

TABLE 17.3
Associations among Functions and Waste Types

Function	Waste									
	OA	OQ	TR	IN	WA	MO	OP	DE	UT	Total
Schedule surgeries								X		1
Request instruments for unscheduled surgeries			X		X	X				3
Count and check instruments at the OR before returning to the sterile unit								X		1
Receive materials at the sterile unit					X	X				2
Clean excess dirty manually	X	X						X		3
Count and check returned instruments			X			X		X		3
Prioritize orders at the sterile unit										0
Wash materials in batches (automatic)	X	X								2
Inspect the quality and number of instruments								X		1
Assemble tray and wrap up	X	X					X	X		4
Search for missing instruments						X			X	2
Sterilize trays in batches	X	X								2
Cool down sterilized trays										0
Store trays on shelves			X	X		X		X		4
Separate trays in carts for delivery at OR				X		X		X		3
Deliver trays to OR			X			X				2
Carry out maintenance of damaged instruments			X			X				2
Carry out maintenance of equipment						X				1
Total	4	4	5	3	1	9	1	8	1	36

Note: overproduction in advance (OA); overproduction quantity (OQ); transportation (TR); inventory (IN); waiting (WA); motion (MO); overprocessing (OP); defects (DE); unutilized talent (UT); and making-do (MD). *Adding-value functions in Italics.*

For example, <schedule surgeries> is a source background function whose output is crucial for the sequencing and prioritization of activities at the sterile unit. However, this function relies on several other functions and factors of considerable complexity (e.g., hospital policies regarding priority clinical procedures and surgeons' schedules with concurrent demands at other hospitals) that were beyond the scope of the model. Thus, the focus was on understanding how <schedule surgeries> interacted with the sterile unit, rather than delving into the details of that function. Similarly, <deliver trays to OR> is a background function that serves as a sink, defining the downstream boundary of the model. Although activities in the OR can influence the sterile unit, a detailed examination of the OR was deemed excessively demanding given the purpose of this study.

There are 13 foreground functions that form the core of the model. For example, <inspect quality and number of instruments> is triggered by the unloading of washed materials, observing the priorities at the moment. This inspection commonly detects missing, damaged, or low-quality materials. If issues are found, the worker will go to the neighbouring storage room to seek a substitute instrument. If a suitable replacement cannot be found, which often happens, the worker assembles the tray as best as possible and sends it to the next stage of the process incomplete. Each tray package includes a checklist on top, indicating any missing, damaged, or low-quality items.

Four functions add value to the patient (22.2% of the total), who is the primary client. They are: <clean excess dirty manually>, <wash materials in batches – automatic>, <assemble tray and wrap up>, and <sterilize trays in batches>. These functions involve physical transformations of the instruments (e.g., removal of blood) and/or alterations to their physical properties. They directly address the patient's requirements by ensuring that instruments are sterilized to prevent contamination and are of high quality to avoid harm during surgery.

The remaining 14 functions are non-adding value functions, comprising 77.8% of the total. These functions add cost but do not directly fulfil patient requirements; for instance, <store trays in shelves> and carry out <maintenance of damaged instruments>. While waste can occur in *both* adding and non-adding value functions, this distinction is useful as the latter usually offer greater potential for improvement and can often be eliminated without compromising system functionality – for example, <maintenance of damaged instruments> ideally should not exist from an efficiency viewpoint. In addition, non-adding value functions are often neglected as targets for process improvement, as they may not be addressed by standardized operating procedures and may occur irregularly – for example, <search for missing instruments at the storage>. Moreover, productivity and quality improvements in adding-value functions are often constrained by technological limitations, making them more challenging to improve. For example, the adding-value function <sterilize trays in batches> is carried out by a machine with a fixed processing time that cannot be shortened without affecting quality.

Results of Step 2: The Variability

In FRAM step 2 (analysis of potential variability), wastes were equated to unwanted output variabilities. Table 17.3 presents the relationship between functions and the lean taxonomy of wastes. Cells marked with an "X" indicate that the waste type manifests as one of the function outputs. It is important to note that overproduction waste occurs only within the adding-value activities, as previously mentioned.

For two waste types – overproduction in advance and waiting – the output variability manifests in terms of timing. Overproduction occurs when the four adding-value functions produce their outputs either too early or too late, resulting from inaccurate production sequencing that does not align with the real demand of the surgical unit. For example, low-priority orders may be processed before high-priority ones. The function <prioritize orders> acts as a resilient compensation for these sequencing issues and is the only function without associated wastes. A handwritten sign in the dirty zone displays current priorities and is updated several times a day by the employee in charge of the storage room, who adjusts it according to the changing demands of the surgical unit. As for waiting, it arises as an output of <request instruments for unscheduled surgeries> since the need for these instruments is often identified too late, leading to surgery delays and increased waiting time for both clinicians and patients.

For the other waste types, output variability occurs in terms of precision. Overproduction in terms of quantity is particularly impactful, as it involves producing more outputs than necessary. This results in a significant number of sterilized instruments (20%–30%, on average) being returned unused from the ORs. Consequently, several functions waste time, machinery, and labour unnecessarily. One contributing factor to this waste is that the composition of the surgical trays is outdated, including instruments that have not been in use in years. Additionally, some of the returned materials are a natural outcome of the necessary slack in surgical procedures, where certain instruments are needed based on patient characteristics and unexpected events.

Motion is the most frequent waste type, primarily due to the confusing material flows in the sterile unit and the limited space for receiving materials from the ORs, which leads to the accumulation of materials in improvised locations. Defective production is the second most frequent waste type, manifesting differently across functions but consistently resulting in outputs that do not meet patient requirements. For instance, the function <schedule surgeries> often schedules surgeries that require the same instruments simultaneously, causing demand to exceed capacity and resulting in delays and cancellations. Three main underlying reasons contribute to this issue: (*i*) the scheduling staff lacks technical knowledge about the necessary supplies for each surgery, and there is no automatic decision-support system; (*ii*) there is no accurate data on inventory levels; and (*iii*) there is no accurate data on the real composition of instruments in each tray.

Results of Step 3: The Aggregated Variability, Actual and Potential

In FRAM step 3 (aggregation of variability), Figure 17.1 presents an instantiation of selected couplings during everyday work, highlighting the systemic impact of

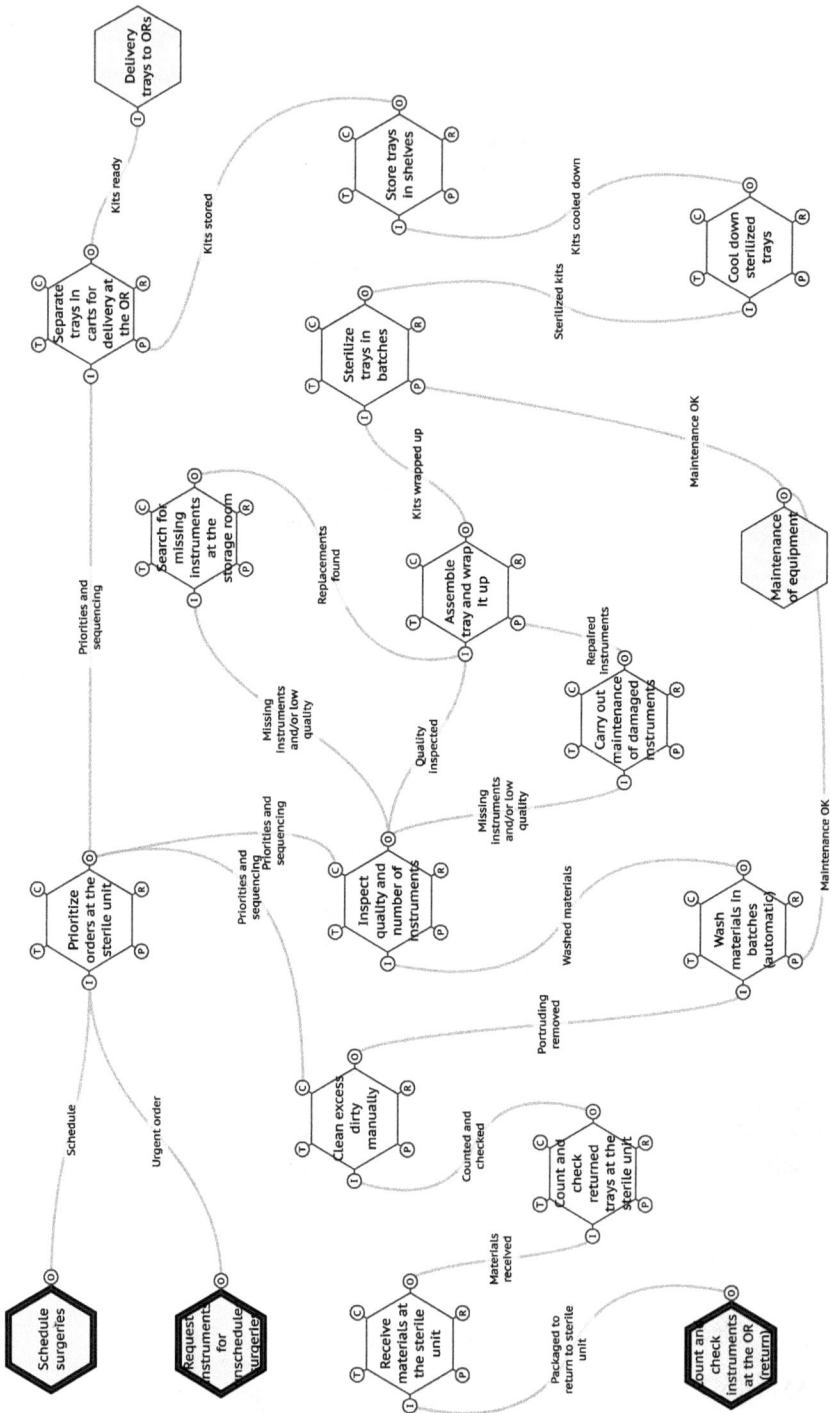

FIGURE 17.1 FRAM instantiation for everyday work at the sterile unit.

the waste associated with overproduction by quantity. Since the trays often contain more instruments than necessary, this overproduction occurs across all adding-value functions. Additionally, the excess instruments require transportation, inspection, and storage, leading to the associated waste types. The aggregation of variability step also reveals feedback loops that contribute to a reinforcing cycle of waste. Wastes originating in the sterile unit (e.g., overproduction and defects) eventually cycle back as instruments return from the ORs.

The relationship between the sterile and surgical units is synergistic, blurring the lines between supplier and customer; however, the sterile unit's role as a client of the surgical unit is largely neglected. This is evident in the unreliable and unrealistic scheduling of surgeries, which places the burden of adaptation on the sterile unit. To some extent, this imbalance stems from power dynamics, as the surgical unit is led by surgeons who bring patients (and revenue) to the hospital, while the sterile unit is managed by registered nurses and technicians providing support services. As a consequence, work-as-done in the sterile unit is largely unknown to doctors and hospital managers, making it difficult for them to understand why the unit does not achieve the expected performance.

RESULTS OF STEP 4: THE CONSEQUENCES OF THE ANALYSIS

FRAM step 4 (consequences of the analysis) was guided by three complementary approaches to address wastes: eliminating unnecessary variability (lean emphasis), coping with residual inevitable variability (resilience emphasis), and enhancing desirable variability (resilience emphasis). The term *emphasis* was chosen in order not to exclude the possibility of both lean and resilience contributing to the three approaches. Although lean production is best known for eliminating unnecessary variability, it also includes mechanisms to cope with residual variability (e.g., safety inventories) and even for increasing desirable variability (e.g., workers' multifunctionality and product variety) (Saurin et al., 2013). In turn, resilience engineering and FRAM literature acknowledge that certain variabilities are intrinsically detrimental to workers' safety and well-being and therefore should be suppressed to the possible extent (Hollnagel, 2012).

Three actions were mostly aligned with traditional lean emphasis: (*i*) revising the composition of the surgical trays; (*ii*) redesigning the layout of all areas in the sterile unit to physically segregate different material flows from their arrival at the unit from the ORs. This segregation prevents the mixing of different flows (e.g., instruments owned by physicians and those owned by the hospital), thereby reducing errors in tray assembly; and (*iii*) updating documents that specify the necessary instruments for each surgical procedure. This, combined with daily demand records for each surgery type, enables estimation of the daily demand for each instrument. Comparing this information with existing inventories helps identify priority instruments that the hospital needs to acquire.

To cope with residual inevitable variability, action (*iv*) involved providing a small inventory of backup instruments that were frequently missing in the tray assembly area. For enhancing desirable variability, actions included (*v*) assessing the level of multifunctionality of each worker in the sterile unit to inform training programs; and (*vi*) initiating a pilot program for autonomous maintenance tasks to prevent the frequent machine breakdowns that had occurred. This last action would also contribute to increasing workers' multifunctionality. Actions (*i*), (*iv*), and (*v*) are discussed in detail below, as they represent the three approaches.

Towards Implementation: Elimination of Unnecessary Variability

Regarding variability elimination, the primary action involved revising the composition of the surgical trays, informed by focus groups described in the data collection section. Overall, this initiative led to the removal of 95 instruments and the addition of 42 instruments. This improvement helped address overproduction in quantity (as unnecessary items would no longer be sterilized), waiting (reducing the likelihood that the surgical team would discover missing instruments during procedures and had to wait for them to be retrieved from the sterile unit), and defects (increasing the chances that trays would contain only the necessary components). This improvement exemplifies the application of lean production principles, particularly pull production, which emphasizes producing according to demand without overproduction (Liker, 2004). Additionally, lean problem-solving principles, such as "go and see for yourself" and "make decisions by consensus" (Liker, 2004), were applied as doctors personally reviewed the composition of trays and met in groups to reach a consensus.

Several functions benefited from this improvement, highlighting its systemic impact. Functions related to the processing, handling, or counting of instruments were directly benefited, leading to expected increases in their quality and productivity. For example, the productivity of <assemble tray and wrap it up> is likely to increase, as workers will no longer spend time assembling trays with unnecessary instruments. The quality of this function may also improve, as assemblies will be completed correctly the first time. Consequently, the function <search for missing instruments at the storage> may become obsolete or needed much less frequently. This reduction would simplify the entire system, as several wasteful interactions would cease to exist.

Towards Implementation: Coping With Residual Variability

To address remaining residual variability, a small inventory of instruments that frequently required corrective maintenance was placed close to the tray assembly workstations. An analysis of records for highly demanded instruments that underwent frequent maintenance helped identify those that would serve as a buffer for the assembly workstations. This buffer consisted of both newly purchased instruments and some of the excess instruments resulting from the earlier rationalization of tray compositions.

This improvement addressed three main waste types: defects (as incomplete trays or trays with damaged instruments would be less likely), motion, and unutilized talent (assembly workers would no longer need to go to the storage room for replacements). However, there was a trade-off: while reducing these wastes, inventory waste increased as a protective measure. This improvement aligns with the concept of

Designing for Resilient Performance, defined as "the use of design principles to support integrated human, technical, and organisational adaptive capabilities" (Disconzi and Saurin, 2022). The buffer of instruments relates to the principle of designing slack resources and strategies. Nevertheless, this buffer is finite and resembles resilience in terms of robustness (Woods, 2015), as it may not be sufficient for extreme conditions. To enhance adaptability, additional resources could include agreements with neighbouring hospitals for the exchange of supplies or flexible instruments that can easily be repurposed when necessary.

It is important to note that, from the perspective of the exemplified resilience improvement, the function <search for missing instruments at the storage> would continue to exist, although more efficiently and anticipated by design. Instead of leaving their workstations to find missing instruments in the storage room (incurring motion waste and hoping to locate the needed instrument), tray assembly workers would have the missing instruments readily available beside their benches. The likelihood of finding the required instrument would also increase, as the inventory was deliberately designed to include the right replacements for frequently damaged instruments.

Towards Implementation: Enhancing Desirable Variability

To support desirable variability, the primary action focused on increasing workers' multifunctionality, addressing several waste types. This measure can reduce defective production by making errors in the assembly of surgical trays less likely. It can also tackle unutilized talent, as multifunctional workers develop a deeper understanding of their roles within the production system, enabling them to offer more impactful ideas for improvement. Finally, it can minimize inventory waste, as work-in-process would be less likely to remain unattended due to a lack of skilled employees available to process and move it down the workflow. Overall, enhancing worker's multifunctionality was intended to improve efficiency while fostering a more adaptable and engaged workforce.

As mentioned in the data collection section, the sterile unit's supervisor carried out an assessment of the multifunctionality of all workers. Results made it clear that some workers were much more multifunctional than others (e.g., there were workers skilled to carry out up to 10 functions while others knew only one function). This imbalance stemmed from individual interests and experiences rather than from deliberate organizational planning. In particular, workers on the night shift were significantly more multifunctional than those on the morning and afternoon shifts, which posed a challenge since demand was lower during nighttime hours.

The multifunctionality assessment set a basis for developing a training program designed to enhance the multifunctionality of workers and address the reliance on a small number of skilled employees for specific tasks. These improvements were anticipated to benefit the Resource aspect of all human functions in the FRAM model, as there would be backup workers available for each function in case the primary worker was unavailable for any reason. This Resource aspect (i.e., multifunctional workers) would be the output of a new function <develop workers' multifunctionality>.

CONCLUSIONS

The models of socio-technical systems generated by the FRAM can be developed for multiple purposes and interpreted through various theoretical lenses. While this flexibility is a strength, it can make applying the FRAM to specific classes of problems challenging, except for safety, which is the focus of a large number of case studies and has a well-established knowledge base. Although the FRAM concepts (e.g., functions, variability, and outputs) are generalizable across domains and problems, their application to contexts such as production waste is not straightforward.

This chapter provides guidance on applying the FRAM in the specific context of production waste, illustrating how outputs and functions can be interpreted as wastes and how lean production and resilience management can complement each other in FRAM step 4 (consequences of the analysis), pursuing efficiency without neglecting protection. It is proposed that lean lenses be used to devise actions that tackle unnecessary variability, while resilience lenses should be employed to manage residual inevitable variability and enhance desirable variability. Complex systems, such as health services, typically require all three approaches. On the one hand, in complex systems, some sub-systems may evolve naturally and disorderly over time, presenting opportunities for lean-oriented initiatives (e.g., optimizing the number and types of instruments in surgical trays). On the other hand, complexity inherently involves residual uncertainty (e.g., demand for surgical instruments and maintenance frequency) and necessitates controllers that match the process variability (e.g., multi-functional workers), thus requiring resilience-oriented improvements. The FRAM offers a framework to analyse these different facets of variability, their implications, and countermeasures. The effectiveness of such analysis is likely to be enhanced by a FRAM modelling team that includes members with expertise in both resilience management and process improvement, such as the industrial engineers involved in the case described in this chapter.

REFERENCES

Bølviken, T., Rooke, J., Koskela, L. The wastes of production in construction – A TFV based taxonomy. In: *22nd Annual Conference of the International Group for Lean Construction*. Oslo, Norway, pp. 23–27, 2014.

Browning, T. R., Heath, R. D. Reconceptualizing the effects of lean on production costs with evidence from the F-22 program. *Journal of Operations Management*, 27(1), 23–44, 2009.

Calabresi, G. *The cost of accidents: a legal and economic analysis*. Yale University Press, 2008.

Card, A. J. The problem with '5 whys'. *BMJ Quality & Safety*, 26(8), 671–677, 2017.

Chiarini, A. Waste savings in patient transportation inside large hospitals using lean thinking tools and logistic solutions. *Leadership in Health Services*, 26(4), 356–367, 2013.

Disconzi, C. M. D. G., Saurin, T. A. Design for resilient performance: Concept and principles. *Applied Ergonomics*, 101, 103707, 2022.

Dos Santos, B. M., Fogliatto, F. S., Zani, C. M., Peres, F. A. P. Approaches to the rationalization of surgical instrument trays: Scoping review and research agenda. *BMC Health Services Research*, 21, 1–15, 2021.

Fireman, M. C. T., Saurin, T. A. Chain of wastes: The moderating role of making-do. In: Tommelein, I.D., Daniel, E. (Eds.), *Proc. 28th Annual Conference of the International Group for Lean Construction (IGLC28)*, 2020. https://doi.org/10.24928/2020/0084, online at iglc.net

Fogliatto, F. S., Anzanello, M. J., Tonetto, L. M., Schneider, D. S., Muller Magalhães, A. M. Lean-healthcare approach to reduce costs in a sterilization plant based on surgical tray rationalization. *Production Planning & Control*, 31(6), 483–495, 2020.

Gayer, B. D., Saurin, T. A., Anzanello, M. The nature and role of informal resilience practices in the performance of lean production systems. *Journal of Manufacturing Technology Management*, 33(6), 1080–1101, 2022.

Guariente, P., Antoniolli, I., Ferreira, L. P., Pereira, T., Silva, F. J. G. Implementing autonomous maintenance in an automotive components manufacturer. *Procedia Manufacturing*, 13, 1128–1134, 2017.

Habibi Rad, M., Mojtahedi, M., Ostwald, M. J. The integration of lean and resilience paradigms: A systematic review identifying current and future research directions. *Sustainability*, 13(16), 8893, 2021.

Hollnagel, E. *FRAM: The functional resonance analysis method: Modelling complex sociotechnical systems*. CRC Press, 2012.

Jasti, N. V. K., Kodali, R. Lean production: literature review and trends. International *Journal of Production Research*, 53(3), 867–885, 2015.

Liker, J. K. *The Toyota way: 14 management principles from the world's greatest manufacturer*. McGraw-Hill Education, 2004.

Nelson, J. Using conceptual depth criteria: addressing the challenge of reaching saturation in qualitative research. *Qualitative Research*, 17(5), 554–570, 2017.

Ohno, T. *Toyota production system: Beyond large scale production*. Productivity Press, 1988.

Patriarca, R., Di Gravio, G., Woltjer, R., Costantino, F., Praetorius, G., Ferreira, P., Hollnagel, E. Framing the FRAM: A literature review on the functional resonance analysis method. *Safety Science*, 129, 104827, 2020.

Patton, M.Q. *Qualitative research & evaluation methods: Integrating theory and practice*. SAGE Publications, 2014.

Potthoff, L., Gunnemann, L. Resilience of lean production systems: A systematic literature review. *Procedia CIRP*, 120, 1315–1320, 2023.

Saurin, T. A., Rooke, J., Koskela, L. (2013). A complex systems theory perspective of lean production. *International Journal of Production Research*, 51(19), 5824–5838.

Scott, J., Marshall, G. *Oxford dictionary of sociology*, (3rd ed), Oxford University Press. 2009.

Soliman, M., Saurin, T.A., Anzanello, M.J. The impacts of lean production on the complexity of socio-technical systems. *International Journal of Production Economics*, 97, 342–357, 2018.

Sujan, M., Lounsbury, O., Pickup, L., Kaya, G. K., Earl, L., McCulloch, P. What kinds of insights do Safety-I and Safety-II approaches provide? A critical reflection on the use of SHERPA and FRAM in healthcare. *Safety Science*, 173, 106450, 2024.

Sujan, M., Pickup, L., de Vos, M. S., Patriarca, R., Konwinski, L., Ross, A., McCulloch, P. (2023). Operationalising FRAM in Healthcare: A critical reflection on practice. *Safety Science*, 158, 105994.

Tortorella, G. L., Fogliatto, F. S., Anzanello, M., Marodin, G. A., Garcia, M., Reis Esteves, R. Making the value flow: Application of value stream mapping in a Brazilian public healthcare organisation. *Total Quality Management & Business Excellence*, 28(13–14), 1544–1558, 2017.

Uhrin, A., Moyano-Fuentes, J., Camara, S.B. Firm risk and self-reference on past performance as main drivers of lean production implementation. *Journal of Manufacturing Technology Management*, 31(3), 458–478, 2020.

Williamsson, A., Dellve, L., Karltun, A. Nurses' use of visual management in hospitals: a longitudinal, quantitative study on its implications on systems performance and working conditions. *Journal of Advanced Nursing*, 75(4), 760–771, 2019.

Woods, D. D. Four concepts for resilience and the implications for the future of resilience engineering. *Reliability Engineering & System Safety*, 141, 5–9, 2015.

18 Maximizing the Impact of the FRAM

From Analysis to Implementation

Mark Sujan and Olivia Lounsbury

CONTEXT: DESIGNING PRACTICAL RECOMMENDATIONS

Before choosing a method, or methods, to study a work setting, you need to have clarity about what you want to achieve, that is, your aims. Typically, researchers and practitioners who adopt the FRAM as their study method want to learn more about how a system works in practice and how the system manages to succeed under varying conditions (Hollnagel et al., 2019). This is the definition of resilience, which provides the theoretical background for the FRAM (Hollnagel, 2010). However, such understanding is rarely an end in itself. Rather, it is the prerequisite and the means for designing and implementing meaningful and sustainable change in practice. Unfortunately, successfully bridging this gap from descriptive analysis of how a system works towards implementing improvement interventions poses significant challenges. This is exacerbated by the lack of guidance in the literature, as many research studies conclude the analysis with recommendations but do not follow these through to implementation and evaluation (Verhagen et al., 2022).

As described in this book, an FRAM analysis typically consists of four steps. In the first FRAM book, the fourth step is described as "Assess the consequences of the analysis", where two specific types of interventions are introduced, namely (a) monitoring and (b) dampening of variability (Hollnagel, 2012). While this is useful in terms of setting a new direction and orienting the mindset away from traditional risk-based thinking, where interventions are described in terms of barriers, this high-level description lacks consideration of the practical challenges around translating such principles into real-world interventions. A corollary to this is that practitioners frequently struggle effectively to communicate the proposed interventions to senior decision-makers, who are often unfamiliar with the theoretical underpinnings of the FRAM.

In this chapter, we offer reflections on how to move from the FRAM analysis to practical implementation in order to maximize the impact of the analysis. The next section describes the clinical scenario we use as an example. The example is concerned with the management of post-surgical patient deterioration. Then, we provide

DOI: 10.1201/9781003518167-23

a brief summary of how the FRAM analysis was undertaken. Following that, we look more closely at the nature of the recommendations that were developed as part of the FRAM analysis. This is helpful to illustrate in the subsequent section what kinds of strategies might be helpful in overcoming barriers to implementation. We conclude the chapter with a summary of common communication pitfalls and how these might be countered.

THE CLINICAL SCENARIO: MANAGEMENT OF POST-SURGICAL PATIENT DETERIORATION

The clinical scenario used in this chapter is the management of post-surgical deterioration. Surgery is a traumatic intervention for a patient's body, and the patient can sometimes get worse after surgery. It is important that healthcare staff recognize and respond quickly to a patient's worsening condition after surgery to ensure their safe recovery. Unfortunately, this does not always happen, and some patients die as a result. The technical term for this is "Failure to rescue" (FTR) (Silber et al., 1992). There is a lot of research on FTR, but how healthcare professionals handle deterioration can vary a lot (Burke et al., 2020). Some of the reasons for this variability identified in the literature include doctors or nurses being new, having too many patients, feeling overly confident, or having trouble communicating. There can also be issues with equipment, delays in getting help or having trouble finding senior doctors who are busy.

Domain Expert

Failure to recognize and respond to acute deterioration of patients, or, FTR, is a well-known and intractable problem, which affects hospitals worldwide (Ghaferi & Dimick, 2015). Silber and colleagues define FTR as the conditional probability of patient death following a surgical complication (Silber et al., 1992). FTR has been proposed as an alternative metric for surgical quality as opposed to mortality rate (Silber et al., 2007), because it is more tightly linked to postsurgical complication rates (Ghaferi et al., 2009; Ghaferi et al., 2011), rather than patient factors. Across surgical settings, FTR rates have been found to range between 8% and 18% (Johnston et al., 2015b; Portuondo et al., 2020). Considering the large number of surgical procedures carried out each year, these figures represent significant, potentially preventable patient harm.

FTR has been linked to lack of clinical experience, high workload, overconfidence, communication problems, equipment and logistical bottlenecks, delayed referrals and transfers, and difficulties in locating senior doctors due to competing priorities (Burke et al., 2020; Callaghan et al., 2017; Donohue & Endacott, 2010; Johnston et al., 2015b; Peebles et al., 2012; Wakeam, Hyder, et al., 2014). Strategies for reducing FTR events were summarized in a recent systematic review (Burke et al., 2020) and include higher nurse staffing levels and a higher percentage of nurses educated to degree level (Blegen et al., 2013; Rafferty et al., 2007). Trigger tools such as the UK National Early Warning Score (NEWS2) are widely used (Royal College of Physicians, 2017) but not universally found to improve outcomes (Bedoya et al., 2019; Donohue & Endacott, 2010). Improvement efforts also include the use of clear standardized escalation and communication protocols, Rapid Response Teams (RRT), and a focus on safety culture (Ghaferi & Dimick, 2015; Johnston et al., 2014; Johnston et al., 2015a; Wakeam, Asafu-Adjei, et al., 2014; Wakeam, Hyder, et al., 2014). While recent US data suggest that top-performing hospitals were able to reduce surgical mortality significantly over the past decade largely by reducing FTR rates (Fry et al., 2020), the management of acute deterioration remains highly variable.

Various strategies for improving FTR rates have been suggested. For example, having more nurses can help because it reduces individual workload. Tools like the National Early Warning Score (NEWS2) can also be used to spot problems early, but research shows they do not always improve outcomes (Bedoya et al., 2019). Other strategies include clear instructions for getting help when needed, specialized teams to respond quickly to emergencies, and a focus on patient safety as a top priority.

The example described in this chapter is based on a study that was carried out on a large surgical emergency unit (SEU) with 54 beds in a teaching hospital in England. SEUs are places where patients who have had emergency surgery are monitored because they are more likely to experience complications. During the day, the SEU has around 35 staff members, including 4 senior surgeons, and a separate team to manage patients before, during, and after surgery. At night, there are fewer doctors – two junior, two mid-level, and one senior.

DATA COLLECTION AND METHODOLOGY

Failure to rescue, along with the associated research, looks at those instances where the management of deterioration has failed. This is the traditional way of looking at a patient safety issue. However, we know from the field of Resilience Engineering that we can – or should – approach safety in another way. Rather than looking at why the management of deterioration sometimes fails, we can study how this activity usually succeeds. This involves studying everyday work or work-as-done (WAD). In our example, FRAM was chosen as the main method to investigate and understand WAD in the management of patient deterioration in order to develop recommendations for improvement interventions that focus on strengthening resilience abilities.

The FRAM analysis was undertaken as part of a larger research project, and the study has been described in detail elsewhere (Sujan et al., 2022). For this reason, only a brief summary is provided here to contextualize the subsequent reflection. Data were collected through semi-structured interviews (n = 31) and workshops (n = 14) with stakeholders working on or with SEU (including nurses, doctors, porters, managers, radiologists, and anaesthetists).

Data were analyzed based on the structure of the FRAM. First, key functions were identified from the interview transcripts. These functions tend to describe what people (rather than technology or organizational functions) do in practice. Interview participants might describe what they do using different terminology and in slightly different ways. The interview transcripts were analyzed thematically. Descriptions of

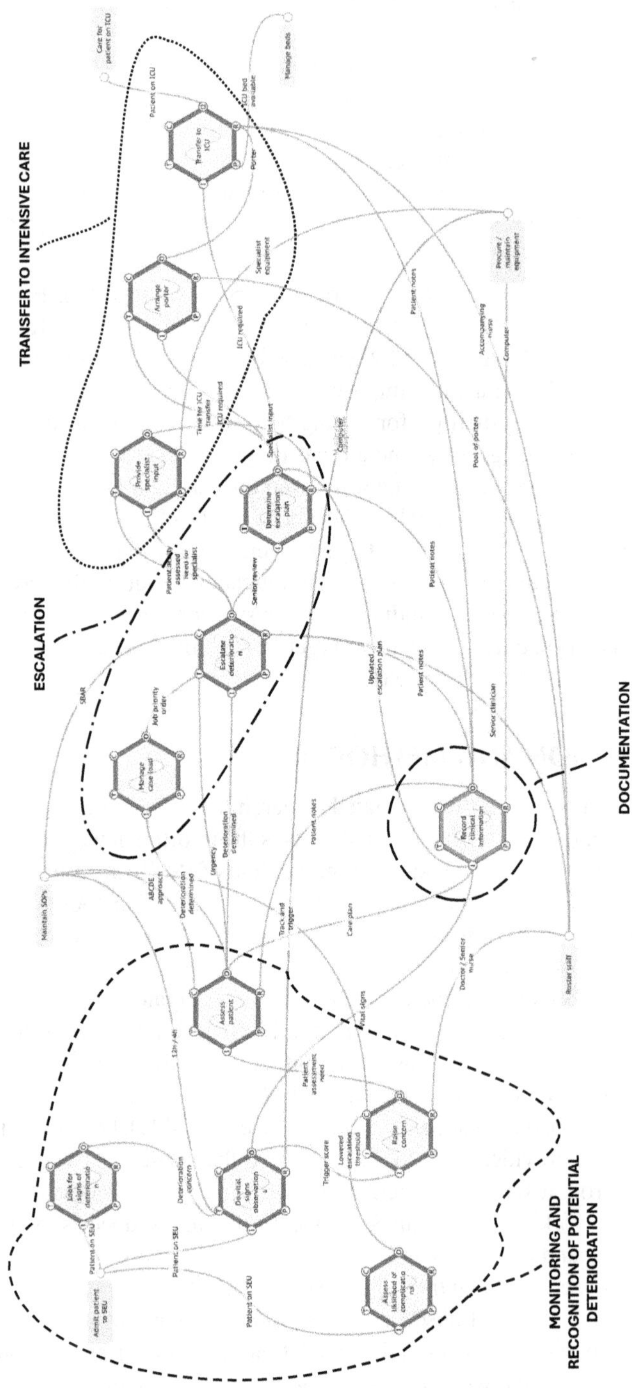

FIGURE 18.1 FRAM model of the management of the deteriorating patient with transfer to intensive care.

functions that were similar or conceptually related were grouped together. For example, one interview participant suggested that the "registrar calls radiology", while another person described that they "need to speak to the [radiology] team to explain the situation". Both instances refer to the situation where the patient's care is escalated and input from radiology is requested. Hence, these were both coded as the function "to escalate deterioration". Then, for each function, couplings and variability were described. During the description of couplings, additional functions were identified and analyzed accordingly. At this stage, a greater emphasis was placed on ensuring organizational and technology functions were identified in addition to human functions. Functions that are carried out within SEU were analyzed in greater depth as foreground functions. Functions outside of SEU were treated as background functions. A graphical representation of the management of the deteriorating patient with transfer to intensive care is shown in Figure 18.1. Four main functional clusters were developed by grouping together functions which relate conceptually to a clinically meaningful broader activity: monitoring and recognition of potential deterioration, documentation, escalation, and transfer to intensive care.

Method Geek

We were interested in performance variability rather than failures. The term variability can be interpreted in different ways. In quality improvement and traditional safety thinking, variability is often regarded as something that is potentially harmful and forms the target of interventions aimed to standardize work. From a resilience engineering perspective, however, variability is perceived as inevitable and often useful (Hollnagel, 2015). This is because people have to manage complexity and uncertainty through dynamic trade-offs, and in this way, they create safe performance (Sujan et al., 2015).

The concept of approximate adjustments is one of the key principles embedded in FRAM (Hollnagel, 2009). We approached variability in the interviews by (a) briefly outlining that the focus was on understanding everyday work rather than whether procedures are followed and (b) prompting interviewees to consider situations of uncertainty and changing contexts.

For example, how might caring for a patient with a complex medical history be different from caring for a patient with a simpler presentation? This was done in a non-normative way, which means we did not assume or suggest that performance variability was bad or undesirable (Furniss et al., 2020). Rather, we encouraged open discussion of how people manage the complexities and uncertainties of their everyday work. We also asked staff to consider how the described variability might affect other parts of the system (functional couplings in FRAM terminology).

Two interview guides were developed, one for staff working on SEU and a second one, slightly modified, for staff working outside but with SEU (e.g., radiology). The interview guide for staff on SEU is shown in Table 18.1. Modifications for staff outside of SEU related to prompts around the interviewee's tasks and their involvement with the management of deterioration, not being based on SEU. The interview guide was used to facilitate the interview, but it was not intended as a set of questions that had to be followed in every interview. The interviewer was free to follow up on specific points of interest opportunistically.

TABLE 18.1

Interview Guide for Interviews With Staff Working in the Surgical Emergency Unit (SEU)

Topic	Prompts
Introduction	Background to the study and the interview. Focus is on everyday work.
Participant background	Interviewee's professional background and current role.
Key activities relating to SEU	Interviewee's main activities working on SEU or interfacing with SEU. What kinds of things do you do on SEU? What other roles do you talk to or work with?
Management of the deteriorating patient	Interviewee's perceptions of how deteriorating patients on SEU are managed, including recognition of deterioration and escalation of care. How do you know if a patient is at risk of deterioration? With whom do you talk about this? What types of information sources do you consider? What do you do if a patient is at risk of deterioration? Do other people ask you for help with managing a patient at risk of deterioration? Do you ask anybody for help?
Variability	Interviewee's perceptions of main types of variability and the reasons for this. Do you always do things in the same way? What are the reasons for doing things differently? What are things you are sometimes struggling with or that make your life harder? How do you deal with these? What makes it work? Where do things start to break down? Why?
Ending	Expression of thanks for contribution Do you have any questions or suggestions?

In our application of the FRAM, Step 4 is the interpretative stage of the analysis (see Sujan et al. (2023) for further discussion on this type of "reflexive" FRAM). This is where we linked the descriptions of performance variability explicitly to the underlying theory of resilience abilities. In this way, we constructed explanations of how performance variability can be an expression of resilient forms of behaviour. In the workshops, participants were then invited to reflect on this and to provide suggestions for how resilience (i.e., the resilience abilities) might be strengthened.

Method Geek

Epistemology refers to our belief in how knowledge comes about or how insights are gained. While the description of the field of epistemology is beyond the scope of the chapter, it is instructive to briefly consider what this might mean for doing an FRAM study. In essence, the FRAM practitioner needs to reflect on whether they believe that their analysis is largely objective and its purpose is to describe an objective reality waiting to be uncovered. This would be indicative of a realist epistemology. This would imply that we focus on the reliability of the analysis as far as different FRAM practitioners should arrive at the same results. In this way, the FRAM model could also be a tool for simulation and quantification of variability and its propagation. Or, in another epistemological stance, the analyst might consider that the FRAM is a method for the practitioner to interrogate a system and its stakeholders in order to gain certain insights. But these insights are actively constructed by the practitioner, and they will also reflect the practitioner's background, experience and focus – hence the need for reflexivity (and hence "reflexive" FRAM). This approach is more aligned with a constructivist epistemology, where we admit the possibility of different versions of reality, each actively created through the interaction of the practitioner with the system as part of the analysis. Neither approach is necessarily superior. However, it is important that the FRAM practitioner is aware of their epistemological stance in order to develop their analysis appropriately and to be able to discern and understand potential limitations adequately.

In the workshops with stakeholders, the scope of the analysis was described, focusing on the concept of variability as embedded in everyday work. Following a summary of the analysis findings, in the first workshop participants were invited to put forward and discuss suggestions for possible interventions in a flexible and unconstrained way. These suggestions were recorded, and after the workshop, we attempted to categorize these according to their impact on resilience abilities. In the second workshop, participants were introduced to the concept of resilience abilities, and they were asked to think specifically about each of the resilience abilities and consider how these might be strengthened. In the final step, we put all of the suggestions together and structured them by their link to resilience abilities. Suggestions that were not aligned with resilience engineering principles or that did not have a clear link to resilience abilities were not included. The findings of the analysis, along with the recommendations made by workshop participants, mapped against the four resilience abilities, are shown in Table 18.2.

TABLE 18.2
Recommendations Based on the FRAM Analysis

Resilience Ability	Capturing What Works	Improvement Suggestions
Monitoring	Looking for red flags for patient deterioration. Understanding colleagues' capabilities and experience. Maintaining a broader departmental view. Actively communicating concerns and observations.	Enhance communication channels, for example, shared workspaces. Leverage technology, for example, clinical prediction tools using machine learning. Establish dedicated monitoring roles and responsibilities. Develop dynamic patient plans that anticipate potential developments and complications.

(Continued)

TABLE 18.2 (Continued)
Recommendations Based on the FRAM Analysis

Resilience Ability	Capturing What Works	Improvement Suggestions
Responding	Collaborative decision-making, for example, sharing tasks across professional groups. Flexible support, for example, stepping up when colleagues are busy. Dynamic and proactive resource allocation, for example, adjusting staffing levels in specific areas to meet changes in demand. Mutual trust and responsiveness, for example, confidence in referring patients to colleagues.	Foster a collaborative culture, for example, reduce hierarchical barriers and empower junior staff. Enhance preparedness and coordination, for example, build on communities of practice and knowledge sharing across departments. Support communication and learning, for example, create safe spaces for discussion and support.
Anticipating	Knowing when peaks are likely to arise in order to support workforce and skill-mix planning	Enhance data insights, for example, data-driven forecasting models. Proactive resource management, for example, implementation of data-driven flexible staffing models.
Learning	Appreciating gaps between work-as-imagined and work-as-done (trade-offs) Collaborative knowledge sharing Psychological safety and trust Role clarity and development	Embed learning into daily operations. Design for adaptability and learning, for example, resilient procedures and work processes. Create opportunities for informal and inter-departmental learning. Develop proactive knowledge exchange.

RESULTS: THE NATURE OF FRAM RECOMMENDATIONS

Step 4 of the FRAM instructs us to consider, among others, interventions to monitor and dampen variability. By linking the application of the FRAM more explicitly to the underpinning resilience theory, we broadened the scope of interventions to consider how resilience abilities, in general, might be strengthened, that is, the ability to monitor, the ability to respond, the ability to anticipate and the ability to learn (Hollnagel, 2010). Interventions based on this type of thinking are intended to support the ability to succeed under varying conditions rather than to narrowly reduce the risk associated with specific hazards (Sujan et al., 2024). But what do such interventions look like in practice?

The FRAM analysis helped us understand and appreciate the variable nature of the work system underlying the management of patient deterioration. An example is provided in Box 18.1. The example concerns the function "do vital signs observations". This should be done at specific times according to a protocol. In practice, the timing of doing vital signs observations can vary significantly. Rather than

---

BOX 18.1   EXAMPLE OF EXAMINATION OF PERFORMANCE VARIABILITY IN DOING VITAL SIGNS OBSERVATIONS

Function: Do vital signs observations

Variability: Timing

Tensions and uncertain performance conditions: Nurses trade off several variables to determine the timing for doing vital signs observations, including suggested observation time as per protocol, patient comfort, patient condition, and their own workload.

Functional coupling: (1) The timeliness of vital signs observations can affect the functions "raise concern" and "determine escalation plan" both positively and negatively. Timely vital signs observations can facilitate prompt escalation, but if concerns based on vital signs observations are raised frequently and unnecessarily, this can negatively affect downstream functions by ultimately resulting in diminished responses ("cry wolf syndrome"). (2) Doctors might create an additional function "assess likelihood of complications" reflect their anticipation of an elevated risk of deterioration. This can change the way vital signs observations are used by (a) increasing the frequency and (b) lowering the escalation threshold for a given trigger score.

---

describing this as a failure (and, hence, something negative), the analysis provided us with a deeper understanding of the complexity of decision-making of people. For example, patients might be asleep, and the nurse does not want to wake the patient, or the nurse might need to attend to other patients whose condition is perceived higher priority. The analysis of functional coupling provided further insights into the system's complexity, which illustrates how people adapt their behaviour. An instance of this is the proactive request by doctors to increase vital signs observation frequency and to lower the escalation threshold (i.e., not waiting for a patient's trigger score to warrant escalation) for patients where they might have intuitively greater concerns about potential deterioration. Such adaptations can provide the resilience required for managing deterioration successfully under changing demands and competing priorities but can also sometimes contribute to deterioration.

Another example is provided in Box 18.2. This example refers to the function "provide specialist input". Frequently specialist input is required for determining the escalation plan, for example, input from radiologists or input from a consultant in intensive care. However, these specialists need to manage their own respective workload, and, hence, prioritization decisions need to be made. The timing of this function directly affects both the downstream function "determine escalation plan", as well as numerous other instances of similar functions across the hospital, which require specialist input as a resource. Interestingly, variability is introduced as part of clinical judgement, in as far as the requesting clinician might follow up the electronic request for specialist input with a phone call or a personal chat; similarly, the specialist (e.g.,

---

BOX 18.2   EXAMPLE OF EXAMINATION OF PERFORMANCE VARIABILITY IN PROVIDING SPECIALIST INPUT

Function: Provide specialist input

Variability: Timing

Tensions and uncertain performance conditions: Collaboration across departmental boundaries is critical for the successful management of deterioration, (e.g., arranging investigations to allow decisions on treatment) and for the onward journey of the patient (e.g., arranging transfer to intensive care or theatre). However, such collaboration requires additional coordination and adjustments. Prioritization decisions are often required, for example, which radiological requests to process first or where to send mobile imaging equipment.

Functional coupling: The timely provision of specialist input is a crucial resource for determining the escalation plan. Timely specialist input can speed up this process, but it also means that there will be delays in responding to other requests. The timing of this function can be affected as part of the function "to escalate deterioration concern" because the requesting clinician can communicate the urgency of the request and can follow up the electronic request with a phone call if required. Similarly, the specialist (e.g., radiologist) can provide the input sooner, for example, findings from a CT scan, if they feel it is important to do so.

---

a radiologist) might call the requesting clinician once the report is available rather than waiting for them to pick it up from the electronic system. In this way, clinicians manage the competing priorities and mismatch in demand and capacity.

We used this analysis of the variability of everyday work to develop suggestions for strengthening resilience in collaboration with staff. There is not necessarily a clear one-to-one mapping from learning about everyday work to specific suggestions for improvement. Learning from everyday work can provide inspiration and areas for exploration, but it is not a formula for generating interventions. This process relies on the facilitation skills of the FRAM practitioner and the collaboration of stakeholders. In this project, we used the learning from everyday work as examples to illustrate to stakeholders the mindset underlying this type of analysis. This was intended to encourage them to start thinking from a resilience engineering perspective. Of course, this still requires active facilitation by the FRAM practitioner as well as a "clean up" of the data following the workshop. In this way, the results are co-created between the FRAM practitioner and the stakeholders.

For example, learning from everyday work about the resilience ability "monitoring" (i.e., knowing what is going on, knowing what to look for) suggested that, among others, staff try to build an awareness of the patients in the department even if they are not assigned to the patient. This is so that in case there is a need for escalation and they need to help out, they are already informed about the patient to a certain extent. It also helps them to know what to look out for. Participants also provided examples of how they attempted to build shared awareness within the team. An instance of this was alluded to above, where a doctor might share their concern about a patient with the nurse so that the nurse can monitor the patient more closely.

These insights from the analysis did not translate immediately into actionable improvements but were fed back to staff during workshops where they were prompted to consider how their resilience abilities could be strengthened. Among the suggestions were, for example, the physical layout of the workspace, including meeting rooms and common rooms to facilitate communication among team members about concerns and to share information. Another suggestion for the creation of shared awareness also addressed the issue of inter-departmental collaboration, which might be improved with the concept of dynamic plans for patient management. Such dynamic plans consider, already upon admission, potential alternative pathways and developments and include professionals from relevant specialities. Further examples are shown in Table 18.2.

DISCUSSION

Conducting an analysis of normal work at all is a significant step forward in a health-care context currently characterized by tremendous resource constraints, which necessitate a near-exclusive focus on what goes wrong. Though the deteriorating patient example provided above was part of a larger research study, these approaches can be incorporated into existing safety investigations as a first step forward. For example, during an investigation of unrecognized patient deterioration, the investigator can prompt the team to consider everyday work and focus on describing and strengthening resilience abilities alongside existing investigation frameworks. Such a strategy is suggested in the recent NHS England Patient Safety Incident Response Framework (PSIRF) (Lounsbury & Sujan, 2023).

The suggestions for improvement resulting from the FRAM analysis outlined above offer promise of strengthening the resilience abilities of the organization by supporting the work people are already doing. However, there are several challenges with moving these improvement suggestions from analysis to practical implementation. The following paragraphs outline common challenges and related recommendations to move forward with implementing FRAM-generated insights.

DIFFICULT TRANSLATION PROCESS (PROTECTING THE INTEGRITY OF FRAM RECOMMENDATIONS)

First, the analysis findings generated require translation into actionable improvements, particularly if the suggestions require organizational support or funding. For example, "Establish dedicated monitoring roles and responsibilities" may be translated into "Hire one nurse responsible solely for monitoring and flagging up patients at risk for deterioration". The naïve act of translating improvement suggestions into actionable points, however, can counter the purpose of the improvement suggestion, which is to support resilient practices. Without careful consideration and familiarity with resilience theory, the action points may paralyze resilience, rather than support it. In the example above, the aim of the "Establish dedicated monitoring roles and responsibilities" improvement suggestion was to maintain a broader departmental view (i.e., share responsibility for monitoring). The following action point (hiring new personnel) may further contribute to the problem of lack of shared awareness by perpetuating silos and rigid job descriptions within the ward. When considering how to implement FRAM-generated improvement suggestions, improvement teams should take heed of the intention of the improvement suggestions and consider how suggested action points support or counter the intent. This can be hampered by a lack of familiarity with the underlying theoretical foundations of the FRAM. Probing questions for considering alignment with resilience theory may include:

- To the analyst: In what ways will this implementation intervention support or hinder the intention of the improvement suggestion?
- To the analyst: Which resilience ability will this improvement intervention support and how?
- To the team(s) doing the work: How will this implementation intervention be normalized into the everyday work from which it was generated?
- To the team(s) doing the work: Is there a clear benefit to incorporating this improvement intervention into daily work, beyond what is already happening?

It should be noted that FRAM can also be used as a tool to proactively understand how certain improvement suggestions or pilot interventions may come to fruition in the future (Damoiseaux-Volman et al., 2021). This can mitigate the difficulty of proactively anticipating the impact of a FRAM-informed intervention before it is implemented.

UNCLEAR LINK, SMALLER PRIORITY

The second challenge associated with implementing improvement suggestions from FRAM is the conceptual distance of the suggestion from a serious risk or harmful event. Improvement suggestions generated from traditional safety investigations are often prioritized to a greater degree by organizational leadership, which in turn reduces barriers to buy-in, implementation, and resourcing. Improvement suggestions generated without a clear, albeit frequently overly simplistic, connection to a high-risk, high-harm event, however, go without that inertia. This may be particularly challenging for improvement suggestions that require significant resources and dedicated support, such as "Implementation of data-driven flexible staffing models". Making the case to translate these improvement suggestions into practice may be accomplished by aligning with institutional priorities, articulating the workload of the current ways in which staff compensate, and retrospectively reviewing previous incidents in which this improvement suggestion may have helped mitigate harm. Probing questions to consider, particularly by those requesting resources for improvement, may include:

- Are there previous examples of investigations related to safety concerns (e.g., patient deterioration) that may align with the improvement suggestions generated from a FRAM analysis of normal work?

- What are the data sources available to me that house staff feedback about workload, safety culture, etc. (e.g., annual surveys and exit interviews)?
- Is there something we can show once the improvement suggestions are implemented that will support organizational reputation?
- Can this improvement suggestion be linked or generalized to problems in other wards or departments?

PERCEIVED LACK OF NOVELTY

There is a risk that, once organizational support for the improvement suggestion has been obtained, staff may respond with "This is what we are already doing". Though this may seem discouraging, this is an indicator that the improvement suggestion could further support trade-offs and adaptations in everyday work. For example, the "Enhance preparedness and coordination" improvement suggestion may indeed already happen, for example, by staff manually pulling together the doctor notes onto one page at the start of each of their shifts, making a note to alert the night shift nurse to patient-specific escalation thresholds, or double documenting their assessments so they are visible to both the medical and nursing teams. While these are the work-arounds that often prevent harm from happening, this individual adaptation work can become duplicative across staff members, can add tremendous, unnecessary burdens to the workforce, and can hinder seamless coordination due to variation in documentation. Gaining staff buy-in to the improvement suggestions is key to ensuring they are actually useful in everyday work. Potential probing questions directed toward those doing the work can include:

- What are your challenges when you are already doing a particular adaptation (e.g., others don't do it the same way, risk of transcription error, and reliance on memory)?
- Under what circumstances might a particular adaptation not always reliably happen (e.g., new nurses may fear escalating concerns and the doctor may be in the operating theatre and thus unavailable)?
- How do you think adaptations could help improve coordination with other departments?

LACK OF COLLECTIVE, CONSCIOUS PRIORITIZATION SYSTEMS

Finally, many of the deteriorating patient improvement suggestions above are contingent on shared awareness and collective, coordinated action. In the current healthcare context, short staffing, ever-changing team members, silos, lack of multidisciplinary education opportunities, fear, mistrust, and a culture of blame serve as deterrents to the team-based culture that is needed to implement many of these improvement suggestions. Overarchingly, prioritization is already happening on a provider-to-provider basis, but there is no coordinated strategy behind it, which further fragments our workplace processes. There is a need for a collective, conscious prioritization system across team members and between teams.

The transition to team-based paradigms starts with modelling this lens at the leadership level and in existing safety investigations. Probing questions for doing so, particularly for those facilitating improvement conversations, may include:

- In what ways could inter-departmental learning benefit both departments (e.g., learning about each department's challenges and reviewing the same data)?
- What information would be helpful to know from specific roles or departments to better manage a situation (e.g., patient deterioration)?
- What are different roles doing independently to build a comprehensive picture of a particular situation (e.g., of a specific patient's case) for themselves? Could any of this be incorporated into team-based conversations?

Though there is a need for a larger, systems approach to developing a collective, conscious prioritization system that enhances adaptive capacity, there may be some steps forward that may be easier to implement, such as:

- Putting data in the guidelines, policies, and procedures to support collective decision-making.
- Specifying how to collectively adjust based on risk in guidelines, policies, and procedures.
- Making the impact of our actions (e.g., unnecessary imaging and antibiotic use) visible to other teams for a mutual social contract.
- Expanding safety investigations to consider all other factors that were being prioritized by the teams at the time of the safety event.

It should be noted that these considerations have been developed based on deep knowledge of healthcare and how it operates. While this is the primary strength of these considerations, and these considerations may extend to other domains, we need to be mindful that we do not suggest a naïve transfer into other domains without understanding the peculiarities of other contexts. For example, we know the history of safety management in aviation is very different from that of healthcare. It would be hurried to transfer learnings from aviation directly to healthcare. Therefore, a similar thought process needs to be leveraged when applying the above considerations to other domains.

COMMUNICATING THE FRAM RESULTS AND ADDRESSING CONCERNS

While the results of the FRAM analysis may be rigorous and understood by those involved in the analysis itself, it is likely that robust improvements will only be actioned if these results are understood by key decision-makers and policy-makers. Anticipating common rebuttals from those unfamiliar with the FRAM and the underlying resilience theory when presented with the FRAM findings is a key step to ensure the message is conveyed. Common rebuttals may include things such as lack of perceived novelty, lack of perceived impact, and lack of perceived distinctiveness (see Table 18.3).

TABLE 18.3
Common Objections to Implementing FRAM Recommendations and How to Counter These

Common Rebuttal	Communication Strategy	FRAM Benefit to Highlight
"This is what we already do".	• Ask whether there are some groups who may benefit from learning "what you already do" (e.g., junior nurses). • Use an analogy of a shop: Most of the time, you are able to navigate quite well using muscle memory, but it becomes challenging when the layout of the shop changes. In complex systems, what is "normally done" is likely to change and methodologies for improvement should reflect that.	• Equitable support between different groups of team members. • Alignment of FRAM with the ever-changing conditions of the workplace.
"I don't understand the need to dedicate resources to support the normal work that is already functioning effectively".	• Quantify and make visible how much work goes into making sure the normal work happens effectively and how much manual or duplicative work could be avoided through better support (e.g., nurses not knowing the appropriate bleep number when trying to reach the doctor for a deteriorating patient, so they instead try every bleep number).	• Reducing the burden on staff.
"These results do not feel unique to FRAM and are instead similar to what we achieve through our existing safety frameworks".	• Ask staff how they feel when engaged in FRAM versus existing safety processes, such as investigations, and communicate findings. • Highlight that, while the intervention itself may look similar (e.g., development of a protocol), the intervention is likely to be more sustainable given the better understanding of the wider context of the work.	• Staff psychological safety when involved in improvement work. • Sustainability of implemented interventions.
"This problem is beyond our control"	• Emphasize what is lost by not at least documenting broader systems issues, specifically acknowledging the risk of losing track and the inability to escalate evidence-based themes to leadership. Systems issues are more likely to get leadership attention and buy-in when there is evidence to show that these problems are recurring, rather than "one offs".	• Detect and escalate interrelated risks between seemingly unrelated areas.

CONCLUSION

This chapter has highlighted the potential of the FRAM to bridge the gap between understanding complex work systems and implementing meaningful interventions. By focusing on work-as-done and integrating resilience engineering principles, FRAM allows practitioners to move beyond traditional risk-based approaches and identify strategies that support adaptive and resilient behaviours at all levels of an organization. However, the challenges of translating FRAM-generated insights into practical, sustainable changes remain significant. Addressing these challenges requires effective communication, alignment with organizational priorities, and collaboration with stakeholders to ensure interventions are contextually relevant and well-integrated. By refining methods for translating analysis into action and fostering a deeper understanding of resilience principles, FRAM can play a critical role in enhancing safety in complex socio-technical systems like healthcare. This chapter illustrates the importance of continued innovation and dialogue in safety science to maximize the practical impact of tools like the FRAM.

REFERENCES

Bedoya, A. D., Clement, M. E., Phelan, M., Steorts, R. C., O'Brien, C., & Goldstein, B. A. (2019). Minimal impact of implemented early warning score and best practice alert for patient deterioration. *Critical Care Medicine*, *47*(1). https://journals.lww.com/ccmjournal/Fulltext/2019/01000/Minimal_Impact_of_Implemented_Early_Warning_Score.7.aspx

Blegen, M. A., Goode, C. J., Park, S. H., Vaughn, T., & Spetz, J. (2013). Baccalaureate education in nursing and patient outcomes. *JONA: The Journal of Nursing Administration*, *43*(2). https://journals.lww.com/jonajournal/Fulltext/2013/02000/Baccalaureate_Education_in_Nursing_and_Patient.8.aspx

Burke, J. R., Downey, C., & Almoudaris, A. M. (2020). Failure to rescue deteriorating patients: A systematic review of root causes and improvement strategies. *Journal of Patient Safety*, https://doi.org/10.1097/pts.0000000000000720

Callaghan, A., Kinsman, L., Cooper, S., & Radomski, N. (2017). The factors that influence junior doctors' capacity to recognise, respond and manage patient deterioration in an acute ward setting: An integrative review. *Australian Critical Care*, *30*(4), 197–209. https://doi.org/10.1016/j.aucc.2016.09.004

Damoiseaux-Volman, B. A., Medlock, S., van der Eijk, M. D., Romijn, J. A., Abu-Hanna, A., & van der Velde, N. (2021). Falls and delirium in older inpatients: Work-as-imagined, work-as-done and preferences for clinical decision support systems. *Safety Science*, *142*, 105355. https://doi.org/10.1016/j.ssci.2021.105355

Donohue, L. A., & Endacott, R. (2010). Track, trigger and teamwork: Communication of deterioration in acute medical and surgical wards. *Intensive and Critical Care Nursing*, *26*(1), 10–17. https://doi.org/10.1016/j.iccn.2009.10.006

Fry, B. T., Smith, M. E., Thumma, J. R., Ghaferi, A. A., & Dimick, J. B. (2020). Ten-year trends in surgical mortality, complications, and failure to rescue in medicare beneficiaries. *Ann Surg*, *271*(5).

Furniss, D., Nelson, D., Habli, I., White, S., Elliott, M., Reynolds, N., & Sujan, M. (2020). Using FRAM to explore sources of performance variability in intravenous infusion administration in ICU: A non-normative approach to systems contradictions. *Applied Ergonomics*, *86*, 103113. https://doi.org/10.1016/j.apergo.2020.103113

Ghaferi, A. A., Birkmeyer, J. D., & Dimick, J. B. (2009). Variation in hospital mortality associated with inpatient surgery. *New England Journal of Medicine*, *361*(14), 1368–1375. https://doi.org/10.1056/NEJMsa0903048

Ghaferi, A. A., Birkmeyer, J. D., & Dimick, J. B. (2011). Hospital volume and failure to rescue with high-risk surgery. *Med Care*, *49*(12), 1076–1081.

Ghaferi, A. A., & Dimick, J. B. (2015). Understanding failure to rescue and improving safety culture. *Ann Surg*, *261*(5), 839.

Hollnagel, E. (2009). *The ETTO principle: Efficiency-thoroughness trade-off*. Ashgate Publishing.

Hollnagel, E. (2010). Prologue: The scope of resilience engineering. In E. Hollnagel, J. Paries, D. D. Woods, & J. Wreathall (Eds.), *Resilience engineering in practice: A guidebook.* Ashgate Publishing.

Hollnagel, E. (2012). *FRAM, the functional resonance analysis method: Modelling complex socio-technical systems.* Ashgate Publishing.

Hollnagel, E. (2015). Why is work-as-imagined different from work-as-done? In R. Wears, E. Hollnagel, & J. Braithwaite (Eds.), *The resilience of everyday clinical work.* Ashgate Publishing.

Hollnagel, E., Sujan, M., & Braithwaite, J. (2019). Resilient health care – Making steady progress. *Safety Science*, *120*, 781–782. https://doi.org/10.1016/j.ssci.2019.07.029

Johnston, M., Arora, S., Anderson, O., King, D., Behar, N., & Darzi, A. (2015a). Escalation of care in surgery: A systematic risk assessment to prevent avoidable harm in hospitalized patients. *Ann Surg*, *261*(5), 831–838.

Johnston, M. J., Arora, S., King, D., Bouras, G., Almoudaris, A. M., Davis, R., & Darzi, A. (2015b). A systematic review to identify the factors that affect failure to rescue and escalation of care in surgery. *Surgery*, *157*(4), 752–763. https://doi.org/10.1016/j.surg.2014.10.017

Johnston, M., Arora, S., King, D., Stroman, L., & Darzi, A. (2014). Escalation of care and failure to rescue: A multicenter, multiprofessional qualitative study. *Surgery*, *155*(6), 989–994. https://doi.org/10.1016/j.surg.2014.01.016

Lounsbury, O., & Sujan, M. (2023). Achieving a restorative just culture through the patient safety incident response framework. *Journal of Patient Safety and Risk Management*, *28*(4), 153–155. https://doi.org/10.1177/25160435231194397

Peebles, E., Subbe, C. P., Hughes, P., & Gemmell, L. (2012). Timing and teamwork—An observational pilot study of patients referred to a rapid response team with the aim of identifying factors amenable to re-design of a rapid response system. *Resuscitation*, *83*(6), 782–787. https://doi.org/10.1016/j.resuscitation.2011.12.019

Portuondo, J. I., Shah, S. R., Raval, M. V., Pan, I. W. E., Zhu, H., Fallon, S. C., Harris, A. H. S., Singh, H., & Massarweh, N. N. (2020). Complications and failure to rescue after inpatient pediatric surgery. *Ann Surg*, https://doi.org/10.1097/SLA.0000000000004463

Rafferty, A. M., Clarke, S. P., Coles, J., Ball, J., James, P., McKee, M., & Aiken, L. H. (2007, 2007/02/01/). Outcomes of variation in hospital nurse staffing in English hospitals: Cross-sectional analysis of survey data and discharge records. *International Journal of Nursing Studies*, *44*(2), 175–182. https://doi.org/10.1016/j.ijnurstu.2006.08.003

Royal College of Physicians. (2017). *National early warning score (NEWS) 2: Standardising the assessment of acute-illness severity in the NHS.* RCP.

Silber, J. H., Romano, P. S., Rosen, A. K., Wang, Y., Even-Shoshan, O., & Volpp, K. G. (2007). Failure-to-rescue: Comparing definitions to measure quality of care. *Med Care*, *45*(10), 918–925. http://www.jstor.org/stable/40221531

Silber, J. H., Williams, S. V., Krakauer, H., & Schwartz, J. S. (1992). Hospital and patient characteristics associated with death after surgery: A study of adverse occurrence and failure to rescue. *Med Care*, *30*(7), 615–629. http://www.jstor.org/stable/3765780

Sujan, M., Bilbro, N., Ross, A., Earl, L., Ibrahim, M., Bond-Smith, G., Ghaferi, A., Pickup, L., & McCulloch, P. (2022). Failure to rescue following emergency surgery: A FRAM analysis of the management of the deteriorating patient. *Applied Ergonomics*, *98*, 103608. https://doi.org/10.1016/j.apergo.2021.103608

Sujan, M., Lounsbury, O., Pickup, L., Kaya, G. K., Earl, L., & McCulloch, P. (2024). What kinds of insights do Safety-I and Safety-II approaches provide? A critical reflection on the use of SHERPA and FRAM in healthcare. *Safety Science*, *173*, 106450. https://doi.org/10.1016/j.ssci.2024.106450

Sujan, M., Pickup, L., de Vos, M. S., Patriarca, R., Konwinski, L., Ross, A., & McCulloch, P. (2023). Operationalising FRAM in healthcare: A critical reflection on practice. *Safety Science*, *158*, 105994. https://doi.org/10.1016/j.ssci.2022.105994

Sujan, M., Spurgeon, P., & Cooke, M. (2015). Translating tensions into safe practices through dynamic trade-offs: The secret second handover. In R. Wears, E. Hollnagel, & J. Braithwaite (Eds.), *The resilience of everday clinical work* (pp. 11–22). Ashgate Publishing.

Verhagen, M. J., de Vos, M. S., Sujan, M., & Hamming, J. F. (2022). The problem with making Safety-II work in healthcare. *BMJ Qual Saf, 31*, 402–408.

Wakeam, E., Asafu-Adjei, D., Ashley, S. W., Cooper, Z., & Weissman, J. S. (2014). The association of intensivists with failure-to-rescue rates in outlier hospitals: Results of a national survey of intensive care unit organizational characteristics. *Journal of Critical Care, 29*(6), 930–935. https://doi.org/10.1016/j.jcrc.2014.06.010

Wakeam, E., Hyder, J. A., Ashley, S. W., & Weissman, J. S. (2014). Barriers and strategies for effective patient rescue: A qualitative study of outliers. *The Joint Commission Journal on Quality and Patient Safety, 40*(11), 503-AP506. https://doi.org/10.1016/S1553-7250(14)40065-5

Section V

Implementation and Reflections on the FRAM

19 FRAM via Software
The FRAM Model Visualizer

Rees Hill

VISUALISING FRAM MODELS

In early 2014, I was working in operational risk management in manufacturing when a friend and colleague told me to add some books to my reading list. We were look-ing for more proactive ways to manage safety and were not content with the current industry standards and largely trailing performance indicators. The books were all by Hollnagel: Safety-I and Safety-II: The Past and Future of Safety Management (Hollnagel, 2014), The ETTO Principle: Efficiency-Thoroughness Trade-Off: Why Things That Go Right Sometimes Go Wrong (Hollnagel, 2009), and FRAM: The Functional Resonance Analysis Method: Modelling Complex Socio-technical Systems (Hollnagel, 2012).

As a systems developer, a statement Hollnagel made in the FRAM book stuck out to me: "The simplicity and the consistency of the method, however, means that it should be straightforward to build supporting software, and probably also to find a way of calculating the magnitude of the Functional resonance". Taking this as a chal-lenge, I set about to find a solution, as I wanted a convenient way to start building my own FRAM models. There were some existing resources on Hollnagel's Functional Resonance website, including spreadsheet templates, graphic templates, and the FRAM Model Builder (FMB), an application that managed the descriptions of the Functions and aspects and produced a text file in a standardised XML format. During the development of the FRAM model, a graphical representation of the text-based descriptions was sometimes created using sticky notes or whiteboards. When a model was finished, it was often drawn in a graphics programme such as Visio. I focused my efforts on taking these text-based descriptions of Functions, created by following the published method, and turning these into a graphics-based view of the FRAM model, bypassing the need to recreate everything manually in a drawing programme.

After two days of coding, I had a working computer programme that took the XML output from the FMB as input and drew hexagon representations of the Functions on the screen. Then, if the Output of one Function had the same name as an aspect of a second Function, the programme drew a straight line between them, showing a "potential coupling". This first version also highlighted the "orphan" aspects if they did not make a coupling, and allowed for the repositioning of the Function's hexagons on-screen. It was February 2014 when I emailed the "proof of concept" programme to Professor Hollnagel with the desire to develop the concept further and coordinate with other efforts. His reply came the next day and included

DOI: 10.1201/9781003518167-25

the following comment, "I think that having a module like this will be very useful for people who try to use the FRAM. Even as it is now, but of course particularly if you can get curved lines, and so on". Thus, a period of rapid continuous improvement development began with Hollnagel making suggestions and testing every development, the first being curved lines. By March 2014 the first public version (0.1.0) was made available on the Functional Resonance website. In April 2014, the public version was updated (0.1.4), adding the ability to create and edit Functions and aspects directly in the FMV software, thus eliminating reliance on the FMB to first "build" the model and produce a data file to be imported. For the first time, the FMV was able to build a FRAM model from scratch, and users could see their model taking shape graphically on screen as they described the Functions.

From the very beginning, the FMV software was developed by following the method strictly as it was written, focusing on step one, to identify and describe the Functions. By describing Functions using the six aspects of Input, Output, Precondition, Resource, Time, and Control, the FMV displays a potential coupling between two Functions when the description of the Output from one Function matches the description of one of the other aspects from the second Function. The question has been asked a number of times: why the FMV software not allow you to click on one Function, then click on another Function, and "draw" a connection between the two. The answer is simply because this does not follow the method. By using a text-based input, the software enforces the method where potential couplings can be discovered, instead of drawing a pre-conceived or "as imagined" diagram that may be more like a process flow. This discovery of potential couplings, or even the discovery that an expected connection is not present, is especially powerful when a collaborative approach is used to build the FRAM model.

Other developments followed, including the graphical distinction between foreground Functions and background Functions, and a "track and playback" feature. Because the concept of an automatic interpretation of the model was not fully developed, it was decided to include a manual recording option that used the FMV graphics as a reference to track and playback a step-by-step analysis of an instantiation. This could show the order that Functions activated and when a potential coupling became an actual coupling, as a way of presenting model interpretation findings.

After testing 12 distinct development versions, the second public version (0.2.0) was released in June 2014 to coincide with the FRAMily meeting held in Gothenburg, Sweden. The FMV was now established as a new standard for creating and visualising FRAM models, and Hollnagel began using the software to teach FRAM courses.

THE CHALLENGES OF QUANTIFICATION

When the public version 0.2.0 was released in 2014, a separate development branch was created with version 1.2.0 to experiment and explore quantification and analysis tools. The first attempt was a "variability tracer" inspired by work from Simon Albery and was based on heat flow modelling that used finite difference equations solved by the Gauss-Seidel iteration method (Albery, 2014). The work also referenced Hollnagel's decision table concept from the Method (Hollnagel, 2012 – Table

8.2) and the first specification for the FRAM Model Interpreter (FMI). The FMV already allowed for the selection of Function types (Human, Technological or Organisational) and some qualification of variability with respect to two phenotypes of time and precision, such as "Too early, On time, Too late, Not at all" and "Precise, Acceptable, Imprecise", respectively, as explained in step two of the method, to characterise the variability. If a Function was tagged with a selection other than "On time" and "Precise" then it was considered to have variability and was displayed with a superimposed curved line. The variability tracer then took this indication of variability and propagated it to downstream Functions with the assumption that a Function dependent on an upstream Function that has variability, is likely to exhibit variability in its own output. This was displayed in the FMV as coloured bars on the Functions. The calculation also included a type of ETTO rule where it was possible to force any Function with propagated variability to be "On time", in which case the variability was manifested as "Imprecise", or if it was forced to be "Precise" then the variability was manifested as "Too late". This was shown to work successfully with the Herald of Free Enterprise example used in the FRAM handbook. The variability tracer allowed for a very simple indication of potential variability propagating through a model, but was not sufficient as an analysis of complicated models because it treated the whole model as a fixed network and did not address the complexity of potential couplings or specific instantiations.

In December 2014, a new version, 1.3.0, began development on a connection between the FMV software and an external analysis tool for calculating conditional probabilities using Bayesian Belief Networks (BBN). The work continued through to version 1.5.0 in December 2018. The software set up starting parameters based on assumptions about the reliability of each Function by type (Human, Technological, or Organisational) considering both exogenous and endogenous variabilities, and analysed the model as a whole, showing the probabilities of each potential coupling becoming an actual coupling, or the probability of each Function producing an acceptable output.

During this time, the use of the FRAM in research papers was increasing, and several researchers were working on their own methods for the quantification and analysis of their FRAM models to apply more rigour to step three of the method, the aggregation of variability. One of the most successful of these endeavours resulted in a new software tool, myFRAM, which was officially developed and released by Riccardo Patriarca and his team from 2015 to 2017 and presented at the 2018 FRAMily meeting held in Cardiff, Wales (Patriarca, Di Gravio, & Costantino, 2017a). myFRAM leverages VBA and utilises Excel spreadsheets to organise the descriptions of the Functions and aspects, making it easier to manage large models. myFRAM has the ability to attach both qualitative and quantitative data and assists in exploring the model in a systematic way, and it was initially developed to operationalise the very first attempt of introducing Monte Carlo simulations within the FRAM (Patriarca, Di Gravio, & Costantino, 2017b). It can connect with other software, such as for statistical analysis or data visualisation, opening the setting to different types of analysis (Python, Matlab, Power BI, etc.). It also produces an output file that is opened in the FMV software for model visualisation and can be initialised from an FMV file, as well.

Notably, the DynaFRAM initiative led by Doug Smith acquired relevance in several use cases in the medical and maritime sectors mainly, too.

In January 2019, another development cycle began to refresh and update the FMV software. Improvements and fixes identified with the 1.X development workstream were typically worked back into the public 0.X versions one step behind. The two workstreams were now to be combined into the same software version. The public functionality would be freely available to any users who installed the software. A suite of more advanced functionality would be available as a paid extension, referred to as the "FMV Pro". The purpose of introducing a licence fee was to help build a fund that could be used to reinvest into continued software development and promotion of the method, as all development on the FMV for the past five years was *pro bono publico*.[1] Version 2.0.2 of the FMV Pro was released publicly at the May 2019 FRAMily meeting in Malaga, Spain. The extended functionality included visually collapsing a group of Functions, exporting and importing Function groups (reproducing what was scientifically discussed earlier in the Abstraction/Agency framework by Patriarca, Bergström, and Di Gravio (2017)), a simplified version of the variability tracer, undo/redo steps, and Metadata with the ability to add and edit any number of key-value pairs for each Function.

The visualisation in FMV has also been used to integrate another family of approaches for quantification, related to the identification of information entropy for the study of random walkers moving in FRAM instantiations, as for the Functional Random Walker (FRW) algorithm developed by Patriarca et al. (2024).

FRAM MODEL INTERPRETER (FMI)

The qualitative analysis of FRAM models and the benefits that the modelling process itself contributed to the understanding of complex systems were firmly established, but the community of FRAM users clearly expressed through research papers and the annual meetings that more quantitative analysis was desired. While there were clearly very good examples for specific applications, there was no default quantitative analysis method in FMV. Hollnagel had corresponded back in 2014 about his intentions to extend FRAM with "some kind of FRAM Model Interpreter (FMI) or FRAM 'inference engine'". In October 2015, Hollnagel wrote a version 0.1 Functional specification for his FMI, which helped inform the variability tracer development. The programming behind the variability tracer was described in a document "Basic qualitative representation of Output variability". Hollnagel responded to this in July 2019 with version 0.2 of the FMI, where he described how a Function's Output may be influenced by its aspects, using decision tables.

Hollnagel began a more concerted effort to develop the FMI in May 2020, during the travel restrictions and lockdown conditions that many of us were experiencing. The basic version (V3) was released to a test audience in June 2020 as a stand-alone application that took an FMV data file as input and gave a text-based interpretation of the FRAM model. By July 2020, the FMI was integrated into version 2.0.7 of the FMV software, such that the interpretation was visualised within the model graphics as it progressed. A text-based report was also available as an output in keeping with the stand-alone FMI. The FMI BasicPlus V4 stand-alone application and the combined FMV/FMI version 2.1.0 application were both released publicly at an FMV/FMI Webinar hosted by Hollnagel in September 2020.

The FMI became the core method for validating that a model is syntactically correct and capable of realising instantiations within its defined bounds and therefore a legitimate system model. It also introduced and defined a number of key concepts in model interpretation. From the very beginning, the FMV identified orphan aspects that did not couple with other Functions, highlighting them in red, and the FMI now further enforced the rule by flagging this as a critical error that would prevent an interpretation. All aspects must couple with at least background Functions as the legitimate method of defining the boundaries and scope of the model. To fix the error, the boundary could be increased by adding additional background Functions or reduced by removing the orphan aspects.

A background Function was already identified as one that had either inputs or outputs but no other aspects. Two special cases of background Function were then identified during the FMI interpretation as needed for the model to be complete. One is an entry Function that only has one or more Outputs that link to Inputs of foreground Functions but has no other aspects. A Function is considered to only activate when it receives an Input, but the entry Function is a special case where it is the most upstream Function that produces the first Output. This would start the interpretation, and the entry Function Output would then couple with the Input of a downstream Function, causing that Function to activate, and so on. A model could have more than one entry Function. The other special case is the exit Function that only has Inputs but no other aspects. This is also referred to as a sink and defines the end of the interpretation. A model could have any number of exit Functions, and the most basic stop condition of an interpretation is when all exit Functions have been activated.

Another critical error that will prevent an FMI interpretation is missing Inputs or Outputs for a foreground Function. If a Function has Time, Control, Precondition, or Resource aspects, then it is by definition a foreground Function, and must therefore also have at least one Input and one Output.

After the FMI has checked and passed the model for critical errors, the interpretation progresses in cycles. The entry Functions activate on Cycle 0 and produce their Outputs. These Outputs are then available to downstream Functions on the next cycle. On Cycle 1, any Functions that received an Input from upstream Functions on the previous cycle then become active. As each downstream Function is activated, it considers each of its aspects, determining if they are actively coupled with other upstream Functions and evaluating if the conditions exist for it to produce its own Output, which is then available to other downstream Functions on the next cycle and so on. In this way, the instantiation progresses throughout the model. The cycle therefore represents the order in which Functions are capable of activating due to the upstream/downstream structure of the model; however, within any given cycle, the Functions are activated asynchronously, or in parallel. In this way, the interpretation could be considered analogous to a Markov chain, where a Function can only consider the state of its aspects as influenced by the previous cycle.

Once a Function receives an Input from an upstream Function on a previous cycle, the default activation conditions for the Function to produce an Output is that all of its other aspects are also present from upstream Functions. This is sufficient for a simple linear model, even one with parallel branches, to progress from entry to exit. If a model is more complicated and includes feedback and loops, as is often the case, then the interpretation will appear to stall or block, as the relative upstream/downstream relationship between Functions can change from one cycle to another. If not all aspects are needed to be present for a Function to produce an Output, meaning that not all potential couplings are always actual couplings at a given cycle in a given instantiation, then this can be adjusted to progress the analysis by using the interpretation profile specific to each Function. Each of the aspect types can be adjusted for the Function to require all, any or none of that aspect type to be allowed to produce an Output.

The key concepts of model interpretation formally introduced by the FMI, and in particular the systematic activation of Functions in cycles, standardised the approach to the interpretation of a model and opened up more possibilities to investigate standard methods of quantification and analysis. In parallel, the visualisation of the FMI within the FMV software allowed the user to see the progression of the interpretation cycles, either one cycle step at a time or as a fully animated interpretation from entry to exit. Errors were highlighted, and loops and blocks could easily be identified graphically to aid the model builder in understanding the potential couplings and instantiations possible within the model and even begin a qualitative what-if type of analysis.

METADATA IN FMV

The release of the combined FMV/FMI sparked another wave of interest in a number of model builders who wanted to add quantification to the systematic method of the FMI, prompting another development cycle of the FMV software.

To encourage experimentation in analysis methods, it was decided to take the predefined parameters that the variability tracer stored within the FMV data file and bring these into the Metadata key-value pairs released in version 2.0.2, so they could be defined by the user and manipulated by the FMI. This was achieved with version 2.1.4 in October 2020 which added another layer to the Metadata Key-Value pair, where an equation could be attached to the pair, such that the equation would be evaluated at the time that a Function was activated by the FMI Cycle, and the Value would be updated with the result. This was initially envisioned for the two phenotypes used by the variability tracer, Timing and Precision, but in theory, any number of Metadata Key-Value pairs could be defined for a Function, opening up the possibility of defining a Function dynamically, as influenced by the couplings made with upstream Functions. The Metadata could also represent any property of the Function, fixed or dynamic, that was then made available to other downstream Functions through the Outputs.

By combining the Metadata equation functionality with the structure of the FMI, the combined FMV/FMI software now had the facility for exploring and utilising these extra dimensions (values) of Functions and to write procedures (Turing algorithms) for how their behaviours emerge from the implications and interactions of these Functional "values". This can also cope with variability in the way the Functions themselves operate, (endogenous variability), as well as the normal variability possible in the aspect couplings (exogenous variability).

When a Function receives an Input during an interpretation, it first considers if the aspect conditions are satisfied before it activates, for example, it may be set to wait until a Precondition aspect is present before activating. If the aspect conditions are fulfilled, then the Metadata is inspected, and any dynamic values are calculated. It is possible to "programme" the Function within the Metadata to further consider the Metadata available to it from the actual couplings that have been realised from upstream Functions and to make a decision if it will produce an Output, and further, what qualities updated in its Metadata will then be available to downstream Functions. In this way, the activation of Functions and propagation of Metadata could be considered stochastic, in that the Output of the Function depends on the present couplings, independent of the model's history. This is significant for the modelling of emergent behaviour in complex systems.

The ability to add, manipulate and output sets of Metadata for the Functions added another dimension to the FRAM analysis, as Hollnagel had pointed out earlier, a Function is more than just an execution node; it also implies a method for executing the instructions from the aspects to produce the Output. The extended Metadata functionality was demonstrated publicly as the FMV Pro version 2.1.5 at the June 2021 FRAMily webinar.

SAFEGUARDING THE SOFTWARE

The principles of open-source software were always important to Hollnagel, stating in 2014 that "My intention has always been to have the basic tools as open source". The FMV software development intentionally used coding that was either open source or could be developed using publicly available development environments and runtimes.

Method Geek

The software was programmed in ActionScript 3 and MXML using the Apache Flex software development kit (SDK). The compiled application used an embedded Adobe AIR runtime, and a significant reason for this choice was the ability to target different operating systems, such as Windows, Apple, and Ubuntu, as this was one of the few choices available at the time. Another advantage was the ability to handle a heavily graphics-centric application. After eight years of software development, this combination was entering the end of its useful life as support for some of the components. It was becoming increasingly difficult to compile new versions to keep pace with the changes to computer operating systems, and particularly difficult to target Apple Mac computers. When version 2.1.5 was presented at the June 2021 FRAMily webinar, a proof of concept for a browser-based version, which would be more operating system agnostic, was also presented. Version 2.2.0 in January 2023 was the final Apache Flex/Adobe Air-based version to be released.

To make the FMV application more accessible and to overcome the challenges of compiling software for specific operating systems, it was decided to explore a web browser-based solution starting in March 2021.

Method Geek

A new standard was WebAssembly (WASM), which allowed for deployment on the web of client-side code that ran inside the existing browser sandbox, eliminating the need to manage separate security policies. It could also run on any operating system that uses a modern browser, including computers and handheld devices. The programming would be done using the Blazor web framework based on HTML, CSS, and C#, with JavaScript interoperability. This was chosen because of the similarities with MXML and ActionScript, such that the structure of the code developed so far could be largely reused. And finally, the integrated development environment (IDE) was Microsoft Visual Studio.

The 2021 concept of the browser-based version contained the most basic functionality, similar to the earliest versions, but the new format offered a simpler and cleaner user interface. This version was tested by Hollnagel and others and was even used for beginner FRAM courses. The functionality was slowly being built up with a more concerted effort starting in 2023 in preparation for the resumption of in-person FRAMily meetings as travel restrictions lifted.

In March 2023 Hollnagel's vision of a truly open-source FRAM software was fully realised, as the FMV software development moved to GitHub under a GNU Affero General Public License. This ensures that the code remains visible and available to anyone with the desire that future developments could build further on the work already accomplished while acknowledging existing copyrights. Using the FMV GitHub open-source project as a starting base would eliminate the need to reinvent core functionality and would ensure that the FRAM software maintained the standards of FRAM building and interpretation that had been developed and tested over the past 10 years. The GitHub open-source project and the FMV Community Edition, with integrated FMI, were announced at the June 2023 FRAMily meeting in Copenhagen, Denmark.

SANDBOX EXPERIMENTATION

To encourage further research and experimentation, especially with regard to quantification and analysis, another fork from the FMV GitHub code was created, while maintaining the FMV Community Edition as a stable public release. The new fork was named the FMV Analysis Sandbox and extends the FMI functionality and utilises the Metadata of the Functions to define and control the instantiations.

The default stop condition for the basic FMI is when all of the exit Functions have been activated. In a complicated model with feedback loops, this stop condition may occur before all of the loops have been tested. Without considering if a given model could be structured differently, it was decided to give more flexibility within the

Sandbox to explore complicated model structures. An override was added where the FMI Cycles could continue past the default stop condition for either a set number of cycles or infinitely until manually stopped. Another option to control the cycles was added to allow an entry Function to activate not only on cycle 0, at the beginning of an interpretation, but on a repeating number of cycles. This now allowed for other types of analysis where instantiations could be repeated automatically or testing for overlapping and interacting data, like one patient after another entering an emergency department.

The FMI interpretation profile allowed for the setting of activation conditions for any given Function, such as if all, any, or none of each type of aspect must be present to produce an Output. The use of Metadata now made it possible to pass on information about the upstream Function through the aspect coupling such that the nature of the coupling could be considered by the downstream Function. A special "Activate" Metadata key was added to the Sandbox that could use an equation to reference the Metadata available from all of the aspects, and if the Function was activated according to the interpretation profile, it could then further calculate if the Function would produce an Output on that cycle. If not, it would hold and would again consider the activate calculation on the next cycle. Further, if the Function had more than one Output and was to activate, it could also direct which Output was produced on that cycle. This created options in modelling instantiations for conditional activation "go or no-go" and conditional pathways "plan A or plan B".

Reusing the colour bars from the original variability tracer outputs, the Sandbox was set up to display the dynamic data from up to six different Metadata values on individual Functions such as colours and text. These change dynamically as the FMI graphics progress.

The FMV Analysis Sandbox version was shared at the June 2024 FRAMily meeting in Lund, Sweden, and was used to present analysis results for some of the research presentations.

WHAT IS POSSIBLE WITH METADATA

To analyse different instantiations of a model, usually, the start conditions or the properties of background Functions are changed to represent the desired parameters. The FMV/FMI and Metadata combination allows for the continual cycling of the Model with changing parameters that can represent successive instantiations. This could be used to test specific "what-if" scenarios or to represent a Monte Carlo method to discover an emergent pattern.

Method Geek

The initial setting or dynamic setting of Metadata can be managed in a number of ways:

- Manual: The data is entered directly into the model and the Interpretation is initiated. Start parameters can then be changed manually and the Interpretation re-run.

- Equations: Conditional statements in the Metadata can change the values based on the cycle count, other counters or other calculations. The model can be set to repeat for a number of cycles or continuously.
- Random: The equations can reference several types of pseudo-random Functions, such as double-precision numbers or integers over a linear range or a defined normal distribution.
- Spreadsheets: The software can read Metadata values from spreadsheets in a CSV format. These can be generated from other software that is performing pre- or post-processing of FRAM models or created specifically as a lookup table or cross-reference source. Because these are user-generated, they can be used to supply random data or specific scenarios.
- URL or API: Metadata can be sourced directly from web resources or web-based Application Programming Interfaces. Easily obtained examples include weather data and aircraft movements.
- User-generated algorithms: The FMV sandbox contains a number of built-in Functions and basic maths Functions available to the user, the most powerful being the conditional statements. A simple demonstration of the available complexity is explained in the "Boil Water" example of Hill's Metadata Approach to Extend the FRAM (Hill & Slater, 2024). In addition to the built-in Functions, because the FMV code is open source, it is possible for advanced users to create their own algorithms for processing Functional Metadata during an FMI Interpretation.
- Direct real-time links and digital twins: Perhaps the most exciting possibility is the ability to link directly to real-time system sensors via direct or indirect digital links. The ability of the FMV/FMI to cycle continuously means that the Metadata and predictions are available continuously. An example of this was in the modelling of the digital twin of the river Usk, which could Monitor the flows and weather data in real time to help improve the passage of migratory fish (Slater & White, 2016).

Because the Metadata is defined by the user, the FMV does not direct any particular method of analysis but rather, by maintaining an open configuration, allows for the possible use of many different forms of interpretation and analysis. This is in keeping with the purpose of the FRAM, which is a method to build a model of how things happen, not to interpret what happens in terms of a pre-existing model. By combining FRAM with the FMV and Metadata, it should be possible to metamodel any complex system and define a meaningful analysis.

RESILIENT SYSTEM MODELLING

Since the publication of Hollnagel's Safety-II in Practice, the Resilience Potentials (Anticipate, Monitor, Respond, and Learn) with their example FRAM models have served as a handy reference for FRAM model builders that wish to look for resilience in their own systems or to build resilience into their systems (Hollnagel, 2017). This is very much in keeping with the fourth step of the method, to propose ways to manage the possible occurrences of uncontrolled performance variability. The Resilience Potentials loop back on each other and create a self-reinforcing system that has the ability to Function as a continuous improvement cycle. If we can Learn from our Response, however reactive, then perhaps we can anticipate what may happen in the future and modify what we Monitor for and react to so that future Responses may be more proactive, preventative, or even predictive.

The ability to loop back and cycle a FRAM model, in a sense using feedback loops to instil a form of feed-forward control, can be modelled using the extended FMV/FMI and Metadata. Using equations to modify the Metadata it is possible to set up a form of machine learning that continually adjusts parameters to better predict outcomes of the model's instantiations. A FRAM-based digital twin created using this approach opens up a wide range of applications, particularly where the systems are too complicated to obtain sensible overviews. Since the FRAM model tracks the Metadata continuously, the observed behaviours can be logged, stored and analysed for trends and patterns. Similarly, historical data can be loaded to pre-learn, process and predict behaviours to enable more intelligent feedforward control and management of the system. In the River Usk example, 10 years' worth of historical river flows and weather data was used to give week-in-advance predictions of expected conditions. In another example, a million days' worth of constructed data was used to train a resilience loop to better predict the number of hospital beds available each day for elective surgeries. But the advantage that an FRAM model has over neural network learning is that instead of a "black box", the analyst can see directly the relations between learned patterns and critical Functions in the FRAM visualisation. "What if" tests of system reactions to challenges and improvements become a lot more meaningful.

FUTURE DEVELOPMENTS

With open-source code and community ownership, it is hoped that the future development and use of FRAM software will be assured. At the time of writing, a number of users are developing FRAM analysis methods for their specific use cases. The GitHub code repositories under the organisational banner of Functional Resonance are overseen by a committee of key FRAM researchers and users. As was the case in the past, it is always the intention that new developments that may be considered consistent with the principles of the FRAM method and of general use to the wider community of FRAM model builders would be built into and released under the FMV Community Edition.

As always, further development, coding, experimentation, research, and peer review are to be strenuously encouraged. This book emphasises that perspective.

APPENDIX 19.A

FMV development timeline. This table provides a structured summary of the FMV development timeline, detailing version numbers, dates, and associated updates.

Date	Version	Features/Updates
Feb 2014	0.0.1	Opens FMB data file (XML), Functions and Couplings (straight lines), Orphans, and Repositioning of Functions with dynamic Couplings.
	0.0.2	Curved lines (using two quadratic Bezier curves), Style property to colour Functions, Save Project/Image, Zoom.
Mar 2014	0.1.0	First public version, automatic resizing of the visualiser window, curve improvements, vector graphics for hexagons, smooth zoom.
	0.1.1	Improved Function label text.
	0.1.2	Embedded fonts for consistent cross-platform appearance.
Apr 2014	0.1.3	Added PDF report output.
	0.1.4	Creation/editing of Functions without relying on FMB.
	0.1.5	Vertical scroll bar for Function description pane, Experiment with "Modern" rendering.
	0.1.6	Manual "Modern" rendering for each Function.
May 2014	0.1.7	Improved performance for moving Functions, added Function types (Human, Technological, and Organisational).
	0.1.8	Selection of variability based on Function type, automatic Foreground/Background identification, and rendering styles.
	0.1.9	Playback Functionality for manual step-through.
June 2014	0.2.0	Bug fixes, public release for Gothenburg FRAMily meeting.
	1.2.0	Experimental quantification methods, added a "Variability Tracer" using heat flow modelling.
Nov 2014	0.3.0	First version is to add globalisation awareness with language selections for on-screen text.
Dec 2014	1.3.0	iDepend connection to BBN analysis.
Mar 2015	0.3.1	Public version for Olten FRAMily meeting.
June 2015	0.3.2	Bug fixes and additional languages
Nov 2015	0.4.0	Additional languages and minor improvements.
	1.4.0	Alignment to 0.4.0 with additional BBN analysis.
May 2016	0.4.1	Public version for Lisbon FRAMily meeting.
Aug 2017	0.4.2	Public version for Cardiff FRAMily meeting.
Sep 2018	0.4.3	New test version with minor updates.
Oct 2018	1.4.3	New extended test version with minor updates.
Dec 2018	1.5.0	End of iDepend BBN connectivity development. First use of grouping, exporting and importing.
Jan 2019	1.5.1-1.5.3	First Metadata version, with improvements and bug fixes from testing feedback.
Feb 2019	1.5.4	Optimisation and performance improvements.
Mar 2019	1.5.5	Updates to specifically target Apple Mac computers in addition to Microsoft Windows and Ubuntu.
	2.0.0	First Pro version with Metadata, combining free/public and Pro versions.
May 2019	2.0.1-2.0.2	Public release for Malaga FRAMily meeting; added undo/redo Functionality.

(Continued)

Date	Version	Features/Updates
June 2019	2.0.3	Improved performance for larger models.
Sep 2019	2.0.4	Additional languages
Mar 2020	2.0.5	Serious bug fix from deleting Functions when editing models.
Jun 2020	2.0.6	Minor but fix from editing certain aspects.
Jul 2020	2.0.7	Integrated FMI (basic V3, basicPlus V1) into FMV Pro.
Sep 2020	2.1.0-2.1.2	Public release with new manual and extended Functionality, with improvements and bug fixes from testing feedback.
Oct 2020	2.1.3-2.1.4	Added "The Lab" for experimental features, including variables and calculations in Metadata.
Mar 2021	3.0.0	First web-based version enters testing.
Apr 2021	2.1.5	Demonstrated extended Pro features and Web concept.
Aug 2021	2.1.6	FMI performance improvement.
Jan 2023	2.2.0	Final Adobe-based version.
Mar 2023	3.0.1	New browser-based interface development moved to GitHub.
Jun 2023	3.0.2 3.0.3	Demonstrated Web version with integrated FMI at Copenhagen FRAMily meeting. Minor improvements and bug fixes.
Aug 2023	3.0.4	Fix globalisation awareness for different number formats and other minor improvements.
Sep 2023	3.0.5	Bug fix on selected couplings text transform.
Jun 2024	3.0.6	Community Edition and Analysis Sandbox used for workshops and presentations during the Lund FRAMily meeting.
Jul 2024	3.0.7	Added reporting similar to original PDF output and added descriptions field, consistent with the 1.X and 2.X versions.
Aug 2024	3.0.8	Changed "Floating" render to be "Greyscale" for publishing use. Added model image to the end of PDF output.
Sep 2024	3.0.9	Added new 2024 icon render. Enabled multiple selections of Functions on the model and on the spreadsheet view.
Nov 2024	3.1.0	Published as the stable production version.

NOTE

1 Voluntary and without payment "for the public good".

REFERENCES

Albery, S. (2014). *An evaluation of the functional resonance analysis method (FRAM) as a practical risk assessment tool within a manufacturing environment*. Royal Melbourne Institute of Technology, School of Health Sciences. Master of OHS Thesis.

Hill, R., Slater, D. (2024). Using a metadata approach to extend the functional resonance analysis method to model quantitatively, emergent behaviours in complex systems. *Systems*, 12, 90. https://doi.org/10.3390/systems12030090

Hollnagel, E. (2009). *The ETTO principle: Efficiency-thoroughness trade-off: Why things that go right sometimes go wrong* (1st ed.). CRC Press. https://doi.org/10.1201/9781315616247

Hollnagel, E. (2012). *FRAM: The functional resonance analysis method: modelling complex socio-technical systems*. Ashgate Publishing.

Hollnagel, E. (2014). *Safety-I and Safety-II: The past and future of safety management* (1st ed.). CRC Press. https://doi.org/10.1201/9781315607511

Hollnagel, E. (2017). *Safety-II in practice: Developing the resilience potentials* (1st ed.). Routledge. https://doi.org/10.4324/9781315201023

Patriarca, R., Bergström, J., Di Gravio, G. (2017). Defining the functional resonance analysis space: Combining abstraction hierarchy and FRAM. *Reliability Engineering & System Safety*, 165, 34–46. https://doi.org/10.1016/j.ress.2017.03.032

Patriarca, R., Di Gravio G., and Costantino, F. (2017a). myFRAM: An open tool support for the Functional resonance analysis method, *2017 2nd International Conference on System Reliability and Safety (ICSRS)*, Milan, Italy, pp. 439–443. https://doi.org/10.1109/ICSRS.2017.8272861

Patriarca, R., Di Gravio, G., Costantino, F. (2017b). A Monte Carlo evolution of the functional resonance analysis method (FRAM) to assess performance variability in complex systems. *Safety Science*, 91, 49–60, https://doi.org/10.1016/j.ssci.2016.07.016

Patriarca, R., Simone, F., Artime, O., Saurin, T., Fogliatto, F., (2024). Conceptualization of a functional random walker for the analysis of socio-technical systems. *Reliability Engineering & System Safety*, 251, 110341. https://doi.org/10.1016/j.ress.2024.110341

Slater, D., White, C. (2016). FINAL REPORT, Developing a Real Time Abstraction & Discharge Permitting Process for Catchment Regulation and Optimised Water Management. InnovateUK File Reference No: 101341.

20 FRAMing Learning
Learning about FRAM through Experience

Pedro Ferreira

CONTEXT: ABOUT LEARNING

Part of the FRAM's uniqueness is its openness to model virtually anything and for whatever purpose. It imposes no strict step-by-step process nor preconditions about what is to be modelled and why. This has been referred to as the FRAM being a *method-sine-model*. The absence of a pre-established model favours a rich learning experience, as modelling activities tend to unfold in a more explorative way. Findings are inevitably bound by the tools being used (and their underlying models), but the FRAM seems to offer valuable means to explore the unknown and alternative perspectives. The FRAM downplays the common saying that "what you get is what you look for" (or "when the tool you have is a hammer all problems look like nails").

On the flip side, such an open nature is often criticised for making the tool excessively exposed to the inevitable human subjectivity. Assumptions and biases are often described as an undesirable influence. They are, however, deeply embedded in every aspect of human performance, even when objectivity is presumed. Above all, biases and assumptions are particularly hard to be perceived by the one who holds them. In the case of the FRAM, its openness simply means that the assumptions at play are those of the person or persons undertaking the modelling activity, as opposed to those normally implicit in the tools' design and underlying models. The point then is not so much to debate whether or not the FRAM becomes more exposed to the influence of human subjectivity than other more normative tools. The focus is rather to pursue a better understanding of how modellers' assumptions are at play in different uses of the FRAM and the extent to which users are explicitly aware of their own assumptions.

Becoming aware of one's own assumptions can be a powerful learning experience. In the case of the FRAM applications, surfacing assumptions improves clarity on the modelling scope and objectives, as much as on the multiple relations (couplings) that are being modelled.

As illustrated by Shorrock (2020) and by Patriarca et al. (2021), the reality of work is unattainable. Whatever the model or representation of work, it only amounts to a particular perspective of that work, or a combination of multiple ones. A FRAM model reflects the perspective of those developing the model and of those providing the data input to the modelling exercise. And no matter how comprehensive the FRAM exercise may be, it will also inevitably leave out many more alternative views about work.

The principles of Work-As-Imagined (WAI) versus Work-As-Done (WAD) can be useful to illustrate the wide spectrum of perspectives that live in between how work is formally described and how it is perceived by those who observe it and even by those who perform it. They are, however, a quite reductionist view of a very dynamic and diverse world of views and understandings about a given work. The FRAM can offer valuable awareness about this diversity and, in doing so, drive a unique sharing and learning experience for all those who engage in the FRAM exercise.

This chapter presents an exploratory approach aimed at better understanding how the openness of the FRAM is perceived by its users and how it is indeed seen as the source of valuable learning and discovery. While not delving too deep into the understanding and awareness of prevailing assumptions and biases, the purpose here is to better understand how people from different industry sectors, different areas of organisations and different professions have come to use the FRAM and how they perceive the value that it has generated. "Value" is here interpreted in terms of the learning experience that emerges, not only for those applying the FRAM as a modelling tool but also for those with whom modellers may come to interact at some stage. Open-ended interviews were conducted with a pool of FRAM users and researchers. Interviewees were selected on the basis of their published or documented work with the FRAM and their experience across different industry sectors.

THE INTERVIEW APPROACH

The overarching goal of the interviews was to capture different "life stories" around the use of the FRAM. The purpose was not to validate specific hypotheses but rather to explore a number of questions related to how the FRAM is perceived by its different users.

These questions can be summarised as follows:

- How is the FRAM perceived by its users?
- How the use of the FRAM changed their views and work?
- How learning through the FRAM perceived?
- What challenges were faced over the years, and how were these dealt with?

A total of seven interviews were conducted with people who have used the FRAM in a more or less regular way and over a period of several years. All interviewees had over 5 years of experience in the application of the FRAM. Particular care was taken to select people with diverse industry and training backgrounds. Most interviewees originated from an engineering discipline and had applied the FRAM across different sectors, namely healthcare, aviation, maritime, and process industries, among others. A certain level of training and/or experience in the domain of human factors was also prevalent. Table 20.1 outlines the key traits of the interviewees.

Interviews had an approximate duration of 45 minutes to one hour. A qualitative analysis of their contents was developed on the principles of thematic analysis. A search for more frequently used words and expressions was also conducted. In the following section, findings are presented around the main themes extracted from the interviews.

TABLE 20.1
Interviewees' Profile to Document the Usage of the FRAM

ID	Background	Years of Experience Using the FRAM (years)	Domains of Application
Int. 1	Mechanical Engineering, Public Quality, and Risk Management	12	Healthcare
Int. 2	Civil and Environmental engineering, Human Factors, and Safety.	6	Oil and Gas, Maritime
Int. 3	Human Factors, Shipping, and Marine Technology	14	Maritime and Shipping
Int. 4	Engineering and Naval Architecture	8	Shipping, Health Care, and Multi-level Governance Structures
Int. 5	Aerospace engineer, Safety, and Resilience Researcher	10	Aviation, Industrial Ops, Healthcare, and Maritime
Int. 6	Aeronautical Engineering, Aerospace Engineering and Safety Engineering	9	Space, Car, Ship, and Chaos systems
Int. 7	Aviation, Human Factors, and Accident Investigations	9	Aviation, healthcare, and Collaborative Robotics

FINDINGS FROM THE INTERVIEWS

Like most explorative approaches, the interviews followed different paths and offered insights on a wide diversity of issues. Rather than aiming to present a thorough and comprehensive account of the interview contents (out of scope here), only the aspects relevant to the scope and objectives of this chapter are highlighted. The questions previously presented were used as guidance. Whenever useful, quotes from interviews are given as an illustration of each point being argued. The use of Method Geek sections indicates a direct quote of sentences taken from the interviews. Occasional usage of Italic format in the main text indicates a quote of shorter expressions from interviews as well.

EXPLORING PERSPECTIVES THROUGH RELATIONS

The idea of "offering a different perspective" on events and operations was quite salient across all interviews. More than pointing towards a different direction, this was described by interviewees as opening up a range of other possibilities.

Method Geek

When a root cause analysis showed that you should train your people or prepare a procedure, then the FRAM showed 20–30 more possibilities (…) a much broader perspective, many more solutions (…) and to discover things you haven't seen.

It's something more oriented perhaps to an increased understanding of how things work in opposition to what most methods or tools, particularly in the safety world, are designed for and used for, which is finding solutions.

One of the key ways through which the FRAM seems to help discovery is through the facilitation of communication with and between "frontline workers" or the people who actually do the work or activities being modelled.

The position of "managers" towards FRAM was described in a very different way. They often require more of a convincing effort. There was a general view that perceptions around responsibility and accountability push managers to seek safety through processes, which often deters many "bottom-up" approaches such as those that the FRAM tends to entail. This also leads people in managing positions to sense and fear any form of variability, which becomes a core issue when applying the FRAM.

Method Geek

The concept of variability is huge; it's the one that most managers struggle with, I think, because they don't want variability (...) if it's there you want to try to constrain it all the time.

Discussions on this topic followed the notion that often those in senior management positions tend to undermine the prevalence of any variability, and if acknowledging it, they expect tools that provide a direct solution to manage it. This challenge was also expressed from the perspective of "control". The fact that embracing the FRAM leads to considering a wide range of aspects that are beyond the control of managers was also mentioned in the course of several interviews.

Method Geek

The FRAM shows you all the things that may somehow be out of your reach. And that can be very scary for a manager who's supposed to show and demonstrate that things are under his control.

I think it is a bit scary for managers because they want to use processes that reassure them that they've covered all bases, whereas FRAM never allows you to do that.

There are going to be some things that you just haven't seen yet, and there are functions there that are very difficult to control. They're actually controlled by someone other than you. They're controlled by the operator.

The FRAM can also facilitate a conversation around difficult points of tension, namely those at which control and the power that sits with it tend to emerge. In one particular context, the FRAM was described as enabling the *understanding of polycentric control*.

On the other hand, when adequate openness was found with management, FRAM was described as providing a tangible way of presenting problems. *You have a platform where you can agree with management.* It may be more difficult to get those in a position of management and responsibility to "sit at the same table", but FRAM seems to enable building from the ground up by first improving on "horizontal" relationships before addressing vertical ones. Across all seven interviews, there were some 35 references to the ideas of conversation (18 ref.), dialogue (13 ref.), and relationship (4 ref.). All of these references were made in the context of the modelling activities and the group or teamwork that is needed to suitably apply the FRAM.

Method Geek

But you give them a way to critically look at what they're actually doing and how rich that is. And it always also gives them a way to go back to their management and disagree with some things and say, hey, this could be done better in the real world.

The functional aspect seems to make it easier for people to situate themselves (their work) in the model and therefore also see how they may or may not relate to other activities. Perhaps more than helping to negotiate through the control and power tensions, the FRAM seems to help develop a higher appreciation for different perspectives and for the diversity of local realities.

Method Geek

So it helped me [the interviewee as modeller] understand what is the system that I actually look at and what the challenges people meet in everyday life are. But it also helped people [participants in the modelling exercise] to explain their work to people making the rules about their work. When we interview a bunch of them from multiple systems and then show them a FRAM map, they get this better situational awareness of the system that helps them understand the complexity and maybe have a little bit more empathy for the frustrations they see on their own day-to-day.

This also comes in line with the idea previously introduced about the FRAM opening up new possibilities. Rather than pointing towards a solution for a given problem, the FRAM can broaden the understanding of problems by exploring multiple perspectives and local realities. This, in return, can strengthen relationships that span across different types of organisational boundaries, as well as hierarchies.

A PARTICIPATIVE AND INTEGRATIVE PROCESS

The engagement with those doing the work, in particular operational persons, was considered a fundamental condition for the application of the FRAM.

Method Geek

FRAM requires a high level of engagement by the operators. If you want to do it well, (…) It requires a lot of engagement.

From my experience, when you use the FRAM, you have this kind of benefit from the people you interact with.

This engagement was described in different ways, but there was a shared perception that a FRAM application ought to be, as much as possible, a group exercise. One of the first indications of this comes from the language that dominates all the interviews. The vast majority of the examples given on work with the FRAM were given in the form of "we", even though some occurrences of "us and them" type of phrases were also registered. An example of an unrealistic scenario was described as *a consulting company coming in and showing people FRAM models without their involvement in it.*

The FRAM was frequently described as a tool that allows for the mapping of knowledge about systems, processes and operations. This knowledge was described as a combination of documented information, interviews with those undertaking the work, and the interpretation of those developing the model. While some stated relying first on the documented information to generate a broader level of understanding of operations, others would tend to prioritise direct contact with operations and work. Despite the different approaches, interviewees unanimously valued the need for an extensive iteration between the different sources of knowledge and their interpretation. This iterative approach enables the integration of multiple contributions and their sources, but also the refinement of the model to be developed.

Method Geek

I think that's another strength of the FRAM (..) it requires you to participate in it so that through your participation, you're constantly exposed to the underlying, fundamentals of it.

While considered as seldom realistic, one ideal scenario was described as the engagement and active participation of those providing the different types of knowledge during the entire modelling exercise. Regardless of whatever degree of participation and integration is made possible throughout the modelling exercise, the validation and calibration of the model through the different sources of knowledge were deemed fundamental for the FRAM.

Method Geek

They're part of creating the model too (…) it's essentially holding a mirror up to them saying this is what you told us and if they tell us, I think this is a little bit off we end up adjusting it.

This was also described as an additional challenge which can become difficult to manage, particularly when dealing, for instance, with highly automated and compartmentalised work. Workers are likely to provide you with very different perspectives, but to some extent, these are inevitably fragmented ones. This makes calibration substantially more difficult. *How do you communicate your model back?* The answer to this could be the case of simply having to navigate through different terminologies. But as conversations and interactions develop around the FRAM exercise, it could emerge that people may share language and terminology whilst having developed a significantly different understanding about them. Subcultures are not uncommon across all kinds of organisations, and sometimes more than the perspective people share about their work, the language they use to describe that perspective can be quite revealing of cultural divides.

WHERE TO BEGIN

For the majority of the interviewees, the first contact with the FRAM came out of a sense of frustration with recurrent shortfalls regarding available tools, in particular how *they force you to think in a certain way*. This comparison with other tools was consistently undertaken in the domain of safety. Indeed, a broad notion of safety appears to remain the main driver for the use of the FRAM. Discussions with interviewees went on to explore the growing perception of recurrent problems for which the approaches they were familiar with did not provide suitable answers. Hence, the use of the FRAM seems to have started off from the question, *what am I not seeing*? Risk and risk assessment were mentioned in six of the interviews but never discussed as a motivation for the use of the FRAM. Most interviewees do not see the FRAM as a tool that, on its own, can suitably address risk assessment needs. Rather, through an FRAM model, an improved approach to risk and risk management can be pursued.

The application of the FRAM remains very much driven by different kinds of research interests and questions. Even in the cases that some interviewees presented and where there was an industry application, the scope remained closely related to the exploration of alternative approaches and solutions. This becomes more evident through several references made to the predisposition towards the learning that must be placed at the start of a FRAM exercise.

Method Geek

So you first need to have in your mind the willingness to learn from people. (…) if you don't come to study with a mind open enough to say <I'm here because I want to learn from people, I want to experiment (…)>

All interviewees highlighted the importance of a "FRAM expert" or a trained person that leads the modelling exercise. This importance was again justified by the absence of an underlying model and a univocal step-by-step process to be followed. Having more than one person in the modelling team with the FRAM experience was considered a valuable contribution, in particular for the interpretation of the different data sources.

Having a clear purpose was considered a foundational step for the application of the FRAM, although different degrees of importance were attributed to the detail that such purposes ought to be given.

Method Geek

The job is to find the purpose of the FRAM because that's the only way you can limit the size of the FRAM. If you don't have a purpose, it will grow and grow and grow and grow.

Establishing a purpose was defined as requiring a broad understanding of work and its context. All interviewees considered that formal aspects of work (i.e., processes and procedures) are important for the definition of purpose. However, complementing and clarifying this knowledge must always be sought through direct contact with workers.

Method Geek

I have to make some hypotheses before, so I can even lead the interviews in an appropriate direction.

From this point onwards, different strategies were outlined by interviewees. There was, however, a common trait across all the interviewees, which pointed towards a more or less systematic, direct and comprehensive engagement with the "reality" to be modelled. When aiming to model larger-scale scenarios or less familiar ones, techniques like hierarchical task analysis were mentioned as providing a helpful way to organise information before approaching FRAM modelling.

Method Geek

And if it was a system that I had absolutely no understanding of, maybe I would have to start with an interview with a person within that system just to get up to speed with the terminology and the general flow of the process. (…) and then use that as a basis to generate interview questions.

When faced with a large-scale modelling scope, most interviewees posited that it is necessary to first focus on a particular and smaller part of the system in question. Even if nothing really prevents large-scale modelling, being too ambitious at this starting stage is likely to bring about challenges when attempting to extract from the model meaningful lessons and guidance towards any system changes. The FRAM provides a wider system perspective that can facilitate the prioritisation of system areas to focus on and even shift the foreground on the model. This can then enhance the understanding of how local actions and changes can impact the system. The *work on the small solutions that could ensure that people are supported to succeed*, whilst perceiving what "sacrificing decisions" or challenges this may bring about in different areas of the system.

ASSUMPTIONS

The influence of different types of assumptions was recognised by all interviewees. The absence of an underlying model renders the FRAM less exposed to design assumptions. The majority of interviewees agreed, however, that this tends to make assumptions from the modellers more prevalent. It is therefore particularly important to explore assumptions.

Method Geek

It's about taking a step back and realising there is something that you're hiding from yourself.

Assumptions were mainly associated with the interpretation of data sources when looking to identify and describe functions. Overall, interviews suggest that assumptions predominantly influence the inevitable "filtering" of what is to become highlighted by the model and what is to be omitted or placed in the background. FRAM is not so much about the stages for building a FRAM model but rather about what data, information and knowledge are placed at its basis and how this is *translated* into FRAM functions and its aspects. Interviewees indicated a general tendency to use open discussions with and between workers so as to determine what is important for them and, as such, ought to be represented in the model.

The model is built *based on your perceptions of what they* (those doing the work) *told you*. Some interviewees referred to this as a kind of *translation* or a *transformation* process. There is a knowledge transformation process that goes from what one interviewee called *primary knowledge*, which would be the perspective conveyed by those who do the work. There is then *secondary knowledge* that refers to the interpretation of the person or group developing the model. From this perspective, there was a general perception that ultimately, the "reality" of work is inaccessible and can only be interpreted by the modeller.

Although not directly related to prevailing assumptions, three interviewees further discussed the importance of having a language and criteria for modelling. This was considered the fundamental basis for the *translation* of knowledge and data collected into functions and their aspects. While some described this translation as a very direct process, some recognised challenges in understanding the language of those providing knowledge and data and creating a suitable representation of that language in the model. The importance of these challenges seems to vary considerably depending on the scope and context of the modelling exercise. Regardless of those context-related variations, functions seem to become symbols of different activities, and the value of FRAM appears to be closely related to the perceptions that are built around these symbols.

The exploratory nature of a FRAM exercise may also amplify the influence of assumptions. In the views of one interviewee, this is, to a great extent, an exploration of one's own assumptions, as much as *those of people who actually do the work, and they don't even realise how many assumptions they have. It's a very rich exploratory work.*

All interviewees considered that the FRAM exercise often surfaces assumptions about different operational aspects. Assumptions emerge not just from the modellers but also from all those participating in the FRAM exercise.

Calibration of the model in direct contact with those who provided the data and knowledge at the source of the model was described as a useful strategy to surface and clarify assumptions. Applying the FRAM as a team exercise was also considered an important contribution, and all interviewees expressed a preference for this approach, as opposed to developing a model on their own.

Method Geek

I also try to make the modelling activity a team thing, so there are a few people who know how to use the FRAM around me, and we might build the FRAM slightly differently from the same data.

When a team exercise is not possible, presenting and explaining the model produced to others was described as a useful source of validation. Once a model is sufficiently developed, *the FRAM allows you to interact with your peers and explore tacit knowledge.* This offers a different perspective that often leads to the review of the model. Discussions with colleagues and workers can often focus on the nature and description of function aspects. For instance, considering a given output as either a control or a precondition could start a reflection on multiple other couplings with many other functions. In the end, interviewees agree that this is not so important as long as a shared understanding is reached regarding how functions and their aspects reflect the understanding of the observed operations. This, of course, further emphasises the critical role of group work and discussions around different stages of an FRAM exercise.

Method Geek

I noticed after having these long discussions with myself and with other FRAM builders who were working on the model that it doesn't really matter as long as you really know what this function is doing in the real world or in your real system.

Assumptions were also considered from the experience with other tools. At least two interviewees highlighted the importance that the FRAM has had in developing a different perspective towards other methods. In particular, one interviewee mentioned that:

Method Geek

Working with FRAM has helped me better understand assumptions underlying other methods. (...) The devil is in the detail. It's what they don't tell you in the assumptions which they make.

Naturally, assumptions and their influence will always be very context-related and every modelling exercise will likely exhibit different degrees and types of influence. All interviewees considered assumptions to be a natural and inevitable prevalence in the application of FRAM, and no particular concerns were expressed in this regard.

Challenges and Benefits of Using **FRAM**

Over the years, the FRAM has often been criticised for lacking clear application guidance and a modelling approach. Most interviewees acknowledged this as an aspect that may have discouraged a wider use of the FRAM, although considering that the "originalities" of the FRAM largely compensate for most alternative tools. One of such originalities was considered to be the much more dynamic nature of a FRAM model. Nevertheless, one interviewee explicitly referred to this as also a source of additional challenges, as what is shown on the model is merely an idea or projection of possible dynamics, as opposed to the actual dynamics of operations and processes. Here lies a certain requisite for imagination that an FRAM model encourages us to explore. While the notion of variability embedded in the FRAM was valued by all interviewees, most also recognised a challenge in capturing the specific dynamic trends that variability may assume.

Method Geek

The model is a picture of all the potential (variability), it's not really a tangible picture. That's overwhelming, and we can't appreciate it by looking at that map.

All interviewees valued the openness of the FRAM, as it offers ample opportunities for application within virtually any context and for any given purpose. While this was valued, it was also mentioned as a source of poorly substantiated or less appropriate application of the FRAM. Because no other guidance is available, adherence to the generic principles for the use of the FRAM becomes critical. In this context, all interviewees considered to be fundamental the integration of people with relevant knowledge and experience in the application of the FRAM within the team developing the modelling exercise.

Several interviewees further added that a clarity of purpose for the application of the FRAM was fundamental. The articulation of a modelling purpose forms the basis for a distinction between a model foreground and its background, supports the definition of criteria and language for the description of functions and their aspects, and also forces careful consideration of the approach and methods to use for the collection of data that will then feed the modelling exercise. Consistency in the definition of criteria and language becomes critical to determining how the understanding of work can be best "translated" into functions and their aspects. Most interviewees considered that the way in which function aspects are defined is sufficiently clear. However, difficulties and *overlaps* often occur when considering how to best represent what is perceived as important or relevant about the work being represented. The "Time" aspect was considered by several interviewees as particularly challenging, not only because it is defined in a more open way but also because *most other aspects also introduce a certain sequencing of functions.*

One interviewee further considered that the absence of a clear purpose is at the source of many applications of the FRAM that, rather than modelling a given perspective of a work situation or a system, end up modelling instantiations of something that is not necessarily clear. Clarity of purpose also supports the recognition of what is often referred to as "the stop rule". On the one hand, *FRAM forces you to go further than most other methods and tools*. Having a declared understanding of what is being pursued helps to recognise a suitable degree of detail in the model developed. On the other hand, this purposefulness drives the exploration of more meaningful questions. For instance, one interviewee gave the example of how it is often tempting to consider the nature of function couplings as a mere coordination issue. Having a clarity of purpose forces discussions around the FRAM modelling to question (i.e., *What does it mean to coordinate?*) this generic view of coordination, namely by looking at a higher granularity of the model being developed. In the end, purposefulness supports the balancing of the modelling exercise by bringing breadth against depth.

Method Geek

I think you have to go down as deep as possible in your granularity on the one hand, but on the other hand, it becomes very complex, and you get these expanding models.

Some of the interviewees identified "the closing of feedback loops" as an indicator of the completeness of a model. Once a model achieves a significant level of development, key processes and elements of "work reality" that provide important inputs to workers start to become apparent. Recognising that these feedback loops are not adequately represented in the model suggests that either more information about work is needed or the model must be revised. This revision could then be at the level of function definition or aspect description, or even of the criteria and language that are at the base of these definitions and descriptions.

Method Geek

Even when interviewing work experts and specialists, there are things that cannot be captured. Once that information is translated into a FRAM model, problems or unclosed loops become a lot more apparent and offer the opportunity to either clarify what was not captured during the interviews or something that is missing.

Sometimes I need something like a dummy function to make the model work and a dummy function to activate it. (...) but it doesn't feel right. Sometimes it makes me go one step back and say this is because I took the wrong direction, and I have to rethink it. I actually don't need a dummy function if I structure my functions right.

Establishing a relation with those providing the information and knowledge about the work was considered a key aspect by all interviewees. *Earning the trust and respect* of those to be interviewed was frequently mentioned as fundamental or *indispensable*. Interviewees had different experiences regarding the building of these relations. When implementing interview approaches, some industry sectors were described as perhaps offering more *resistant* than others. Several possible reasons for this were put forward, namely the prevalence of a more hierarchical structure of organisations, more conservative cultures, strong and rigid regulations, among others.

The FRAM provides a background on systems' understanding that can then render the use of other analytical tools much more effective. The FRAM can offer an important basis for the improvement of causality analysis, risk assessments, or the pursuit of other probabilistic reasonings. At least two interviewees advocated that such approaches are often flawed precisely because they lack a system understanding, such as the one that the FRAM can help develop. The language that emerges during interviews and the one then used to describe functions was considered to have a significant influence across the different steps of a modelling exercise. On the one hand, several interviewees acknowledged that challenges could emerge when relating certain context-related language and terminology with the description of functions and their aspects. Ensuring that workers recognise their "reality" in the functions produced may often require a number of iterations around consulting with those workers and function revision. On the other hand, the FRAM terminology was described as accessible for those related to operations, namely because it is *process-based*. This suggests the pursuit of a compromise between modellers and workers around an acceptable degree of precision in the language and a level of detail in the model.

LEARNING EXPERIENCES

In general, all interviewees valued the learning from the experience of developing the model much more than the model itself. This was described as *the journey to arrive at the model*. Once again, the dialogue between those developing the model and those contributing with the knowledge (i.e., workers) was placed at the core of this learning.

Method Geek

The collaborative aspect of building FRAM models is that the users or the operators end up feeling involved and engaged and they end up they absorb some of the FRAM knowledge from you.

There were numerous references to the learning value of engaging with people, establishing a dialogue, and processing all the outcomes from interviews, prior to addressing the actual development of the model.

Method Geek

The result is the written report after you have conducted interviews and after you have talked with people. In a sense it is the journey to get to the model that gives you the result.

Developing a system model that enables some degree of understanding of its dynamics was considered one of the most original and transformative aspects of the FRAM. All interviewees placed this idea more or less at the core of the learning that FRAM can foster.

Method Geek

The idea of system modelling kind of changed my own thinking and the way I think about almost everything in in my life. It gave me a way to understand the world differently that I'm kind of thankful for.

Having a broader system view enables a better understanding of dependencies and dynamics. One of the most highlighted learnings by interviewees was the ability to develop a much more comprehensive perspective on how a person's work or a given operation is situated in a wider system frame. To some extent, the functional model and the conversations that emerge around it, facilitate the expansion of the boundaries on a mental model. This was also considered to contribute to a better appreciation of the flexibility and adaptability that people bring to work and the value this may have towards system performance.

As models become more detailed and system understanding improves, there is an opportunity to capture certain *functional unbalances*, like when certain functions end up showing a much bigger number of couplings than others. This opens the opportunity to discuss what could change to reduce pressure on these functions. Simplifying or reshaping dependencies emerges from the enhanced understanding by both the modellers and the workers. *You see the complexity; you see possibilities to make it more simple.*

An important learning from FRAM was associated with the realisation of things that may go unappreciated but yet can be fundamental for work and operations. This was also referred to as the surfacing of tacit knowledge about work. *They usually are very surprised to see that their task is so complicated and very chaotic sometimes.* The FRAM offers the opportunity to question assumptions around how work takes place both by the modellers and the workers. *The biggest lesson of FRAM is about the Implicit secrets, the many things that are not written in the textbook.* The *questioning of normal* opens up a wide range of opportunities for change. This makes a FRAM exercise much more about learning "how workers do", rather than about "what they do".

Interviewees also made references to the better understanding of priorities and trade-offs that can be achieved. Through the FRAM, both modellers and workers often perceive multiple different aspects of work and operations that are concurrent or take place in tight sequences (i.e., tight couplings). Discussions around the FRAM modelling provide a basis for a better and shared understanding of the challenges and for an agreement on what should be prioritised.

Method Geek

If we only have the time to choose five [aspects of work or process needs to devote time and attention to], it's these five things. And they agree on it across professionals and across frontline and management.

The non-linear nature of the FRAM models contrasts with more conventional representations of work, which tend to convey a siloed and narrow view of work and processes. Important learning was considered to emerge from the insight the FRAM can provide into how the work of some people, teams or departments relate to others and how they could impact each other.

Method Geek

The work between the departments is much more interconnected, so it really matters what I do within my department. It could affect another department and vice versa.

There is a prevalent idea that an FRAM model is never quite finished. There is often a continuous or iterative pursuit towards improvement and recalibration of the model. This emphasises the idea that the FRAM builds on multiple perspectives, as over time, more than operational changes, the understanding of those operations changes, and perspectives tend to shift from both the modellers and the workers.

Method Geek

It's just capturing what I observe on that day. Then maybe I observe the model in one month and I should accept that it's different, but not because people behave differently but because I have a different understanding of the system.

In the case of human interactions with technology and when considering the introduction of new technology, the FRAM highlights how *this is far more than simply replacing one agent with another*. Several interviewees described the learning value of the FRAM as the ability to illustrate how such transformations must always consider the restructuring of the whole system. To an extent, the FRAM enables learning about future operations through a simulation of different possible scenarios, based on the instantiation of an imagined functional organisation. The comparison of different organisations and scenarios, sometimes in terms of "the before and after" the introduction of new technologies, enables a more explicit exploration of changes. The engagement of workers in this comparison process offers valuable opportunities for the understanding of challenges when faced with new technologies. *Functions tend to allow for a very direct transfer into software design.* The FRAM expands on what typically engineering diagrams and process flows provide, not only by introducing a more dynamic perspective but also a multi-dimensional one. *It offers a lot more about emerging possibilities, as opposed to sequential logics.*

REFLECTIONS FROM THE INTERVIEWS

The previous section outlined the extracted themes from the interviews, aiming to, in line with the objectives, explore learning and assumptions around the use of FRAM. This section undertakes a more integrated reflection that builds on these themes, as well as on the author's own experience with the FRAM.

HOSTING CONVERSATIONS

The dominating feature of the FRAM appears to be the ability to host conversations that, with different degrees of systematicity and iteration, develop a shared understanding of operations. The system knowledge and understanding that is achieved is valued for its focus on relations between different operational states, as well as for the potential it offers to explore non-linear and dynamic aspects. The FRAM does not build on the precision nor accuracy of models, but rather it opens up opportunities for further exploration and for the question of assumptions about operations. Discussions and values of the FRAM were much more focused on the process of gathering data and information than on what is achieved as a model. The model was often viewed as a guide for further iteration around the perspectives about work and a way of gauging the extent to which modelling purposes are being pursued.

ENGAGEMENT WITH WORK AND PEOPLE

The need for a strong engagement with different organisational levels was recognised by all interviewees as the basis for any FRAM exercise, in particular with workers providing information on the operations to be modelled. The importance of this direct relation was seen as triggering valuable reflections around implicit issues that tend to be poorly addressed. Even before a comprehensive FRAM model is achieved, interviewees recognised the potential for learning opportunities of unique value for both workers and the modellers. This feeds the development of the model but also creates grounds for a relevant iteration and instantiation of the model. It also becomes useful in identifying the need or the value for any further analysis, namely by resorting to quantitative methods.

FRAM AND SAFETY

The contribution of FRAM towards safety was considerably associated with the discovery of safety-related challenges, namely those emerging from relations that remain poorly known and understood. The FRAM bears a broad notion of safety at its core, and contributions to this end were more frequently coming from process improvements or even the quality of processes and a better understanding of relations between people and technology. This generated a diversity of perspectives on the advantages and challenges of applying the FRAM, but the basis was always considered to be the gathering of knowledge and understanding about relations between different operational aspects, conditions, or stages.

This is in line with the mindset of problem exploration that seems to dominate the use of the FRAM. Several other aspects seem to be related to this orientation:

- Most interviewees initially engaged with FRAM out of a perception of shortfalls in available tools or a sense of insufficient knowledge available regarding certain safety-related aspects. Such perceptions seem to have then triggered the exploration of alternatives.

- Keeping in mind that research activities tend to be more prone to the questioning of available knowledge, the close relationship with academia and the research community could also further support the continued pursuit of problem understanding and exploration.

Conversely, the fact that a FRAM analysis does not offer certainties and conclusive solutions may discourage a wider application in industry, where the search for solutions is naturally more pressing. There were nevertheless some examples of industry implementation and even of its integration into processes, namely those devoted to event investigations. Indeed, investigations seem to remain an important source of FRAM applications.

FRAMing Interviews

Keeping with the core theme of this discussion, an attempt was made to represent the learnings from the interviews as a FRAM model. As illustrated in the previous section, this model inevitably reflects the perspectives of the interviewees (as the providers of information), as much as it does those of the interviewer (the modeller in this case). The purpose is therefore to feed further reflections, as opposed to providing any definite conclusion.

This modelling exercise started from the main topics that were extracted from the interviews, although subsections do not necessarily match a function in the model. The author's reflection and contribution were mainly around the description of functions and their aspects. While inputs and experiences from the interviews were a continuous influence, the author's own views and experience with FRAM are also important contributors to this process.

Domain Expert

Table 20.2 provides a description of the functions in the model produced. For practical reasons, only the relevant function aspects are shown.

TABLE 20.2
Tabular View of the FRAM Model Developed to Represent the Learnings from the Interviews

Name of Function	Defining Modelling Purpose
Description	Clarifying questions to be explored through FRAM
Input	Perceived Difficulties and Variabilities
Output	Open Questions
	Scope and Objectives

(Continued)

TABLE 20.2 (Continued)
Tabular View of the FRAM Model Developed to Represent the Learnings from the Interviews

Name of Function	**Hosting Conversations**
Description	Dialogues around work and related issues
Input	Open Questions
Output	Shared Perspectives
Precondition	Openness
Resource	Wider Understanding
	Functional Description
Control	Time and Space for Conversation

Name of Function	**Engaging With Workers**
Description	Creating space and time for a relation with others
Input	Openness
Output	Time and Space for Conversation
Control	Scope and Objectives

Name of Function	**Building Trust**
Description	Developing trust
Input	Time and Space for Conversation
Output	Openness

Name of Function	**Exploring Perspectives**
Description	Embracing diversity of views about work
Input	Shared Perspectives
	Open Questions
	Shifting Understanding
Output	Shared Contributions

Name of Function	**Integrating Contributions**
Description	Bringing together information gathered and different views about that information.
Input	Shared Perspectives
	Shared Contributions
Output	Shifting Understanding
Resource	Perceived Assumptions

Name of Function	**Sharing Understanding**
Description	Reflecting on and articulating a wider understanding
Input	Shifting Understanding

(Continued)

TABLE 20.2 (Continued)
Tabular View of the FRAM Model Developed to Represent the Learnings from the Interviews

Name of Function	Sharing Understanding
	Shared Contributions
Output	Wider Understanding

Name of Function	Surfacing Assumptions
Description	Perceiving what was taken for granted about work and workers.
Input	Shifting Understanding
Output	Perceived Assumptions
Resource	Shared Contributions
	Perceived Difficulties and Variabilities
	Functional Description

Name of Function	Translating Information
Description	Producing the description of functions and their aspects
Input	Wider Understanding
Output	Functional Description
Control	Perceived Assumptions

Name of Function	Experiencing Challenges
Description	The issues and problems that have led to engaging with FRAM
Input	Functional Description
Output	Perceived Difficulties and Variabilities
Resource	Perceived Assumptions

The colours in the model illustrate different kinds of processes that emerged as critical for how FRAM tends to support learning. Functions in medium grey (i.e., <Defining modelling purpose>, <Experiencing challenges>) were perceived as initiating functions, whereas the three additional colours highlight different feedback loops that illustrate the iterative nature of an FRAM exercise.

These three loops can be briefly described as follows (see Figure 20.1):

- Engaging with work and the workers (lightest grey, functions on the left: <Engage with workers>, <Building trust>) Developing trust-based relations with those providing information to the modelling was frequently highlighted as a cornerstone for the value of the whole exercise. This requires putting aside (even if momentarily) the modelling purposes and their related questions in order to create adequate time and space for listening and understanding the persons being interviewed. This should generate openness from both the interviewer and the interviewees.

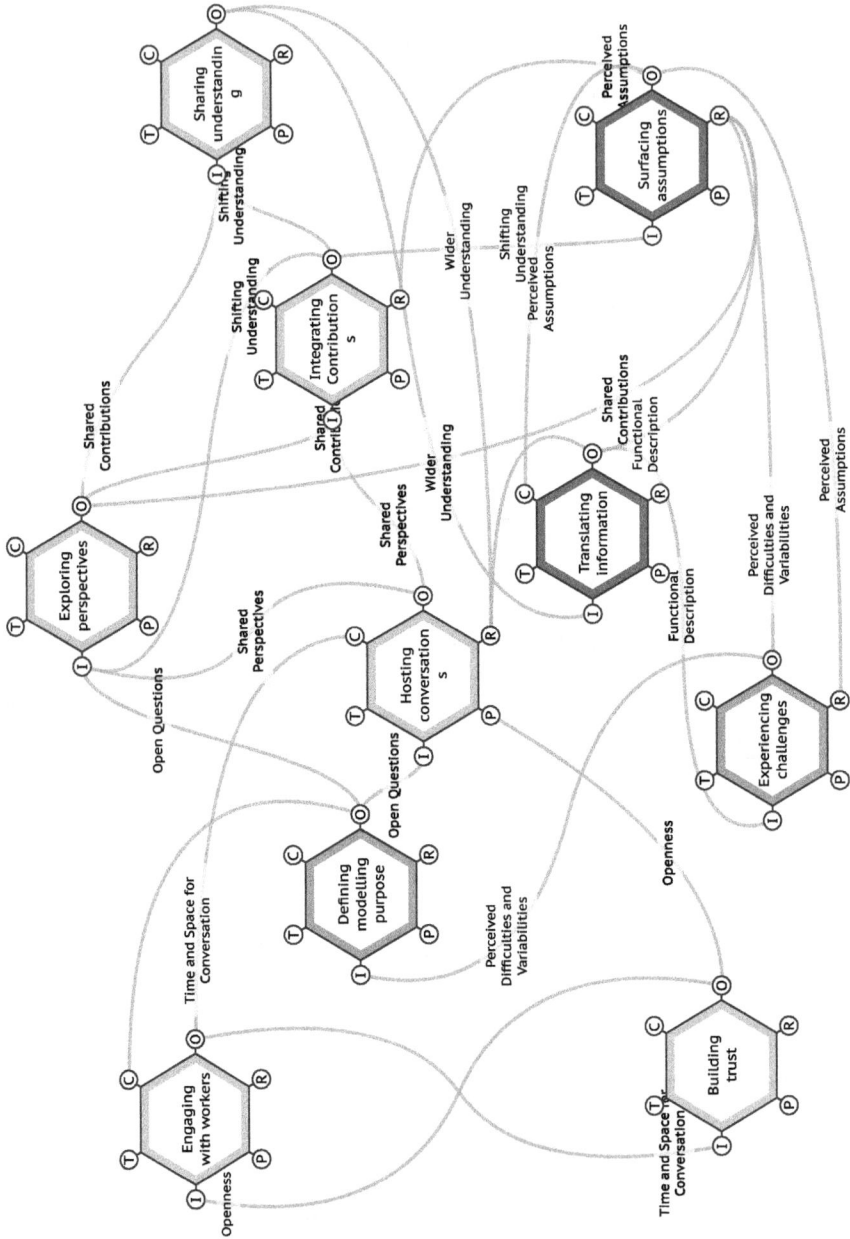

FIGURE 20.1   FRAM model of the learning that emerged from the interviews about the FRAM.

- Surfacing and articulating own assumptions (black, functions on the bottom right: <Translating information>, <Surfacing assumptions>): The previous process tends to expand boundaries and understanding of work, some of which will very likely come to challenge the modeller's perceptions about work. Making sense of this diversity and exploring the way in which one's own assumptions are being challenged, requires a systemisation of the information gathered. The visualisation of how this systematic approach is progressively building up helps refocus attention and further clarify issues and challenges.
- Building on relations with workers (lightest grey, functions on the top right: <Exploring perspectives>, <Hosting conversations>, <Integrating contributions>, <Sharing understanding>): The trust-based relations enable the exploration of open questions and issues within a conversational environment that embraces multiple perspectives. As more of these conversations are experienced by the modeller, reflections (and perhaps even more conversations) emerge around the overlaps, boundaries, and complementarities between the multiple shared perspectives.

These three loops do not seem to take place in any sequential way but rather overlap or alternate with each other, as the modeller perceives and reassesses the way in which the modelling exercise is progressing. This is why there is no strict source or sink function in this model. Focus on each of these loop shifts as discovery is experienced. It could also be argued that the loop "Building on relations with workers" tends to consume the most time and effort. This would mean that, to a great extent, the "Engaging with work and the workers" and "Surfacing and articulating own assumptions" loops often anchor on the way in which the "Building on relations with workers" process unfolds.

The following types of questions from the modeller seem to underline the alternation between these loops:

- What are we trying to achieve?
- What is being told about work?
- How do we understand what is being told about work?
- Where are the contradictions and conflicts?
- What complements on what?
- What does this say about our own understanding of work?
- What are we missing?
- What did we find new about work?
- How does our understanding make sense to others?

CONCLUSIONS

In its early stages, applications of the FRAM consisted of complicated paper-and-pencil exercises. These exercises often discouraged FRAM users and made it impractical for most purposes other than those driven by some research intention. The use of spreadsheets and later the availability of the FMV softened many of these barriers, and a very positive increase in the number of FRAM applications was registered over the years. This also facilitated the combination of the FRAM with other types of tools, namely those pursuing the quantification of various risk-related aspects.

The FRAM is not seen as a method that provides closure to problems but rather one that opens up multiple opportunities for better understanding and tackling problems. While this tends to be valued as part of the FRAM's uniqueness, it may also be the source of challenges towards a more significant industry application of the FRAM. The pressures for problem-solving that often emerge in business and industry life do not seem to agree with the FRAM. The open nature of the FRAM seems to prevail as a core value, precisely because it offers the opportunity to explore alternative perspectives and learn about what is being modelled. The learning opportunities emerge from the dialogue that challenges different views and understandings of work.

The FRAM is, in essence, a compilation of different contributions towards an improved depth and breadth and shared perspective of systems. While much of this can be traced in a FRAM model and its instantiations, important parts are reliant on how the dialogues around that model are taken forward and how the improved work understanding is reported. The FRAM is not only a collection of multiple perspectives about work, but also the value that can be extracted from it, which lies considerably on the new perspectives that may continuously emerge from reflections around the model and its possible revisions.

The FRAM ends up being more than just the sum of different perspectives (those of the modellers and of the workers). There are things that no one actually realises until they are confronted with the gaps during conversations and iterations of the model. The modelling developed based on the interview process here presented offers additional opportunities for reflection. The feedback loops illustrated by the model, as well as the whole model, highlight the continuous nature of any meaningful learning process, and the iterative way in which a FRAM exercise seems to pursue that continuity of learning.

While there is certainly no recipe for conducting an FRAM exercise, there appear to be a number of driving trends that can lead to valuable learning opportunities for both those providing input to the modelling and those developing the model. The model presented earlier about the interviews and the set of questions derived from it, could offer guidance on better understanding these trends. Ultimately, it is up to the modellers to be as aware as possible of their own assumptions and question them at every step of the way.

REFERENCES

Patriarca R., Falegnami A., Costantino F., Di Gravio G., De Nicola A., Villani M.L. (2021). WAx: An Integrated Conceptual Framework for the Analysis of Cyber-Socio-Technical Systems. *Safety Science*, 136, 105142.

Shorrock S. (2020). Proxies for Work-as-Done: 1. Work-as-Imagined. Retrieved on 10 July, 2024 from https://humanisticsystems.com/2020/10/28/proxies-for-work-as-done-1-work-as-imagined/

Section VI

Final Thoughts

21 To FRAM or Not to FRAM?

Riccardo Patriarca and Gesa Praetorius

As we reach the end of this journey, a question naturally arises: should one embrace the FRAM or look elsewhere for methodological guidance?[1]

Throughout these chapters, the FRAM has been presented as a flexible lens to understand complex systems and as a rigorous framework designed to deal with the intricate interplay among specific system activities. A conscious FRAM analyst knows that the method urges us to explore performance variability as an inherent characteristic of real operations, rather than an exception or anomaly. In doing so, it challenges traditional cause-effect thinking, urging a shift towards a more nuanced, emergent perspective on system behaviour. In short, it suggests we abandon Newtonian reasoning when coming to analyse, design, and eventually manage complex socio-technical systems.

Yet, the FRAM is not a universal cure-all. While it offers a means to analyse complexity without oversimplifying it into tidy, deterministic models, its successful application requires thoughtful preparation, self-acceptance of ignorance about domain knowledge, and willingness to acquire it the hard way, that is, in the field and through the eyes of those working within the system we seek to understand. The FRAM encourages us to acknowledge that not all phenomena can be easily understood through simplistic models with arrows indicating cause-effect links, and not all events have a comfortable, identifiable root cause. This can be unsettling for some while encouraging to others.

Yet, for certain use cases, it is also less beneficial, or even unnecessary.

The choice to deploy the FRAM hinges on the scope of its analysis, the nature of the system, and the resources available to the analyst(s). If a system is relatively stable, well-understood, mostly technological and with limited human–system interaction, possibly even amenable to quantification, the FRAM's added complexity may not prove necessary. For a system that is dynamic and interactive, encompassing human operators and constrained by loose (or, conversely, too tight) procedural aspects, usually the subtle daily variations in performance defy linear explanation. This is when an FRAM investigation can offer critical insights.

Such insights come at a cost, though. The efforts to develop a FRAM study refer to the need to have access to experts within the system under study and to the need to condense dozens of hours of interviews and observations into a set of functions and attempt to reconcile some otherwise hidden adaptive behaviours. Delving into the nitty-gritty of work inherently means there are no easy answers, no definitive endpoint, no fixed level of granularity, and no predetermined conclusion to the analysis.

DOI: 10.1201/9781003518167-28

Interpretation, iterative refinement, and close collaboration between those within the system (sharp- or blunt-end operators) and the analyst are fundamental to fully saturating the FRAM model and its instantiations. This dynamic interplay reveals not only what has transpired and what is currently occurring within the system but also highlights potential trajectories of variability in everyday work. Such an approach ensures that the model remains deeply grounded in the lived realities of the system, offering a comprehensive lens to explore its adaptive capacities and emergent behaviours.

This inherent complexity is what makes the FRAM valuable as a debriefing tool or as a common floor where expectations at the blunt end, and practices at the sharp end can be discussed. By serving as a shared reference point for different work varieties (cf. Work-As-X, WAx model by Patriarca et al. (2021)), the FRAM supports diverse aims, including improved ergonomics, enhanced efficiency, and collaborative co-design.

Adopting the FRAM successfully implies embracing a particular mindset: one that values understanding over mere explanation, contextual insight over simple metrics, and dynamic relationships over static constructs. This direction can improve the understanding of what makes a complex system resilient and what supports adaptive behaviours and help to identify what hidden potential for both success and failure – key objectives in modern safety, reliability, and risk assessment – in everyday work. The value of this direction spans also beyond these usual fields, reaching ergonomics, human performance management, quality and even productivity, as discussed in the "Synesis" book by Hollnagel (2022), too.

The decision to engage with the FRAM should not be viewed as a simple choice between old and new approaches. Instead, it prompts us to reconsider what it means to "model" in the context of complex adaptive socio-technical systems. Traditional approaches have long offered the comfort of numbers, yet FRAM reminds us that not all meaningful aspects of performance can be neatly measured. Attempts to quantify emergent behaviours may miss the subtle interdependencies that truly shape outcomes. Conversely, purely qualitative descriptions, though rich in insight, may feel lacking in the power to show recurrent patterns supporting robust comparisons or trend analyses.

This tension opens a space for innovation. Innovation has been addressed in various chapters of this book, where hybrid strategies have been portrayed, coupling the FRAM's nuanced conceptual modelling with carefully selected indicators or simulation tools (i.e., Discrete Event Simulation, AI for modelling learning processes, and fuzzy assessment, among others). This trend reflects what was already noticed in the literature (Patriarca et al., 2020).

In walking down this path, there is, however, an implicit trap of oversimplification. Analysts shall remain vigilant against it, ensuring that the drive for numbers does not erode the inherent complexity of operations.

An analytical translation of the FRAM essentials has been proposed in the introductory chapter of this book. The formulation has been developed purposefully in a way that can accommodate both qualitative and quantitative perspectives, leaving to the conscious analyst the final decision about which one to embrace. By thoughtfully balancing the strengths of qualitative insight and judicious quantification, we might better capture the resonances that shape real-world performance.

"To FRAM or not to FRAM" as such is not a binary decision. It is a matter of selecting the right tool for a valuable scope. The ultimate measure of any analytical method lies in its capacity to reveal what matters: to open our eyes to what we could not previously see and to guide us towards meaningful, sustainable improvements. The FRAM, when wisely chosen and carefully applied, can do precisely that. Some analysts even argue that the value of the FRAM lies more in the journey to obtain the model than in the model itself. From our personal experience, we can attest that in some cases we have not even shown the entire "forest" of hexagons produced by the intensive data-gathering process. Instead, we focused on the insights drawn from the model, which ultimately provided real value to the stakeholders involved in the analysis.

In essence, "to FRAM or not to FRAM" becomes less about choosing a method and more about forging a path towards a more informed, context-sensitive form of analysis. One that marries the rigour of modelling with a sober understanding that root causes or numbers alone cannot fully contain the complexity we seek to understand.

NOTE

1 The title "To FRAM or Not to FRAM" draws its inspiration from a vibrant (from the authors' perspective, at least, who were also the ones delivering it) joint lecture delivered in 2020, where international PhD students and safety professionals gathered to explore the depths of systems thinking in modern safety science.

REFERENCES

Hollnagel, E. (2022). *Synesis: The Unification of Productivity, Quality, Safety and Reliability*. Taylor and Francis.

Patriarca, R., et al. (2020). Framing the FRAM: A Literature Review on the Functional Resonance Analysis Method, *Safety Science*, 129, 104827. https://doi.org/10.1016/j.ssci.2020.104827

Patriarca, R., et al. (2021). WAx: An Integrated Conceptual Framework for the Analysis of Cyber-Socio-Technical Systems, *Safety Science*, 136, 105142. https://doi.org/10.1016/j.ssci.2020.105142

Glossary

Riccardo Patriarca and Erik Hollnagel

Welcome to the glossary of this book, conveniently tucked away at the very end of the book – just where you'd go first if you needed a quick explanation of all those technical terms that have been used so far. Here, we have compiled a comprehensive list of the key concepts and jargon about the FRAM. Whether you are a seasoned practitioner seeking clarity, or a curious newcomer wondering why we keep talking about "couplings" and "resonance," this section aims to untangle the key terminology around the method. The content started from an initial collection available on the FRAM website, but it has evolved over time and specialised for the content of this book.

> …who doesn't love learning what a term means after they've already read it 20 or more times?

ABOUT PRINCIPLES

Approximate adjustments (principle): When working conditions are turbulent and uncertain or when time or resources are limited, it is necessary to adjust performance to match the current conditions. Accordingly, performance adjustments are necessary on virtually all occasions, but they are inherently approximate rather than perfect. The approximations are, however, good enough under most conditions to ensure successful performance.

Emergence (principle): Variations in normal performance are rarely significant enough on their own to cause an accident or even to constitute a malfunction. However, when variability from multiple functions combines in unexpected ways, it can lead to disproportionately large consequences, resulting in a non-linear effect. Both failures and normal performance are emergent phenomena rather than resultant ones because neither can be solely attributed to or explained by the (mal)functions of specific components or parts.

Equivalence of successes and failures (principle): For each action, the choice of what to do is determined by many different things, including competence, understanding of the situation, experience, habit, demands, available resources, and expectations about how the situation may develop – not least about what others may do. If the expected outcome is obtained, the next action is taken, and so on. But if the outcome is unexpected, then the preceding action is re-evaluated and classified as wrong rather than right, as an error or as a mistake, using the common but fallacious post hoc ergo propter hoc argument. With hindsight, it is pointed out what should have been done if only people had made the necessary effort at the time. The whole argument is, however, unreasonable because the action was chosen based on the expected rather than the actual outcome. Failures and successes are equivalent in the sense that we can only say whether the preceding action was right or wrong after the outcome is known. That only changes the judgement of the action but not the action itself.

Functional resonance (principle): Functional resonance is defined as the detectable signal that emerges from the unintended interaction of the everyday variability of multiple signals. The signals are usually subliminal, both the "target" signal and the combination of the remaining signals that constitute the noise. However, the variability of the signals is subject to certain regularities that are characteristic of different types of functions, hence not random or stochastic. Since the resonance effects are a consequence of the ways in which the system functions, the phenomenon is called functional resonance rather than stochastic resonance.

On the other hand, stochastic resonance refers to the increased sensitivity of a system or device to disturbance. Unlike classical resonance, where the response builds up over time, stochastic resonance produces a non-linear outcome, meaning the output is not directly proportional to the input, and it can occur – sometimes instantaneously. More broadly, in physical systems, resonance describes the phenomenon where a system oscillates with greater amplitude at specific frequencies, known as its resonant (or resonance) frequencies. At these frequencies, even small, repeated external forces can generate large oscillations, potentially causing significant damage or even the destruction of the system.

ABOUT FUNCTIONS

Aspect (of a FRAM function): Each FRAM function is described by six aspectsInput, Output, Precondition, Resource, Control, and Time. An aspect should be described if relevant by the analyst and if there is sufficient information on it. For foreground functions, it is necessary to describe at least one Input and one Output. For background functions that represent a source (entry function), it is sufficient to describe the Output. Similarly, it is sufficient to describe the Input for background functions that represent a sink (exit function). Note, however, that if only Inputs and Outputs are defined for all functions, the FRAM model regresses to a simple flowchart or network.

Background function: Foreground and background refer to the relative importance of a function in the model. The background functions denote what affects the foreground functions being studied, that is, the context or working environment. A designated background function may become a foreground function if the focus of the analysis changes.

A background function can have one or more Outputs but cannot have any other defined aspects (Entry background function or source function). Background functions are typically used to denote the sources for Input, Precondition, Resource, Control, and/or Time aspects of downstream functions. Alternatively, background functions are also used to set the final scope of the process. In that case, they are characterised by receiving an

Input-only (Exit background function or sink function). A background entry function may, therefore, be seen as a placeholder for an upstream/downstream function that does not need to be specified further. Since background functions do not need to be described in the same details as foreground functions, they provide a convenient way of setting the boundaries of the analysis.

Boundary: The semi-explicit stop rule of the FRAM is that the analysis should continue until there is no unexplained (or unexplainable) variability of functions, that is, the analysis has reached a set of functions which can be assumed to be stable rather than variable. More practical approaches to setting the boundary are addressed towards recognising a relevant (yet manageable) purpose of the analysis, as prescribed by the building Step 0.

Control (as aspect): Control is what supervises or regulates a function so that it results in the desired Output. Control can be a plan, a schedule, a procedure, a set of guidelines or instructions, a program (an algorithm), a "measure and correct" functionality, as well as a type of social control (i.e., external as for the expectations of the management, or internal, such as self-regulation).

Coupling: Coupling describes the degree to which subsystems, functions, and components are connected or depend upon each other; the degree of coupling can range from loose to tight. The FRAM makes a distinction between the potential couplings that are defined by a FRAM model and the actual couplings that can realistically be assumed to exist for a given set of conditions (an instantiation). While the actual couplings will always be a subset of the potential couplings, there is usually value in studying their differences for capturing different varieties of work.

Downstream functions: A FRAM model describes the functions and their potential couplings rather than how they are organised for specific conditions. It is, therefore, not possible to say with certainty whether one function will always be carried out prior to or after another function. In an instantiation of the model, detailed information about a specific situation or scenario is used to create an instance or a specific example of the model. This establishes a temporal organisation of the functions as they are likely to unfold (or become activated) in the scenario, and it is only possible to consider functions in their temporal and causal relations when an instantiation has been produced. Functions that – in the instantiation – happen after other functions and which, therefore, may be affected by them are called downstream functions. The notion of downstream function is clearly a relative one rather than an absolute one.

Foreground functions: Foreground and background refer to the relative importance of a function in the model. The foreground functions denote what is being analysed or assessed, that is, the focus of the investigation. A designated foreground function may become a background function if the focus of the analysis changes.

Function: In the FRAM, a function represents the means that are necessary to achieve a goal. More generally, a function refers to the activities – or set of activities – that are required to produce a certain outcome. A function describes what people – individually or collectively – have to do in order to achieve a specific aim. A function can also refer to what an organisation doesfor example, the function of an air operator is to move passengers safely and efficiently from X to Y. A function can finally refer to what a technological system does either by itself (an automated function) or in collaboration with one or more humans (an interactive function or coagency). The description of a Function should be a verb if it is a single word or begin with a verb if it is a short sentence.

Input (as aspect): The Input to a function is traditionally defined as that which is used or transformed by the function to produce the Output. The Input can represent matter, energy, or information. This definition corresponds to the normal use of the term in flowcharts, process maps, logical circuits, etc. There is, however, another meaning that is just as important for the FRAM, namely the Input as that which activates or starts a function. The Input in this sense may be a clearance or an instruction to begin doing something, which in turn requires that the input is detected and recognised by the function. While this nominally can be seen as being data, it is more important that the Input serves as a signal that a function can begin. Technically speaking, the Input represents a change in the state of the environment, just as if the Input was matter or energy. The description of an Input should be a noun, if it is a single word, or begin with a noun if it is a short sentence. If written with an initial lowercase letter (i.e., *input*), the term refers to any type of input for a function, including Input, Precondition, Resource, Control, or Time.

Instantiation: An instantiation of a FRAM model is a "map" of how a set of functions are mutually coupled under given conditions (favourable or unfavourable) or for a given timeframe. The set may include all the functions in a model or a subset thereof. The couplings that are represented by a specific instantiation can be seen as representing the order or sequence in which the functions were triggered for the given conditions. The sequence may include parallel "paths" and "iterations". If the model is used to represent an event that has taken place, such as an accident investigation, the instantiation will typically cover the time frame of the event and represent the couplings that did exist at the time. An instantiation then only covers a specific functional propagation (an actuated scenario) as a portion of the possibilities provided by a FRAM model. For a risk assessment, it is more appropriate to work with a set of instantiations, where each instantiation shows the coupling between upstream and downstream functions at a given time or for given conditions. When an instantiation is calculated, it allows converting the FRAM model into a graph, with functions as nodes and edges as couplings.

Model: A FRAM model describes all the system's functions, that is, both foreground and background. The potential couplings among functions are defined by how the aspects of the functions are described. The FRAM model, however, does not describe the actual couplings that may exist under given conditions (seeInstantiation). To represent the result of performance variability of a function – or of several functions – therefore requires an instantiation. A graphical representation of a FRAM model will be a set of hexagons, where each hexagon stands for a function, and with all potential couplings being represented. Because the couplings in a FRAM model are potential rather than actual, a FRAM model is not a process model or even a net (graph). Conversely, an instantiation of a FRAM model may be isofunctional (i.e., having the same function) to a process model or a network.

Output (as aspect): The Output from a function is the result of what the function does, for instance by processing the Input. The Output can therefore represent matter, energy, or information – the latter as a command issued or the outcome of a decision. The Output can be seen as representing a change of state – of the system or of one or more output parameters. The Output can also represent the signal that starts a downstream function. The description of an Output should be a noun, if it is a single word, or begin with a noun if it is a short sentence.

Performance variability: The FRAM principles suggest that performance is always variable in practice. The performance variability of upstream functions may affect the performance variability of downstream functions and thereby lead to non-linear effects (functional resonance).

Phenotypes of variability: Due to resource constraints, time pressure, and operating conditions, performance variability is inherent and inevitable in socio-technical systems. The phenotypes of variability help identify how these variations occur and how they may combine to generate potentially emergent outcomes. Typical examples of phenotypes are timing, duration, distance, direction, magnitude, speed, precision, pressure, quantity, sequence, etc. The list is not exhaustive as any system may have its own relevant measures.

Pre-condition (as aspect): A function in many cases does not begin before one or more Preconditions have been established. The presence of Preconditions can be understood as a system state that must be true, or as conditions that must be verified before a function is carried out. Although a Precondition is a state that must be true before a function is carried out, it does not itself constitute a signal that starts the function. Input, on the other hand, can activate a function. This simple rule can be used to determine whether something should be described as an Input or as a Precondition. The description of a Precondition should be a noun, if it is a single word, or begin with a noun if it is a short sentence.

Resource (as aspect): A Resource is something that is needed or consumed while a function is carried out. A Resource can represent matter, energy, information, competence, software, tools, manpower, etc. Time can, in principle, also be considered as a Resource, but since Time has a special status, it is treated as a separate aspect. Since some Resources are consumed while the function is carried out and others are not, it is useful to distinguish between non-execution conditioned-Resources (simply called Resources) on the one hand and Execution-condition-Resources on the other. The difference is that while a Resource is consumed by a function, there will be less of it as time goes by. Contrarily an Execution condition is not consumed but only needs to be available or exist while a function is active. The description of a Resource (or an Execution condition) should be a noun, if it is a single word, or begin with a noun if it is a short sentence. Note that there still is a difference between a Precondition and an Execution Condition (in essence a particular type of Resource) being that the former is only required before the function starts, but not while it is carried out.

Time (as aspect): The Time aspect of a function represents the various ways in which Time can affect how a function is carried out. Time, or rather temporal relations, can be seen as a form of Control, but one in which temporal relations are involved. One example of that is when Time represents the sequencing conditions. A function may, for instance, have to be carried out (or be completed) before another function, after another function, or overlapping with – parallel to – another function. Time may also relate to a function alone, seen in relation to either clock time or elapsed time. Time can also be seen as representing a Resource, such as when something must be completed before a certain point in time, or within a certain duration (like when a report must be produced in less than a week). Time can, of course, also be seen as a Precondition, for example, that a function must not begin before midnight, or that it must not begin before other functions have been completed. Yet rather than having Time as a part of either of the three aspects of a function – or indeed, the four since it conceivably could also be considered as an Input – it seems reasonable to recognise its special status by having it as an aspect in its own right. Ultimately, the purpose of the analysis or specific coupling should prevail in determining whether something is called a Resource, a Precondition, or a Time aspect. The description of a Time aspect should be a noun, if it is a single word, or begin with a noun if it is a short sentence.

Upstream functions: An upstream function is defined in a similar way as a downstream function (q.v.) as a function that – in a given instantiation – happens before other functions, and which therefore may affect them. The notion of upstream function is relative rather than absolute.

EXTENSIONS TO THE FRAM (SELECTED)

Abstraction/Agency (framework): This framework examines how individual components or actors (the agents) operate across various levels of abstraction—from detailed, task-specific activities to broader system-wide functions. The framework helps to understand interactions at different levels (inter-level) and among different agents (inter-agents), as well as conducting analysis in depth (intra-level, inter-agent). It extends the Abstraction/decomposition framework originally described by Rasmussen[1] to the idea of agency and co-agency in order to support a consistent analysis of contributions to emergent behaviours and overall system performance.

Functional random walker: In traditional network theory, a random walk is a dynamical model used to simulate the evolution of a walker that moves between adjacent nodes, respecting the directionality of the links. For the simulation to run, time is assumed as discrete and at each time step, a movement of the walker occurs. This notion needs to be adapted to suit the notion of FRAM instantiationsthe notion of "functional random walker" represents the evolution of a random dynamic process that is forced to adhere to the functional and logical constraints imposed by a FRAM model structure, that is Input/Output/Resource/Control/Precondition/Time couplings. In a functional random walker, the movements of the walker, which represents an information agent, are constrained to reflect the specific instantiations being analysed. With dedicated metrics (e.g. probability of visiting nodes and mean passage time), it is also possible to move the unit of the analysis from a single instantiation to larger models.

Functional signature: Functional signature is a recorded account of the functional activity and individual function variability over time. The signature is thus a variable that can be traced over time for any instantiation and helps to understand the dynamic nature of the system. They can usually be created by logging the functional activities of each function over time, in addition to the usual static visualisations.

Interdependence analysis: Interdependence analysis adds a dimension of investigation to FRAM couplings, referring to how system functions are interconnected through their dependencies. It utilises the principles of Observability, Directability, and Predictability (ODP) as a breakdown of interdependence, the latter being defined as a capacity generated by the upstream agents and a dependency of that capacity received by a downstream agent between any two consecutive functions performed by two different agents. It identifies how variability in one function can influence others in particular terms of system readability (Observability and Predictability) and system control (Directability). By mapping the six aspects of each function – Input, Output, Precondition, Resource, Control, and Time – with this additional layer of analysis reveals how performance variability propagates through couplings in terms of multi-agent coordination

Metadata: it refers to supplementary information associated with elements of a FRAM model, such as functions, aspects, and their interdependencies. Metadata includes details like descriptions, annotations, labels, or numerical attributes that provide additional context beyond the primary modelling data.

OTHER FOUNDATIONAL ASPECTS (SELECTED)

Activity: the description of work as carried out in practice (from 1950s French ergonomics, activité), distinct to task, which corresponds to a formal description of work (from the 1950s study by Ombredanne & Faverge, tâche). Task corresponds to work-as-imagined (WAI) and activities to work-as-done (WAD).

Data collection: Virtually any kind of data and information can be potentially useful for a modelling exercise. If not explicitly useful for the description of functions and their aspects, any insights tend to improve background understanding, which in return can also improve functional description. There are, however, important data elements and sources, which if over-looked, could compromise the modelling exercise. The FRAM is very suit-able for capturing the WAD, that is, it allows being in contact with the work reality and those who perform it, a fundamental cornerstone for any model-ling exercise. On-sight observations and interviews in all their shapes and formats are quite useful, if not unavoidable. A FRAM model based only on documented information and archival data would be dangerous unless, of course, you are planning to build a model of documentary knowledge only. If that's not the case, you should take a step back and rethink your approach.

Non-linearity: FRAM is often said to differ from most other modelling tools for enabling non-linear representations of work and operations. But it is all up to the modellers. Understanding and embracing the non-linear nature of most things in life must start with you, and with that, there's little that FRAM nor any other tool can do. If your heart and mind are in the right place, then FRAM can highlight important aspects of the non-linear nature that characterises every work and operation. Feedback loops, multi-layered operations and other interdependency-related aspects can emerge as mani-festations of non-linearity.

Task: The task corresponds to the formal description of work, opposite to the activ-ity, which corresponds to the description of how the task is carried out in practice. See activity.

Work variety (varieties of work): This concept highlights the differences between how work is perceived, planned, prescribed, disclosed (etc.), and how it is actually performed in practice. This idea dates back to 1950s French ergo-nomics, where it was early noted that there is often a gap between theoreti-cal models of work and reality. Such a gap is relevant within the scope of the FRAM, as this latter can be a valid method to explore such nuances. A consistent FRAM analysis is usually meant to delve into the nitty-gritty of work-as-done (WAD) and explore the activities, rather than the task (work-as-imagined). Other varieties can be observed, as discussed extensively by Shorrock (2016) and formalised in the WAx framework by Patriarca et al. (2021). It is usually beneficial to compare various FRAM instantiations for capturing the complexities, and trade-offs of different work varieties and gain a larger understanding of system dynamics, and emergent system behaviours.

NOTE

1   J. Rasmussen, "The role of hierarchical knowledge representation in decision making and system management," in IEEE Transactions on Systems, Man, and Cybernetics, vol. SMC-15, no. 2, pp. 234-243, March-April 1985, doi10.1109/TSMC.1985.6313353.

Index

Pages in **bold** refer to tables.

For Product Safety Concerns and Information please contact our EU
representative GPSR@taylorandfrancis.com
Taylor & Francis Verlag GmbH, Kaufingerstraße 24, 80331 München, Germany

www.ingramcontent.com/pod-product-compliance
Lightning Source LLC
Chambersburg PA
CBHW060425220326
41598CB00021BA/2293